Energy Transition and Environmental Sustainability

Energy Transition and Environmental Sustainability

Editors

Prafula Pearce
Tina Soliman Hunter

MDPI • Basel • Beijing • Wuhan • Barcelona • Belgrade • Manchester • Tokyo • Cluj • Tianjin

MDPI

Editors
Prafula Pearce
Edith Cowan University
Joondalup
Australia

Tina Soliman Hunter
University of Aberdeen
Scotland
UK

Editorial Office
MDPI
St. Alban-Anlage 66
4052 Basel, Switzerland

This is a reprint of articles from the Special Issue published online in the open access journal *Energies* (ISSN 1996-1073) (available at: https://www.mdpi.com/journal/energies/special_issues/energy_transition_environmental_sustainability).

For citation purposes, cite each article independently as indicated on the article page online and as indicated below:

LastName, A.A.; LastName, B.B.; LastName, C.C. Article Title. *Journal Name* **Year**, *Volume Number*, Page Range.

ISBN 978-3-0365-7166-9 (Hbk)
ISBN 978-3-0365-7167-6 (PDF)

Contents

About the Editors

Prafula Pearce

Prafula Pearce is an Associate Professor in Law, a member of the Research and Research Training Committee, and ECU Tax Clinic Coordinator in the School of Business and Law at Edith Cowan University, Perth, Australia. She has over 30 years of experience that includes academic and commercial, and her research focus is around regulatory measures to promote sustainable energy use.

Tina Soliman Hunter

Professor Tina Soliman Hunter is a Professor of Energy and Resources Law at Macquarie University, and the Director of the Centre for Energy and Natural Resrouces Innovation and Transformation (CENRIT) at Macquarie University. She was also the former Director of the Aberdeen University Centre for Energy Law (AUCEL). Tina's current research interests include: mitigating the environmental impacts of resource extraction, the role of science in the regulation of resource extraction activities, offshore wind licensing and regulation, shale gas extraction in Australia, energy security, Arctic petroleum law, regulation, and environment, multidisciplinary approaches to resource regulation, State control in hydrocarbon extraction, the new hydrogen economy, and Indigenous and religious concepts in natural resources law.

energies

Editorial

Special Issue "Energy Transition and Environmental Sustainability"

Prafula Pearce

School of Business and Law, Edith Cowan University, Joondalup, WA 6027, Australia; p.pearce@ecu.edu.au

This Special Issue on "Energy Transition and Environmental Sustainability" includes thirteen papers on policies including: the challenges of the United Nations Sustainable Development Goals regarding energy transition and legal reforms in Taiwan and Japan [1] successful energy transition toward solar PV in South Korea [2]; transition from diesel buses to hybrid-driven (HEV) and electricity-driven buses (BEV) for public transport in Central Europe [3]; vehicle transition and the development of electric car production in three regions, the United States, the European Union and Japan [4]; affordable and environmentally friendly cooling solutions for buildings in Pakistan [5]; development of projects to replace fossil fuels with renewable energies, mainly hydropower in Ecuador [6]; the Role of Electrification in the Decarbonization of Central-Western Europe [7]; energy technology innovation through the application of new technologies in oil resource development [8]; the cost of a Wastewater Treatment Plant in Poland and the Impact of Atmospheric Precipitation [9]; analysis of the design of new buildings respecting the "zero-energy and low carbon emission" concept in tropical climatic regions [10]; climate change litigation risks for Australian energy companies and investors from a policy and governance perspective [11]; a demand and supply management study of a Childcare Facility in Tokyo and the need to shift from conventional power generation to renewable energy sources [12]; implementation of new policies supporting renewable natural gas production from organic wastes in Canada [13].

A brief summary of the content associated with each of the selected papers belonging to this Special Issues is included below:

In 'National Energy Transition Framework toward SDG7 with Legal Reforms and Policy Bundles: The Case of Taiwan and Its Comparison with Japan', the authors Mu-Xing Lin, Hwa Meei Liou and Kuei Tien Chou [1] construct an analysis structure for national energy transition to analyse the current situation within Taiwan's electricity sector reforms, while providing evidence of the national experience of electrical industry reforms as an international reference. This study also compares the differences between the seventh Sustainable Development Goal relationship and national energy transitions in Taiwan and Japan, based on the similar initiative of the revised Electricity Act within the policy bundle.

In their paper, 'Which Institutional Conditions Lead to a Successful Local Energy Transition? Applying Fuzzy-Set Qualitative Comparative Analysis to Solar PV Cases in South Korea' [2], the authors Youhyun Lee, Bomi Kim and Heeju Hwang use a fuzzy-set qualitative comparative analysis to measure the success of a local energy transition. Their study provides insights on energy transition for developing or newly industrialized countries.

As diesel is the most used fuel for buses and other urban transport vehicles in European countries, in their paper, 'Environmental Sustainability of the Vehicle Fleet Change in Public City Transport of Selected City in Central Europe' [3], the authors Vladimír Konečný, Jozef Gnap, Tomáš Settey, František Petro, Tomáš Skrúcaný and Tomasz Figlus explore the extent to which air pollution can be reduced if the urban public transport fleet is renewed in the city of Žilina.

In 'Perspectives for Mitigation of CO_2 Emission due to Development of Electromobility in Several Countries', the authors Karol Tucki, Olga Orynycz and Mateusz Mitoraj-

Citation: Pearce, P. Special Issue "Energy Transition and Environmental Sustainability". *Energies* **2023**, *16*, 2675. https://doi.org/10.3390/en16062675

Received: 17 November 2022
Accepted: 26 November 2022
Published: 13 March 2023

Wojtanek [4] use the creep trend method using historical data for the years 2007–2017 for the analysis of the development of electric car production in three regions, the United States, the European Union and Japan, and project the reduction in emissions of over 14,908,000 thousand tonnes of CO_2 in the European Union, 3,786,000 thousand tonnes of CO_2 in United States, and 111,683 thousand tonnes of CO_2 in Japan. Their study concluded that electricity-powered cars along with appropriate choices of energetic resources as well as electricity distribution management will play an important role in achieving a sustainable energy economy.

In 'Energetic, Economic and Environmental (3E) Assessment and Design of Solar-Powered HVAC Systems in Pakistan', the authors Sajid Mehmood, Serguey A. Maximov, Hannah Chalmers and Daniel Friedrich [5] explored cooling systems for industrial buildings in Lahore, Pakistan. They evaluated two systems, namely, electrically driven water-cooled vapour compression systems and thermally (solar) driven vapour absorption cooling systems, and concluded that electrically driven vapour compression-based cooling systems have much higher running costs and are potentially hazardous for the environment but have lower capital costs. On the other hand, solar thermal systems have lower running costs and emissions but require further reductions in the capital costs or government subsidies to make them viable.

In 'Low-Carbon Energy Governance: Scenarios to Accelerate the Change in the Energy Matrix in Ecuador' [6], the authors Flavio R. Arroyo M. and Luis J. Miguel used the system dynamics methodology in their study to model supply, demand and CO_2 emissions scenarios for Ecuador for the year 2030. Since oil remained the most important source of energy, their proposal for energy policies aimed at mitigating emissions by replacing fossil fuels with renewable energies, mainly hydropower.

In 'The Role of Electrification in the Decarbonization of Central-Western Europe', the authors Gauthier de Maere d'Aertrycke, Yves Smeers, Hugues de Peufeilhoux and Pierre-Laurent Lucille [7] used the "variational scenario" analysis, which showed that tilting the central role of electricity to a mix of electricity and green gas offers several advantages in terms of efficiency, flexibility of investment strategies, and robustness with respect to major uncertainties.

The authors HaeOk Choi and Hwanll Park, in 'Oil is the New Data: Energy Technology Innovation in Digital Oil Fields' [8], explained the evolution of Digital Oil Fields (DOFs) over the course of 10 years and showed that DOF technology together with device-related technologies is developing through convergence and close links with other industries, specifically the equipment, parts, and material industries.

In 'The Impact of Atmospheric Precipitation on Wastewater Volume Flowing into the Wastewater Treatment Plant in Nowy Targ (Poland) in Terms of Treatment Costs', the authors Piotr Bugajski, Elwira Nowobilska-Majewska and Michał Majewski [9] determined the costs resulting from the treatment of accidental (rain) water entering the analysed sewerage system in that region. Their research provides important information for sewage network operators to replace the combined sewage system in Nowy Targ with a distributed sewerage system.

In 'Transition to Zero Energy and Low Carbon Emission in Residential Buildings Located in Tropical and Temperate Climates', the authors Modeste Kameni Nematchoua, José A. Orosa, Paola Ricciardi, Esther Obonyo, Eric Jean Roy Sambatra and Sigrid Reiter [10] analysed the design of new buildings with respect to the "zero-energy and low carbon emission" concept in tropical climatic regions in comparison with temperate zones. The results showed that the renovation of existing residential buildings facilitates a reduction of up to 35% of energy demand and a great quantity of CO_2 emissions in both climate zones; however, the investment rate linked to the construction of zero-energy buildings in tropical zones is 12 times lower than in temperate zones, and the payback was double.

The editor of this journal, Associate Professor Prafula Pearce, explored whether company directors can and, in some cases, should be considering the impact of climate change litigation risks on their business, or else risk breaching their obligation to exercise care and diligence under the Corporation Act 2001 (Cth, Australia) in the paper entitled 'Duty to Address Climate Change Litigation Risks for Australian Energy Companies—Policy and Governance Issues' [11].

In 'Solar Energy Demand-to-Supply Management by the On-Demand Cumulative-Control Method: Case of a Childcare Facility in Tokyo', the authors Hiromasa Ijuin, Satoshi Yamada, Tetsuo Yamada, Masato Takanokura and Masayuki Matsui [12] used actual power data from a childcare facility in Tokyo and used a demand-to-supply management method for a solar power generation system by using the on-demand cumulative-control method.

Finally, the authors Omid Norouzi and Animesh Dutta, in 'The Current Status and Future Potential of Biogas Production from Canada's Organic Fraction Municipal Solid Waste' [13], explored findings from data gathered from published papers, the Canadian Biogas Association, Canada's national statistical agency, and energy companies' websites to gain insight into the current status of anaerobic digestion plants in recovering energy and resources from organic wastes.

It is my hope that you will enjoy reading these studies and find them to be worthy of serious consideration by policy makers around the world.

Conflicts of Interest: The authors declare no conflict of interest.

References

1. Lin, M.-X.; Liou, H.M.; Chou, K.T. National Energy Transition Framework toward SDG7 with Legal Reforms and Policy Bundles: The Case of Taiwan and Its Comparison with Japan. *Energies* **2020**, *13*, 1387. [CrossRef]
2. Lee, Y.; Kim, B.; Hwang, A.H. Which Institutional Conditions Lead to a Successful Local Energy Transition? Applying Fuzzy-Set Qualitative Comparative Analysis to Solar PV Cases in South Korea. *Energies* **2020**, *13*, 3696. [CrossRef]
3. Konečný, V.; Gnap, J.; Settey, T.; Petro, F.; Skrúcaný, T.; Figlus, T. Environmental Sustainability of the Vehicle Fleet Change in Public City Transport of Selected City in Central Europe. *Energies* **2020**, *13*, 3869. [CrossRef]
4. Tucki, K.; Orynycz, O.; Mitoraj-Wojtanek, M. Perspectives for Mitigation of CO_2 Emission due to Development of Electromobility in Several Countries. *Energies* **2020**, *13*, 4127. [CrossRef]
5. Mehmood, S.; Maximov, S.A.; Chalmers, H.; Friedrich, D. Energetic, Economic and Environmental (3E) Assessment and Design of Solar-Powered HVAC Systems in Pakistan. *Energies* **2020**, *13*, 4333. [CrossRef]
6. Mehmood, S.; Maximov, S.A.; Chalmers, H.; Friedrich, D. Low-Carbon Energy Governance: Scenarios to Accelerate the Change in the Energy Matrix in Ecuador. *Energies* **2020**, *13*, 4731.
7. de Maere d'Aertrycke, G.; Smeers, Y.; de Peufeilhoux, H.; Lucille, P.L. The Role of Electrification in the Decarbonization of Central-Western Europe. *Energies* **2020**, *13*, 4919. [CrossRef]
8. Choi, H.; Park, H. "Oil is the New Data": Energy Technology Innovation in Digital Oil Fields. *Energies* **2020**, *13*, 5547. [CrossRef]
9. Bugajski, P.; Nowobilska-Majewska, E.; Majewski, M. The Impact of Atmospheric Precipitation on Wastewater Volume Flowing into the Wastewater Treatment Plant in Nowy Targ (Poland) in Terms of Treatment Costs. *Energies* **2021**, *14*, 3806. [CrossRef]
10. Nematchoua, M.; Orosa, J.; Ricciardi, P.; Obonyo, E.; Sambatra, E.; Reiter, S. Transition to Zero Energy and Low Carbon Emission in Residential Buildings Located in Tropical and Temperate Climates. *Energies* **2021**, *14*, 4253. [CrossRef]
11. Pearce, P. Duty to Address Climate Change Litigation Risks for Australian Energy Companies—Policy and Governance Issues. *Energies* **2021**, *14*, 7838. [CrossRef]
12. Norouzi, O.; Dutta, A. Solar Energy Demand-to-Supply Management by the On-Demand Cumulative-Control Method: Case of a Childcare Facility in Tokyo. *Energies* **2022**, *15*, 475. [CrossRef]
13. Hiromasa, I.; Satoshi, Y.; Tetsuo, Y.; Masato, T.; Masayuki, M. The Current Status and Future Potential of Biogas Production from Canada's Organic Fraction Municipal Solid Waste. *Energies* **2022**, *15*, 4608.

energies

Article

Solar Energy Demand-to-Supply Management by the On-Demand Cumulative-Control Method: Case of a Childcare Facility in Tokyo

Hiromasa Ijuin [1], Satoshi Yamada [1], Tetsuo Yamada [1,*], Masato Takanokura [2] and Masayuki Matsui [2]

1 Department of Informatics, The University of Electro-Communications, 1-5-1 Chofugaoka, Chofu-shi 182-8585, Japan; hju1n@yahoo.co.jp (H.I.); satoshi.yamada@uec.ac.jp (S.Y.)
2 Department of Industrial Engineering and Management, Engineering Research Institute, Kanagawa University, 3-27-1 Rokkakubashi, Kanagawa-ku, Yokohama-shi 221-8686, Japan; takanokura@kanagawa-u.ac.jp (M.T.); matsui@kanagawa-u.ac.jp (M.M.)
* Correspondence: tyamada@uec.ac.jp; Tel.: +81-(42)-443-5269

Abstract: In recent years, environmental and energy issues relating to global warming have become more serious, and there is a need to shift from conventional power generation, which emits an abundance of carbon dioxide, to renewable energy sources without emissions, such as solar and wind. However, solar power generation, which is one of the renewable energies, changes dynamically, depending on real time weather conditions. Thus, power supplied mainly by solar power generation is often unstable, and an appropriate on-demand energy management for demand-to-supply is required to ensure a stable power supply. Demand-to-supply management methods include inventory management analysis and on-demand inventory management analysis. The cumulative-control method has been used as one of the production management methods to visually manage inventory status in factories and warehouses, while the on-demand cumulative-control method is an extension of inventory management analysis. This study models a demand-to-supply management method for a solar power generation system by using the on-demand cumulative-control method in an actual case. First, a demand-to-supply management method is modeled by an on-demand cumulative-control method, using actual power data from a childcare facility in Tokyo. Next, the on-demand cumulative-control method is adopted to the case without batteries, and the amount of electricity to be purchased is estimated. Finally, the effectiveness of the maximum battery capacity and the amount of the initial charge are examined and discussed by sensitivity analysis.

Keywords: renewable energy; electricity storage management; power demand and supply forecast; initial charge; dynamic inventory control

Citation: Ijuin, H.; Yamada, S.; Yamada, T.; Takanokura, M.; Matsui, M. Solar Energy Demand-to-Supply Management by the On-Demand Cumulative-Control Method: Case of a Childcare Facility in Tokyo. *Energies* **2022**, *15*, 4608. https://doi.org/10.3390/en15134608

Academic Editor: Peter D. Lund

Received: 20 April 2022
Accepted: 18 June 2022
Published: 23 June 2022

1. Introduction

In recent years, global warming has become increasingly serious. The 26th session of the Conference of the Parties to the United Nations Framework Convention on Climate Change (COP26) was held in Glasgow, where more than 130 leaders showed an effort to counter global climate change (UNFCCC) [1]. U.S. President Joe Biden set a 50–52% reduction target for GHG emissions by 2030 at the climate summit [2]. According to BBC news [2], Japan has the fifth largest greenhouse gas (GHG) emissions in the world. The Japanese Prime Minister, Yoshihide Suga, has established a plan for a carbon-free society, aiming to reduce greenhouse gas emissions to zero by 2050 [3]. Encouraging the decarbonization of the energy industry is essential to achieving this aim, as this industry contributes a substantial fraction of global carbon dioxide emissions, a major greenhouse gas. Thus, the Ministry of the Environment (2020) [4] in Japan has also demanded more effective measures for the entire energy sector. The introduction of carbon-free renewable power generation (e.g., solar and wind) is also being actively implemented [3,5,6].

Solar power is dependent on sunlight; therefore, it has been stated that, provided the sun is present, solar power resource depletion should not be of concern. While many manufacturing companies and service organizations are beginning to use solar power for factory and facility operations, achieving stable solar power generation has been problematic. This is because solar power is weather dependent, making it difficult to balance supply and demand [7]. Thus, batteries and power purchasing from electric power companies are often used to meet the gap between energy demand and supply. Figure 1 shows a renewable energy system based on the relationship among supply, demand, and inventory. In supply, private power generation from solar power and power purchased from electrical companies are inflow to a battery. The outflows from the battery are connected to a demand facility as the power consumption in the facility. There are energy transition issues for solar power demand-to-supply as follows: When should we switch power purchasing to private power generation or private power generation to power purchasing? Additionally, how much energy should we store in batteries at each moment?

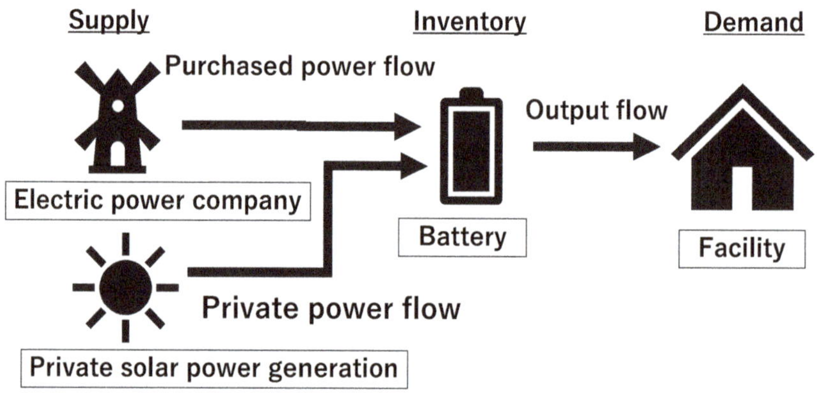

Figure 1. A renewable energy system based on the relationship among supply, demand, and inventory.

Regarding renewable energy for hardware and mechanism improvement, Agyekum et al. (2021) experimentally investigated actual solar power generation [8–10] and emulated the wave energy [11]. Agyekum et al. (2021) [8] aimed to be a lower temperature of the solar photovoltaics (PV) through double-sided water cooling to increase power generation efficiency. However, they did not focus on the power consumption. Agyekum et al. (2021) [9] investigated the effect of a combination of active and passive cooling mechanisms for the solar PV module. They combined aluminum fins, water, and ultrasonic humidifier to cool solar PV panels under real weather conditions. However, they did not mention the balance of solar power generation between supply and demand. Agyekum et al. (2021) [10] conducted an experiment on the performance enhancement of a solar PV from exergy, energy, and economy. They improved the performance by the cooling of a PV panel though they did not mention the balance of solar power generation between supply and demand. Agyekum et al. (2021) [11] proposed a test bench device to emulate a wave energy converter. They developed a prototype of a wave energy converter. Nevertheless, they did not compare the actual data to the simulation data. Furthermore, PraveenKumar et al. (2021) [12] conducted an experiment effectively for the passive cooling mechanism on the temperature change inside the solar cell and power generation performance. They prevented the solar cell's temperature from heating up, and they improved its performance by using this cooling mechanism. However, they did not consider the opportunity loss of energy storage due to full charging.

From the viewpoint of production research, this uncertain balance requires highly accurate demand forecasting and robust supply planning; this is also referred to as the demand-to-supply management method (DSMM) [13,14]. They used a cumulative control method [15], which has long been used in process management and production planning for the intuitive quantification of the subsequent production amount. That production amount was based on the expression of the cumulative amount and the on-demand cumulative-control method (OCCM) [14]; this is the development method of DSMM.

Table 1 shows a literature review on renewable energy. In the research on energy demand-to-supply management, Wichmann et al. (2019) [16] approached the energy-oriented general lot-sizing and scheduling problems by combining them with decisions on utilizing energy storage for a single machine shop. They modeled production planning to minimize total costs related to production and energy, accommodating for energy trading. Moon and Park (2013) [17] attempted to minimize the total production cost with energy management by the flexible job-shop scheduling problem. However, their models did not include renewable energy management such as wind and solar. Uhlemair et al. (2014) [18] investigated capacity planning for a biogas plant and the course of the district heating network in Germany bioenergy villages. They generated a linear model to economically optimize the production and distribution systems of bioenergy villages. The sole power supply to their model was the biogas plant; as such, they did not consider procurement from other power generation companies, such as lignite power generation or natural gas.

Power generation from renewable energies such as solar has multiple issues. One of them is supply uncertainty because renewable energy generation is greatly affected by the weather. In order to overcome the supply uncertainty, there are three main approaches as follows: introducing a battery to store electricity; forecasting supply and demand; and using a stochastic model.

In terms of renewable energy with batteries, Trappey et al. (2013) [19] developed a hierarchical cost learning model for wind energy. Although the model had an improved fit between a hierarchical model and actual data, power supply and demand management were not considered. Pham et al. (2019) [20] developed a multi-site production and micro-grid planning model for net-zero energy operations. This approach was a two-stage optimization programming, such as the scheduling of production for electricity demand. In addition, the sizing and sitting of the microgrid systems were optimized. However, the study did not forecast supply and demand related to renewable power generation and consumption using an actual facility.

Regarding forecasting, Jahanpour et al. (2016) [21] proposed a collaborative platform for communities in the energy distribution network. This collaborative platform addressed the stochastic nature of electricity demand, the dynamics in power generation over time, and uncertainty through collaboration between energy providers to build a sustainable energy distribution network. However, their management model did not include energy supply from a battery. Rentizelas et al. (2012) [22] formulated the national electricity generation system in Greece. They analyzed and applied an optimization method to determine the optimal generating mix that minimized generation costs, while operating within system constraints and incorporating the uncertainty of emission allowance prices. However, their model operated using only annual energy supply and demand management, and they did not address daily supply and demand.

With respect to a stochastic model, Xydis (2013) [23] analyzed wind resources on Kythira Island, proposing an evaluation methodology and an investment tool. They estimated the wind resources and costs using the probability density function and cumulative distribution for variation in wind velocity. Furthermore, they proposed planning wind farms on Kythira Island. However, their model did not consider energy demand. Takanokura et al. (2014) [24] proposed energy management using a newsboy problem via storage systems for smart cities; however, their study did not account for demand-to-supply management using actual power generation. Santana-Vieraa et al. (2015) [25] proposed the implementation of a demand response program in large manufacturing facilities featuring distributed wind and solar energy using a stochastic programming model. The model allowed the manufacturer to meet curtailment requirements without causing major electricity shortages that could adversely affect the normal production schedule. However, they did not deal with the change of supply and demand for each time zone.

Therefore, the proposed model in this study only consider the above three main methods: battery, forecasting, and a stochastic model. Thus, the originality and the novelty of this paper are finding that solar energy management issues have a similar structure with production management in overcoming uncertainty, and that the on-demand DSMM and OCCM methods can be applied to a solar power generation system. Furthermore, we demonstrated that both methods were applicable to the case of solar power generation system. One of the advantages of this method was that it enabled us to predict and to decide the amount of electricity to be purchased in the next period by changing storage capacity in real time in accordance with the amount of electricity consumed by using DSMM.

This study applies DSMM [13,26] to a solar power generation system associated with a renewable energy consulting company in Japan, Ecolomy Co., Ltd. (Tokyo, Japan) [27]. The total amount of generated renewable energy and purchased power is set as the input flow, the total power consumption is considered the output flow, and the remaining battery storage capacity is the inventory amount. These values are utilized as a case study for the demand-to-supply management of renewable power generation using data from private power generation and the power consumption grid in an actual childcare facility in Tokyo, Japan. This paper demonstrates that power demands could be met using a storage battery if solar power was unable to generate power. After that, the battery capacity and the storage required for the optimal operation of the solar system could be determined using on-demand and demand-to-supply management methods as it applied to actual childcare facility cases. Thus, the following research questions (RQs) were posed:

- RQ1 What are the managerial issues including capital investment in a transition to solar power generation in the targeted renewable energy company?
- RQ2 How was the demand-to-supply balance between the amount of electricity consumption and the amount of on-site solar power generation at the target facility?
- RQ3 How to operate a battery more effectively by applying a demand-to-supply management developed in production and logistics? Additionally, how was the moving base storage capacity for on-demand DSMM?
- RQ4 What was the effect of the day of the week, time of day, and weather on the demand-to-supply management in solar-power generation at the target facility?
- RQ5 How effective are the capacity of the battery and the initial storage in the battery? Moreover, what is the energy storage opportunity loss at that time?

The remainder of this manuscript has been structured into four sections. Section 2 describes the model and the formulation of DSMM using the Cumulative-Control Method and OCCM. Section 3 presents the results of DSMM using the Cumulative-Control Method and OCCM, respectively. Section 4 conducts a sensitivity analysis for power storage capacity and initial storage to know the effects of designed storage capacities, obtains feedback from a partner renewable energy company, and discusses the results of the RQs. Finally, Section 5 concludes this study and develops future works.

Table 1. Literature review on renewable energy.

	Type of Renewable Energy				Demand-to-Supply		Overcoming Uncertainties				
	Solar	**Wind**	**Bio-Gas**	**Biomass**	**Supply**	**Demand**	**Battery Capacity**	**Stochastic Model**	**Forecast-ing**	**Real Data**	**Methods**
Wichmann et al. (2019) [16]						√					Energy-oriented lot-sizing and scheduling
Moon and Park (2013) [17]					√	√	√				Mixed integer programming
Trappey et al. (2013) [19]		√					√		√	√	Hierarchical learning model
Xydis (2013) [23]		√			√			√		√	Weibull and Rayleigh probability density function
Jahanpour et al. (2016) [21]		√			√	√		√	√	√	Trigonometric regression model
Santana-Vieraa et al. (2015) [25]	√	√			√	√		√		√	Stochastic programming
Takanokura et al. (2014) [24]	√				√	√	√	√	√		Newsboy problem
Pham et al. (2019) [20]	√	√			√	√	√	√			Stochastic planning model
Uhlemair et al. (2014) [18]			√		√	√				√	Mixed integer linear programming
Rentizelas et al. (2012) [22]	√	√		√	√	√		√	√	√	Forward-sweeping linear programming
This study	√				√	√	√	√	√	√	On-demand cumulative-control method

2. Methods

In order to apply the DSMM to renewable energy, Section 2 describes the model and the formulation of DSMM, using the cumulative-control method and OCCM. Section 2.1 shows the procedure performed for the demand-to-supply management analysis of renewable power generation. Section 2.2 models a DSMM for renewable energy. The cumulative-control method is explained in Section 2.3. In Section 2.4, the on-demand DSMM model is applied to solar power generation.

2.1. Procedure

Figure 2 shows the procedure performed for the demand-to-supply management analysis of renewable power generation. The notations used in this study are shown in Table A1 in Appendix A.

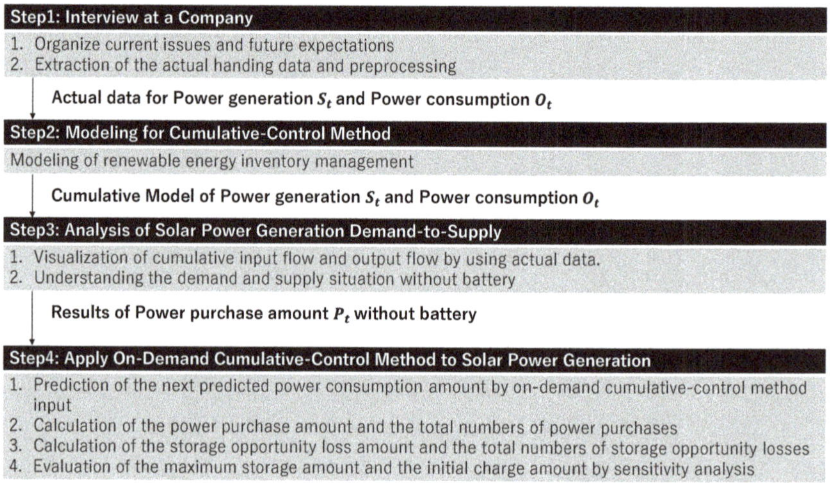

Figure 2. The procedure for the demand-to-supply management analysis of renewable power generation.

In Step 1, interviews are conducted with Ecolomy Co., Ltd., which operates a renewable energy-related business, to identify current issues and future expectations for the analysis when the actual data is examined. In Step 2, the DSMM model is applied to actual private power generation and consumption data.

In Step 3, the cumulative inflow and outflow are visualized using actual data and confirmed for the demand-to-supply situation in a case without a storage battery. In Step 4, the on-demand DSMM [26] is applied to the model created in Step 2. Specifically, the next forecasted power consumption amount X_{t+1} is determined, and the moving base storage capacity N_t is obtained. Moreover, the next power purchase amount P_{t+1} after the parameter $\overline{\beta}_{t+1}$ is updated using Matsui logic [13]. The inventory distribution $F(L_t)$ used in the N_t calculation is derived from actual storage data. Furthermore, the next power purchase amount P_{t+1} and the total numbers of power purchases TP_{count} are determined. The storage opportunity loss amount L_{loss_t} and the total numbers of storage opportunity losses TL_{count} are then calculated to evaluate the initial storage capacity L_{init} and the maximum storage capacity L_{max} of the storage battery by sensitivity analysis.

2.2. DSMM Model

Figure 3 shows a DSMM for renewable energy based on the relationship among supply, demand, and inventory. In this study, the DSMM is applied using the remaining level of the storage battery as the inventory; the time until the stored electrical power is released is regarded as the lead time; the amount of private power generation and power purchased from electrical companies is the inflow, so that the sum of this electric power over time period t is set as the inflow amount I_t. The outflow amount O_t is used as the power consumption of the facility, the maximum storage capacity of the battery is L_{max}, which means the warehouse capacity in the production system, and the initial charge amount is L_{init}, which corresponds to the initial inventory amount in the production system.

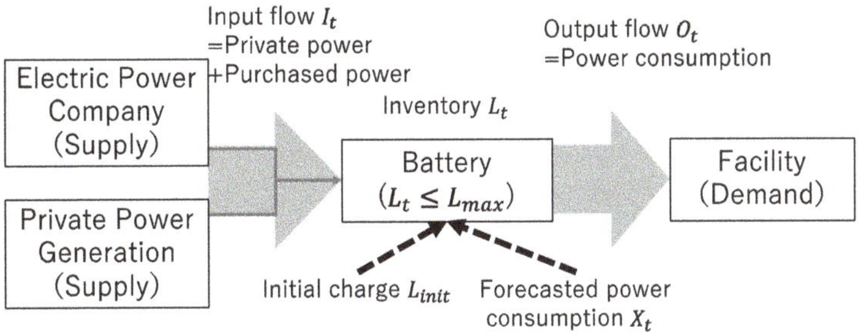

Figure 3. Demand-to-supply management model for renewable energy.

2.3. Cumulative-Control Method

The relationship among supply, demand, and inventory is modeled to apply the DSMM to the amount of private energy generation and consumption obtained from the actual facility. The DSMM is often used in process and progress management for the quantitative analysis of work and inventory [13]. The supply chain refers to the products in progress or inventory of the process.

In the DSMM, cumulative input and output are plotted on a two-dimensional graph, with the period and the quantity set as the horizontal and vertical axes, respectively. The difference values in the vertical represents the numbers of flows, and that in the horizontal means the lead time.

2.4. Application of On-Demand DSMM to Solar Power Generation

The cumulative inflow and outflow are obtained from the recorded data to grasp the demand-to-supply situation for solar power generation without storage batteries. Cumulative inflow refers to the total amount of private power generation corresponding to the amount of electrical power from solar power generation in the facility at period t, S_t. Cumulative outflow refers to the total amount of power consumption corresponding to the amount of power consumption in the facility at period t, O_t.

Further, the cumulative power purchase amount is determined to understand the demand-to-supply situation when no storage batteries are present. The power purchase amount at period t is represented as $P_t = \max(0, O_t - S_t)$. This means the amount of electricity purchased from electric power companies when the private power generation is insufficient.

Unlike the conventional DSMM, the on-demand DSMM [26] changes the base supply in response to the reconfiguration of demand structure. Thus, it enables on-demand changes in the next input amount. The procedure for applying the on-demand DSMM is shown in Figure 4.

In Step (a), the following parameters were set to their initial values: coefficient α $(0 \leq \alpha \leq 1)$ used in the exponential smoothing method, the initial value $\overline{\beta}_0$ $(0 \leq \overline{\beta}_0 \leq 1)$ of the input amount determination parameter updated by Matsui logic, the cumulative probability distribution $F(L_t)$ for the frequency of the remaining storage capacity, the initial storage capacity L_{init} $(0 \leq L_{init} \leq L_{max})$, and the maximum storage capacity L_{max} $(0 \leq L_{max})$.

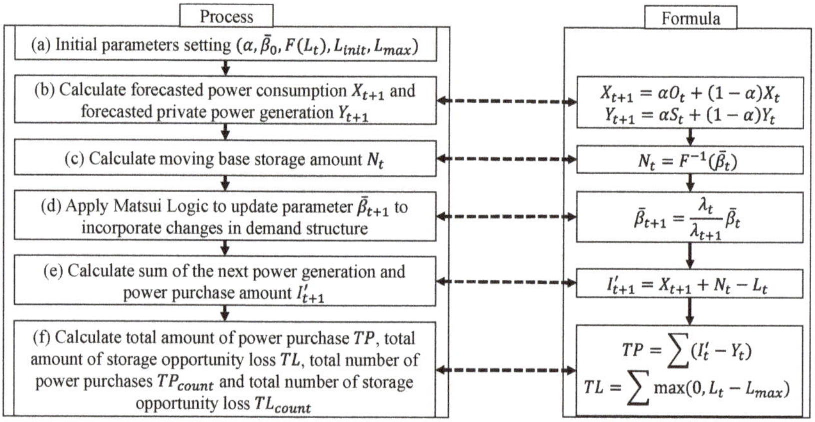

Figure 4. Procedure for applying the on-demand DSMM to renewable power generation.

In Step (b), the amounts of forecasted power consumption X_{t+1} and forecasted private power generation Y_{t+1}. were calculated using the exponential smoothing method [28] using Equations (1) and (2). The storage opportunity loss amount L_{loss_t} was determined using the forecasted power consumption amount, the next input amount I'_t, and the forecasted private power generation.

$$X_{t+1} = \alpha O_t + (1-\alpha)X_t \tag{1}$$

$$Y_{t+1} = \alpha S_t + (1-\alpha)Y_t \tag{2}$$

In Step (c), the moving base storage capacity N_t was calculated. Insufficient inventory could lead to out-of-stock losses, whereas excess inventory could lead to capacity pressure and increased storage costs, so an appropriately sized inventory is desirable. To that end, Matsui et al. (2005) [26] used the newsvendor problem to determine the base supply chain that minimized costs due to excess inventory or insufficient stock. In this study, the newsvendor problem was used to calculate N_t.

The calculation of N_t required solving the newsvendor problem of cost minimization, expressed by the total cost function $C(N_t)$ in Equation (3):

$$C(N_t) = \beta_1 N_t + \beta_2 (N_t - L_t)^+ + \beta_3 (L_t - N_t)^+ \tag{3}$$

Here, L_t is the remaining storage capacity in period t, β_1 is the cost coefficient of maintaining storage, β_2 is the cost coefficient of insufficient storage, β_3 is the cost coefficient of excess storage [26], and $(a)^+ = max(a, 0)$. At this time, the solution N_t is equal to L_t, which satisfies Equation (4) using the cumulative probability distribution $F(L_t)$ of the frequency of stored electricity.

$$F(L_t) = \overline{\beta}_t \tag{4}$$

where $\overline{\beta}_t$ is determined by Equation (5), which consists only of the cost coefficients β_1, β_2, and β_3.

$$\overline{\beta}_t = \frac{\beta_3 - \beta_1}{\beta_2 + \beta_3} \tag{5}$$

However, $\overline{\beta}_t$ cannot be updated in the calculation of Equation (5), and thus, cannot follow fluctuations in demand. In the on-demand DSMM, $\overline{\beta}_t$ is updated every period according to the demand fluctuations using Matsui logic [13] described in Step (d).

Figure 5 shows the calculation procedure for the cumulative probability distribution $F(L_t)$ of the remaining storage capacity L_t and the moving base inventory N_t.

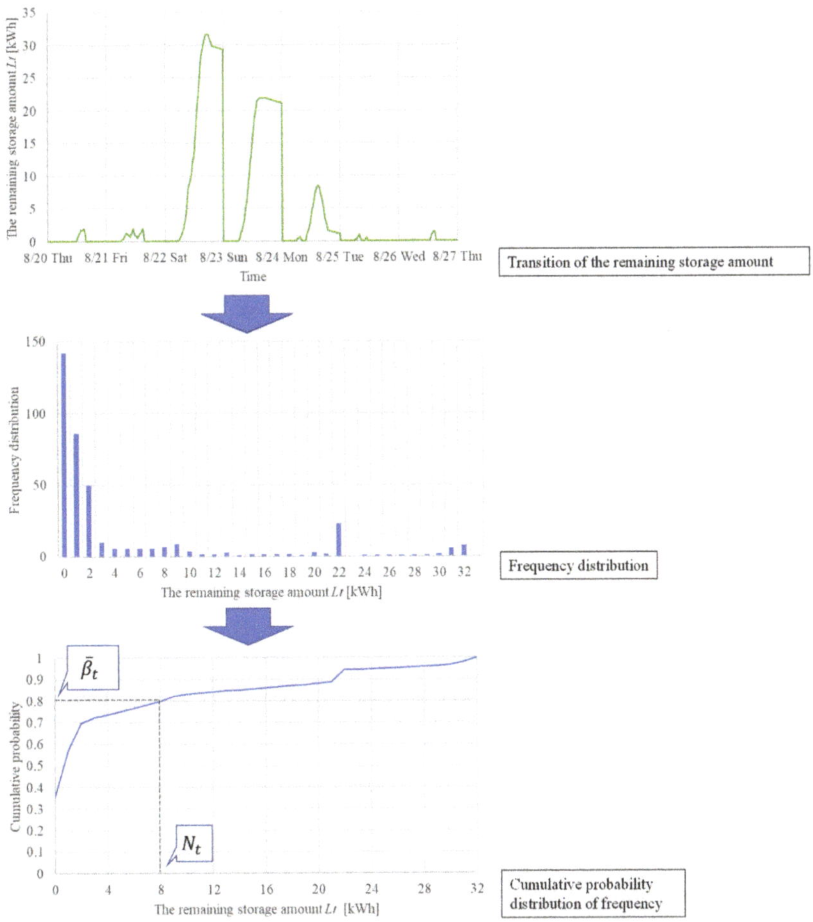

Figure 5. The calculation procedure for the cumulative probability distribution $F(L_t)$ with the remaining storage capacity L_t and the moving base inventory N_t.

The data period was limited to create the frequency distribution of L_t to obtain the cumulative probability distribution $F(L_t)$. Subsequently, the cumulative probability was derived from $F(L_t)$ and converted into a cumulative probability distribution. The cumulative probability distribution $F(L_t)$ of the remaining storage capacity L_t was determined, and the intersection between $F(L_t)$ and $\overline{\beta_t}$ was used to obtain the moving base inventory N_t.

In this study, six types of $F(L_t)$ were created: five from the electricity storage data from 5:00–19:00 on weekdays (20 August, 21 August, 24 August, 25 August, and 26 August) and one created from all electricity storage data from the period of 5:00–19:00.

In Step (d), $\overline{\beta}_t$ was updated using Matsui logic. Matsui et al. (2005) [26] updated parameters to incorporate changes in demand structure, adjust the base inventory to the momentary demand, and determine the subsequent input amount to maintain the corresponding base inventory. It was assumed that the parameter $\overline{\beta}_t$ over period t expressed in Equation (5) canceled the changes with the demand ratio λ_t shown in Equation (6):

$$\lambda_{t+1}\,\overline{\beta}_{t+1} = \lambda_t\,\overline{\beta}_t = \cdots = \lambda_0\,\overline{\beta}_0 \tag{6}$$

Equation (7) is modified from Equation (6) to obtain the next-period parameter $\overline{\beta}_{t+1}$, known as the Matsui logic parameter (Matsui et al., 2009) [13].

$$\overline{\beta}_{t+1} = \frac{\lambda_t}{\lambda_{t+1}} \overline{\beta}_t \tag{7}$$

From this, if the initial value of $\overline{\beta}_0$ is given, the cost for the next period will be updated continuously in response to fluctuations in demand, and the moving base storage capacity N_t can be calculated.

Here, the demand λ_t and forecasted demand λ_{t+1} were calculated using Equations (8) and (9), respectively.

$$\lambda_t = O_t \tag{8}$$

$$\lambda_{t+1} = X_{t+1} \tag{9}$$

For example, when $O_t = 10$, $X_{t+1} = 15$ and $\overline{\beta}_t = 0.3$, the parameters are updated, as expressed in Equation (10).

$$\overline{\beta}_{t+1} = \frac{\lambda_t}{\lambda_{t+1}} \overline{\beta}_t = \frac{O_t}{X_{t+1}} \overline{\beta_t} = \frac{10}{15} \times 0.3 = 0.2 \tag{10}$$

Furthermore, $\overline{\beta}_{t+1}$ must be $0 < \overline{\beta}_{t+1} \leq 1$ to calculate the base supply chain. Furthermore, the updates become more gradual as $\overline{\beta}_{t+1}$ approaches zero, and $\overline{\beta}_{t+1}$ does not move from around zero. Therefore, 0.1 was set to a lower limit of $\overline{\beta}_{t+1}$. Therefore, the parameter $\overline{\beta}_{t+1}$ was modified every time it was updated so that $0.1 < \overline{\beta}_{t+1} \leq 1$.

Finally, in Step (e), the next input amount I'_t to be input into the next period to bring the supply chain closer to the base supply chain, was calculated. This was the final output in the on-demand DSMM and the supply–chain management equation (Matsui et al., 2005 [26]; Equation (11)) was used.

$$I'_{t+1} = X_{t+1} + N_t - L_t \tag{11}$$

This next input amount I'_t must be controllable. Hence, in this study, the next power purchase amount P_{t+1} that was controllable in the inflow amount I_t was set as the next input amount I'_t. In Step (f), the next power purchase amount P_{t+1} was determined by Equation (12). A lower total sum $\sum P_{t+1}$ of P_{t+1} or total number of power purchases TP_{count} was thought to result in more favorable supply–demand management.

$$P_{t+1} = I'_{t+1} - Y_{t+1} \tag{12}$$

Furthermore, the storage opportunity loss amount L_{loss_t} when the opportunity to store electricity was lost due to a full charge was considered and determined by Equation (13). A lower total sum of storage opportunity loss $\sum L_{loss_t}$ or total number of storage opportunity losses TL_{count} was thought to result in more favorable supply–demand management.

$$L_{loss_t} = \max(0,\ L_t - L_{max}) \tag{13}$$

3. Results

Based on the procedure and model in Section 2, Section 3 demonstrates how to apply the proposed demand-to-supply management for renewable power generation to a case study. First, Section 3.1 organizes the company hearing from a company developing a renewable energy-related business named Ecolomy Co., Ltd. Next, the assumption of inflow and outflow for the power is explained in Section 3.2. Then, Section 3.3 analyzes demand-to-supply of solar power generation in a child care facility. Finally, the on-demand cumulative-control method is applied to solar power generation in Scenario 3.4.

3.1. Result of Step 1: Company Hearing

A hearing was conducted with Ecolomy Co., Ltd., [27], which is developing a renewable energy-related business, to discuss their current and future issues. The company was asked to provide the power consumption data of the facility that was operating both the solar and private power generation. After pre-processing, analyses using the DSMM were performed.

From the hearings with Ecolomy Co., Ltd., we were able to extract the following current issues: introduction costs are unrealistic if the storage battery has a large capacity; large variations are observed if one household attempts to make demand forecasts, leading to difficulty in achieving accurate predictions; attempts to accommodate electricity through a widespread power grid incur high transport charges so remote accommodation of electricity would be costly; and securing space for large-scale renewable energy power generation is difficult in Tokyo, a major electricity consumer.

We extracted opinions through the hearings, with the following particular exceptions: the proposal of a model that would require economical battery storage capacity, applications that would lead to a network of local production for intra-community consumption (i.e., microgrids), and the construction of a model that enabled stable power generation with other forms of renewable energy, such as wind and biomass.

In this study, we focused on a proposed model that could obtain an economically viable battery storage capacity and pursued this focus through the on-demand DSMM. Ecolomy Co., Ltd. provided private power generation and consumption data from actual solar power generation at a given facility. Three meetings with the company were held in October, November, and December 2020 to further understand the data.

Table 2 shows the details of the facility that provided the private power generation and consumption data. The facility was a children's daycare in Tokyo. The data during business hours for one week were obtained, with the target facility having a renewable power generator output of 10 [kW] in Tokyo.

Table 2. Source data on private power generation and consumption amount and their corresponding details.

Facility Attributes	Childcare Facility
Business hours	7:00–19:00 on weekdays and Saturdays
Data period	20 August–26 August in a given year
Panel power generation output	10 kW

Furthermore, power consumption from the facility was recorded as wattmeter log data at 15-min intervals, whereas the private power generation data were recorded as the total power generation amount per day. Hence, the time intervals needed to be arranged for analysis. First, the daily total of incoming solar radiation in Tokyo was obtained from Japan Meteorological Agency (2020) [29], and the total private power generation for one day was partitioned hourly with the corresponding fraction total solar radiation. Further, the hourly private power generation was divided by 1/4 to convert the data to 15-min intervals.

3.2. Result of Step 2: Modeling for Cumulative-Control Method

The variables used in the DSMM of renewable energy were implemented in this study. Private power generation, one of the components for the inflow amount I_t, which was pre-processed from daily private power generation data to 15-min increment data. The power consumption amount, which corresponds to the outflow amount O_t, required no further adjustment as the facility data provided by Ecolomy Co., Ltd. were wattmeter log data recorded every 15 min.

3.3. Result of Step 3: Analysis of Solar Power Generation Demand-to-Supply

Figure 6 shows a cumulative graph of observed power consumption, private power generation, and estimated power purchase amount for a given week in summer. It is assumed that there are no storage batteries. In this case, the cumulative power consumption amounts for a target period were 167.01 [kWh], and the cumulative private power generation was 153.24 [kWh], with a relatively slight difference of 13.77 [kWh].

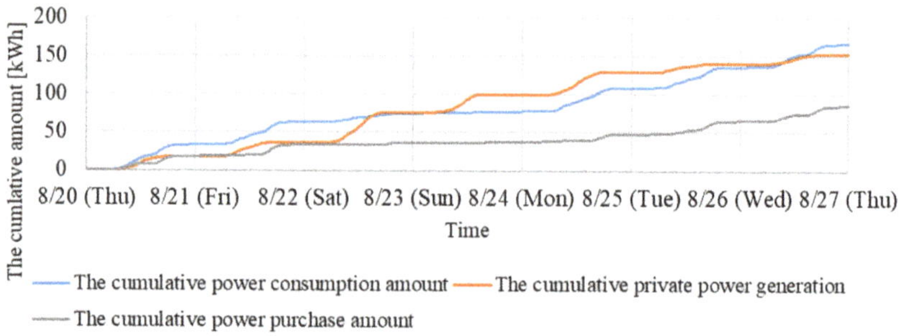

Figure 6. Cumulative graph displaying the amounts of observed power consumption and private power generation, and estimated power purchases across the time period analyzed.

Furthermore, the cumulative power purchase amount in 1 week was estimated as 85.83 [kWh], which corresponds to 51.4% of the cumulative power consumption. The total numbers of power purchases TP_{count} was 482, and the purchase of power occurred at a rate of 71.7% of the total measurement period, 672 periods. These results indicated that 72.06 [kWh] of electric power, 47.0% of the electric power generated by private power generation, was discarded without being used. Therefore, it turns out that such cases enabled us to reduce this cumulative power purchase amount by introducing a storage battery and more suitably managing demand-to-supply incompatibilities.

Moreover, cloudy and rainy weather meant that actual data were lower than the measured values of incoming solar radiation strength and duration. Therefore, potentially a further reduction for the cumulative power purchase amount would be expected.

3.4. Result of Step 4: Apply on-Demand Cumulative-Control Method to Solar Power Generation

The input data, including measurements, and output data of the model are related with the resulted main figures in Figure A1 in the Appendix A. The right-hand side of each equation in that figure is the input data, including the measurements, while left-hand side means the output data of the model.

In Step (a) the initial settings, five types of the cumulative probability distribution $F(L_t)$ of the remaining storage amount frequency were created from 5:00 to 19:00 on weekdays 20 August, 21 August, 24 August, 25 August, and 26 August 2020 and one from the total electricity storage data from the target period. Figure 7 shows the results for each $F(L_t)$. It is found that the maximum value of the remaining storage amount occurred on 24 August in Figure 7, as one of the best optimal weather conditions (e.g., sunshine hours, total insolation) were experienced on this day, and it allowed electricity storage to be increased for all the time. Furthermore, the frequency was widely distributed, and the cumulative probability distribution gradually rose. Therefore, the subsequent input amount could be determined.

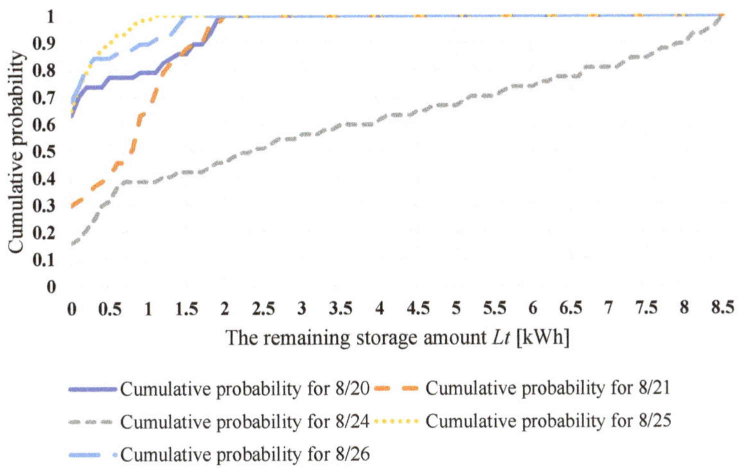

Figure 7. Cumulative probability distribution of the remaining storage amount from 20 August to 26 August.

In Step (b), forecasted power consumption X_{t+1} and forecasted private power generation Y_{t+1} on 20 August were calculated using the exponential smoothing method for each $\alpha = 0.2$, 0.5, and 0.8, and the results are shown in Figure 8. The results indicate that accuracy increases with α. A large difference between the forecasted and the measured values is observed at $\alpha = 0.2$; however, making α too large increases the weight of the prior measured periods and results in the predicted values lagging behind.

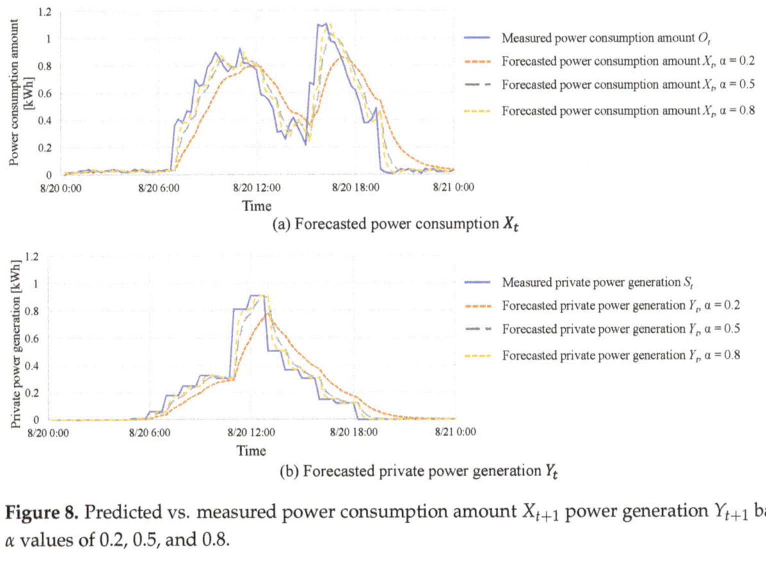

Figure 8. Predicted vs. measured power consumption amount X_{t+1} power generation Y_{t+1} based on α values of 0.2, 0.5, and 0.8.

Based on the above analysis, all subsequent calculations will use the forecasted power consumption amount X_{t+1} and private power generation Y_{t+1} calculated with an α value of 0.5.

In Step (c), the moving base storage amount N_t was obtained. Figure 9 shows the storage amount L_t ($L_{init} = 0$, $L_{max} = 5$) controlled by the moving base storage amount, obtained from the cumulative probability distribution $F(L_t)$ of the remaining storage amount L_t on 25 August, which had poor weather with clouds and occasional rain. This pattern was observed because $N_t = 0$ when an increase in storage amount was detected, making it possible to refrain from purchasing electric power, However, it was also found that management of N_t. was difficult on 24 August when private power generation was large due to favorable conditions and holidays (22 August and 23 August) when the power consumption was small.

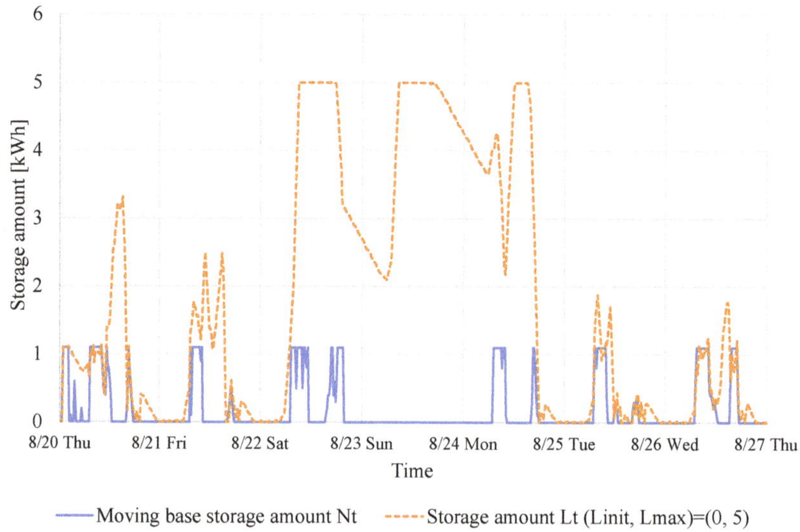

Figure 9. Moving base storage amount N_t and storage amount L_t ($L_{init} = 0$ kWh, $L_{max} = 5$ kWh) over the period analyzed.

Figure 10 shows the updated result of $\overline{\beta}_t$ due to Matsui logic when $\overline{\beta}_0 = 0$ (from step (d)). Further, $\overline{\beta}_t$ increased during the morning hours from 06:00 to 12:00 and in the hours around 16:00, with the exception of Sundays when the facility was closed. The possible reason is that these were the hours when the power consumption amount of the facility began to increase. Figure 11 shows shifts in the power consumption rates, which were often higher than the private power generation during these hours and are thus critical hours in terms of demand-to-supply management and the difference between private power generation and power consumption.

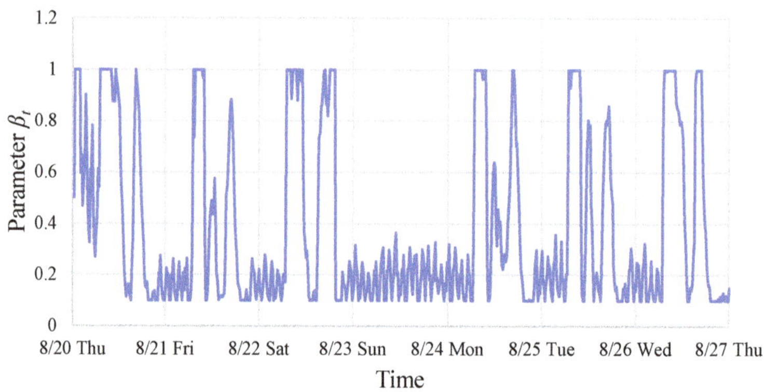

Figure 10. Updated result of $\overline{\beta}_t$ by Matsui logic ($\overline{\beta}_0 = 0.5$).

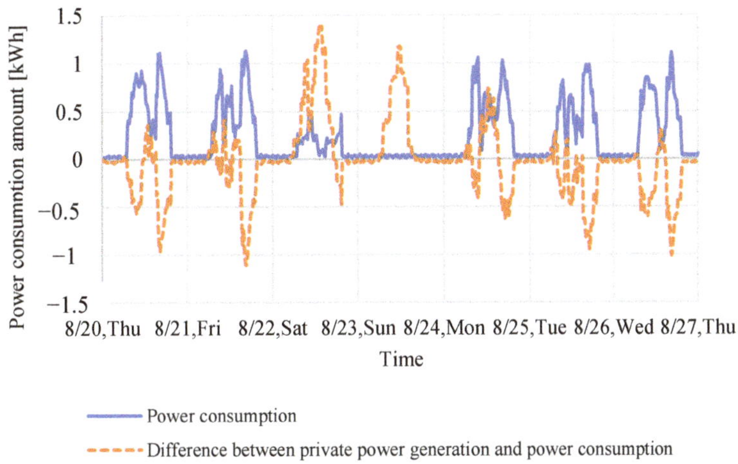

Figure 11. Power consumption and difference between private power generation and power consumption.

For all cumulative probability distributions $F(L_t)$, we conducted sensitivity analysis of the initial inventory amount L_{init} and the maximum storage amount L_{max} by evaluating the total power purchase amount TP, number of power purchases TP_{count}, storage opportunity loss amount TL, and the number of storage opportunity losses (TL_{count}, at Step (d)).

4. Impact of the Proposed Method

From Section 3, the solar energy at the facility surveyed was not used effectively because the timing of supply and demand was different. Therefore, it is desirable to introduce a storage battery to generate and store electricity. However, the larger capacity of the storage batteries brings a higher installation cost.

This section analyzes the impact of the proposed method in the designed system. In Section 4.1, sensitivity analysis is conducted for the initial inventory and maximum storage capacity, with the total power purchase amount, the total number of power purchase, and the amount of storage opportunity loss as evaluation targets. Section 4.2 receives feedback from Ecolomy Co., Ltd. about the results of Section 4 and evaluates the application of this study based on this feedback. Finally, Section 4.3 discusses answers of the RQs.

4.1. Maximum and Initial Storage Capacities by Sensitivity Analysis

A sensitivity analysis is carried out to assess the value of the maximum and initial storage capacity of the batteries.

Figure 12a shows the sensitivity analysis results of the total power purchase amount TP in the cumulative probability distribution $F(L_t)$ of the remaining storage amount L_t on 25 August. TP at this time was the smallest compared to the other cumulative probability distributions when the maximum storage amount $L_{max} = 45$ [kWh] and 50 [kWh] and the initial inventory amount $L_{init} = 30$ [kWh]; however, L_{init} needed to be purchased at $t = 0$. Furthermore, increases in L_{max} and L_{init} by 5 [kWh] each often resulted in a TP reduction by 5 [kWh] or less each time. Since TP can be difficult to be decreased, even by L_{max} or L_{init}, intentionally increasing these values should be avoided. If demand-to-supply management was performed with a $TP \leq 30$ [kWh], then combinations of $L_{max} = 40$ [kWh] and $L_{init} = 0$ [kWh], or $L_{max} = 20$ [kWh] and $L_{init} = 20$ [kWh] could be conceivable. Though the introduction costs of storage batteries are very high, it is demonstrated that a large initial inventory with a small battery was effective such as $L_{max} = 20$ [kWh] and $L_{init} = 20$ [kWh].

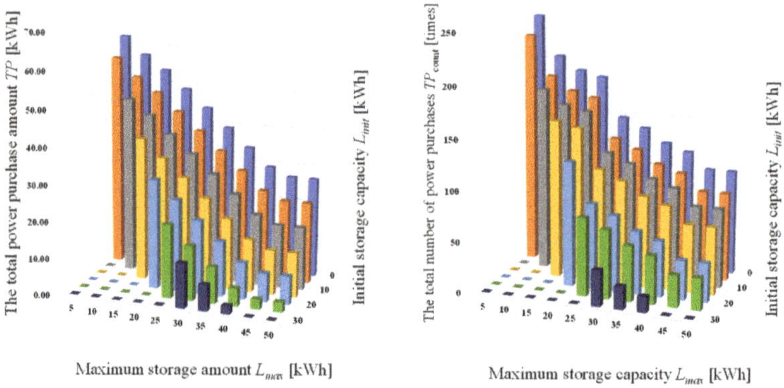

(a) Total amount of power purchase TP (b) Total number of power purchases TP_{count}

Figure 12. Sensitivity analysis for (**a**) total power purchase amount TP and (**b**) total number of power purchases count TP_{count}.

Figure 12b show the results of the sensitivity analysis of the total number of power purchases TP_{count}, with the cumulative probability distribution $F(L_t)$ of the frequency of the remaining storage amount L_t on 25 August.

These results indicate that the change in the total power purchase amount TP corresponded to the change in TP_{count}, but the TP did not always match the changes in TP_{count} when 5 [kWh] of change was observed. For example, changing from a maximum storage amount L_{max} from 25 to 30 [kWh], while maintaining initial inventory L_{init} at 15 [kWh], resulted in a decrease in TP_{count} by 9 [times]. Alternatively, changing L_{max} from 30 to 35 [kWh], while maintaining $L_{init} = 15$ [kWh], resulted in a decrease of TP_{count} by 13 [times]. The reason for this difference was likely because increasing L_{max} could bring more storage amount on Saturday and Sunday. Thus, decreasing the timing at which the storage decreased to and became zero and the power purchase occurred was slower when changing from 30 to 35 [kWh] than from 25 to 30 [kWh].

Furthermore, differences in the amount of change of L_{init} were observed depending on the difference in the periods of high power consumption from 20 to 21 August. Figure 13 shows the changes in the power purchase amount when $L_{max} = 10$ [kWh], and L_{init} values are 0, 5, or 10 [kWh]. The changes in the total power purchase capacity TP_{count} is also added. It was shown that unifying the amount of change in the power purchase amount was difficult as some factors, such as power consumption amount and private power generation, are difficult to control as constant.

Figure 14a shows the results of the sensitivity analysis for the total storage opportunity loss amount TL when a cumulative probability distribution $F(L_t)$ of the remaining storage amount L_t on 25 August was used. TL reached its minimum during this period, when compared to the other cumulative probability distributions.

There were no changes due to the initial inventory amount L_{init} when $L_{init} \leq 25$ kWh, and changes in the maximum storage amount L_{max} resulted in decreases of the storage opportunity loss amount by the same amount. However, TL increased at $L_{init} = 30$ [kWh] compared to the range of the initial inventory amount $L_{init} \leq 25$ [kWh] because battery consumption peaked at 27.9 [kWh] on 22 August when the weather was poor, as shown in Table A2 in Appendix A [29]. As a result, L_{init} was not entirely consumed by 22 August, and the battery was fully charged at a faster rate than in $L_{init} \leq 25$ [kWh], when the weather improved and consumption decreased. Further, $L_{init} \geq 30$ [kWh] was unfavorable when conducting demand-to-supply management from Thursday. It could also be observed that L_{init} could be decreased as the starting date of demand-to-supply management approached the day when power consumption decreased and private power generation was high.

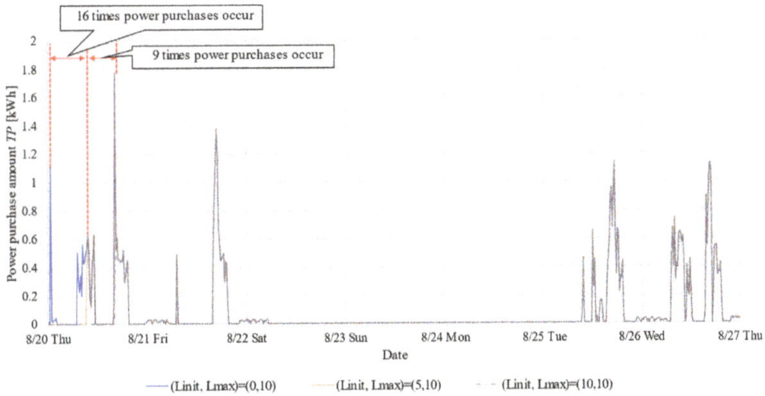

Figure 13. Changes in the power purchase amount at $L_{max} = 10$ [kWh] and $L_{init} = 0\,,5\,,10$ [kWh].

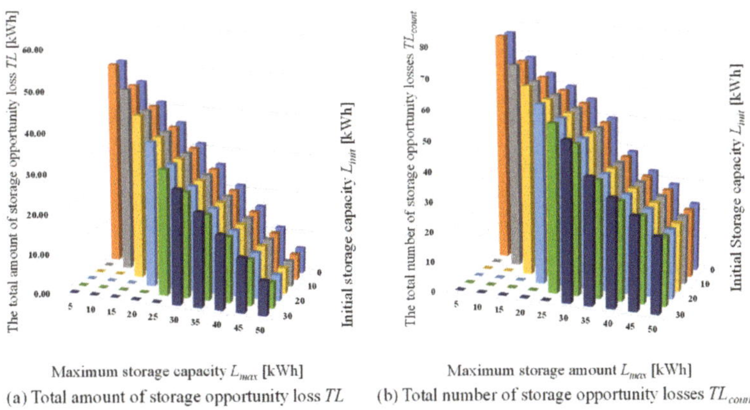

Figure 14. Sensitivity analysis for total amount of (**a**) storage opportunity TL and (**b**) total number of storage opportunity losses TL_{count}.

Figure 14b shows the results of the sensitivity analysis for the total number of storage opportunity losses TL_{count}. Similar to the total number of power purchases TP_{count}, the changes in the total storage opportunity loss amount TL generally corresponded with those in the total number of storage opportunity losses TL_{count}. However, decreases in TL were inconsistent for locations where L_{init} or L_{max} changes of 5 [kWh] were observed in the inventory amount. The reason for this is that the storage amount reached L_{max} on 22 August, when the private power generation greatly exceeded power consumption. Thus, a large amount of power was stored, and the timing at which the storage opportunity loss occurred varied according to each L_{max}.

Figure 15 shows the changes in power purchase amount when $L_{init} = 0$ [kWh] and the maximum storage amount $L_{max} = 0$, 5, or 10 [kWh]. The changes in TL_{count} are also added.

4.2. Feedback

We received feedback from the staff of Ecolomy Co., Ltd. regarding the analysis results relating to the application of the on-demand DSMM to solar power generation. The applicability of this study is investigated based on this feedback:

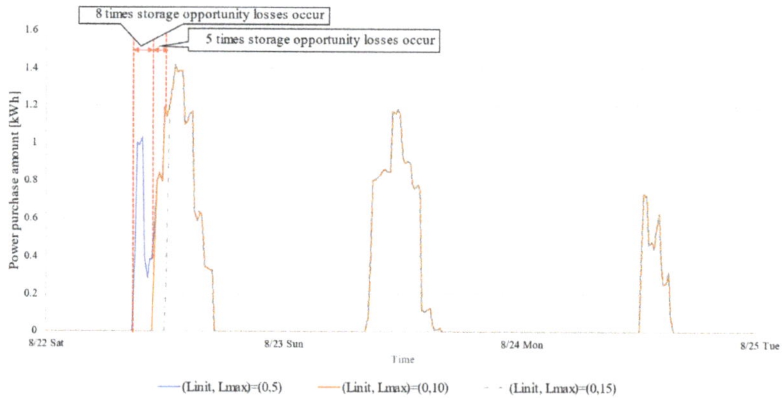

Figure 15. Changes in the power purchase amount when the initial inventory amount $L_{init} = 0$ [kWh] and the maximum storage capacity $L_{max} = 0,\ 5,\ 10$ [kWh].

- We received a positive opinion regarding the sensitivity analysis results of the total power purchase amount *TP*. They stated that "it is great that even a 20 [kWh] storage battery can achieve the same effect as a 40 [kWh] storage battery depending on how it is used." This opinion indicates that the modeling of demand-to-supply management of solar power generation using the on-demand demand-to-supply management method was useful;
- We also received a comment stating that "there is a sufficient amount of solar energy falling on the earth. However, we have heard that the technology for fully utilizing this is insufficient, and a storage battery with infinite capacity would be desirable if feasible." The importance of demand-to-supply management that seeks a realistic storage capacity or initial storage amount was also recognized once again.

4.3. Discussion

This section summarizes the findings of the results in Sections 3 and 4 along with RQs in Section 1.

RQ1. What are the managerial issues including capital investment in a transition to solar power generation in the targeted renewable energy company?

As per the answers to RQ1 based on the company interviews in step 1, the main managerial issue was the high initial introduction costs of a solar energy system with a larger capacity storage battery. The second issue was a too large variance to make proper predictions of electric power demand on a household basis. Therefore, it turned out that a methodology for dealing with the uncertainty of solar power generation under limited battery capacity, such as the DSMM and OCCM, is required.

RQ2. How was the demand-to-supply balance between the amount of electricity consumption and the amount of on-site solar power generation at the target facility?

As per the answer to RQ2 based on the results in steps 2 and 3, the difference in the cumulative amounts of power demand and private power supply during the week analyzed at the facility in Tokyo was only 8%. This fact means that appropriate demand-to-supply management could be employed to cover the private consumption with private power generation from solar power sources. However, 47% of the cumulative power consumption was discarded due to a dynamic gap between power demand and supply. As a result, it indicated that private power generation was not fully utilized, therefore, battery storage is helpful to satisfy the dynamic gap.

RQ3. How to operate a battery more effectively by applying a demand-to-supply management developed in production and logistics? Additionally, how was the moving base storage capacity for on-demand DSMM?

They are answered from the results in step 4. The morning from 06:00 to 12:00 and the early evening around 16:00 were the hours when the power consumption amount of the facility began to increase. Therefore, in order to operate the battery more effectively, the private power generation would be charged to the battery storage in the hours from 13:00 to 15:00, before the early evening.

Additionally, it is found that the moving base storage capacity N_t became 0 for saving the power purchase when an increase in storage amount was detected.

RQ4. What was the effect of the day of the week, time of day, and weather on the demand-to-supply management in solar-power generation at the target facility?

Regarding the day of the week: on weekdays, the power consumption amount was larger than the private power generation based on the results in Section 3.3. The average power purchase amount was 30 [kWh] and the average of private power generation was 17 [kWh] on weekdays. However, on Saturday and Sunday, the power consumption amount was lower than the private power generation. For the time of day, power consumption exceeded that of private power generation during the morning hours of 6:00–12:00 and the afternoon hours around 16:00 for all days except Sunday and holidays, when the facility was closed. With respect to the weather, the power generation efficiency actually depended on the weather in the week. The amount of solar power supply per day on cloudy and sunny days was twice and 4 times higher than that on rainy days, respectively.

RQ5. How effective are the capacity of the battery and the initial storage in the battery? Moreover, what is the energy storage opportunity loss at that time?

Since the introduction costs of storage batteries are very high, a large initial inventory with a small battery was effective such as L_{max} = 20 [kWh] and L_{init} = 20 [kWh] from the results in Section 4. At that time, the total storage opportunity loss amount TL become 46.36 [kWh].

The above discussions demonstrate that the proposed on-demand DSMM and OCCM could be useful and applied for demand-to-supply management of solar power generation and self-consumption.

5. Conclusions

This study applied the demand-to-supply management method (DSMM) and its development method of the on-demand DSMM to the solar energy at a facility.

The effective maximum and initial storage amount could be obtained simultaneously through the on-demand DSMM. Therefore, it demonstrates that the on-demand DSMM could be applied to demand-to-supply management of renewable power generation. Moreover, it was also shown that power consumption at times when renewable power was unable to generate could be covered by introducing a storage battery, and the battery capacity and storage required for smooth operation of the renewable system could be calculated using the on-demand DSMM. The main conclusions are as follows.

- The proposed method enables us to pre-process the private power generation data and align them to the same timescale as recorded for power consumption;
- The difference in the cumulative power consumption and the private power generation amounts during the course of the week analyzed was minimal. Results show that appropriate demand-to-supply management could be employed to cover private consumption with private power generation from PV sources;
- The cumulative amount of power purchased during this one week, however, corresponded to approximately half of the cumulative power consumption. It was also shown that the total number of power purchases comprised ~70%, indicating that private power generation was not fully utilized;

- Power consumption exceeded that of private power generation during the morning hours of 6:00–12:00, and the afternoon hours around 16:00 for all days except Sunday, when the facility was closed. It was also shown from shifts in the updating of $\overline{\beta}_t$ by Matsui logic that these were crucial hours for demand-to-supply management;
- It was shown that in the moving base storage amount obtained from the cumulative probability distribution of the remaining storage amount on a cloudy day, it was difficult to manage days where the daylight hours and total solar radiation were favorable or holidays where power consumption was minimal;
- A sensitivity analysis of the total power purchase amount showed that an effective combination of the initial and the maximum storage amount could be obtained simultaneously through the on-demand DSMM;
- The amount of changes in the total number of power purchases was inconsistent;
- It was not preferable to have a certain amount of initial inventory when setting the supply–demand management start day as Thursday because the storage opportunity loss would arrive earlier due to a full charge;
- With regard to the storage opportunity loss amount, it was found that the total number of storage opportunity losses did not necessarily correspond to times when there was a change of 5 [kWh];

These results indicate that the on-demand DSMM could be applied to, and useful for, demand-to-supply management of solar power generation. It was also shown that power consumption at times when solar was unable to generate could be covered by a storage battery, and the battery capacity and the storage required for optimal operation of the solar system could be calculated using the on-demand DSMM.

Future studies should include the extension of the target period of the data, management of the considered storage amount for seasonal fluctuations to further increase accuracy, proposal of a demand-to-supply management method that can accommodate for a wider range of demand situations by extending target facilities, and use of a moving average that can smooth data against power consumption of highly variable data. Additionally, future expectations include the following: the proposal of a model that can obtain economical storage battery capacity; application to a network of local production for local consumption within a community, called a microgrid; and building a model that enables stable power generation in combination with other renewable energy sources.

Author Contributions: S.Y. and T.Y. conceptualized the goals and the aims of this study. S.Y., M.M. and T.Y. designed the methodology. S.Y. wrote the original draft assisted by H.I., H.I. and T.Y. shaped this manuscript based on the original draft. M.T. and M.M. provided their related studies. T.Y. acquired funds. S.Y. applied formulations assisted by M.M., S.Y. programmed and validated the formulation, and visualized the results with H.I. and T.Y., S.Y. conducted the numerical experiment, and H.I., T.Y. and M.M. reviewed it. T.Y. managed this project and supervised the overall content. All authors have read and agreed to the published version of the manuscript.

Funding: This study was partially supported by the Japan Society for the Promotion of Science (JSPS), KAKENHI, Grant-in-Aid for Scientific Research (A), JP18H03824, from 2018 to 2023.

Institutional Review Board Statement: Not applicable.

Informed Consent Statement: Not applicable.

Data Availability Statement: Not applicable.

Acknowledgments: We hereby express our sincere thanks to Ecolomy Co., Ltd. who generously provided their data for this study.

Conflicts of Interest: The authors declare no conflict of interest.

Appendix A

Table A1. The notations used in this study.

Variables	This Study
I_t:	Sum of private power generation and purchased electrical power in period t [KWh]
I'_t:	Sum of the next private power generation and the next power purchase amount in period t [kWh]
O_t:	Power consumption amount in period t [kWh]
L_t:	Remaining storage capacity in period t [kWh]
N_t:	Moving base storage capacity in period t [kWh]
X_t:	Forecasted power consumption amount in period t [kWh]
λ_t:	Demand rate at period t
$\overline{\beta}_t$:	Input amount determination parameter in period t
S_t:	Private power generation in period t [kWh]
P_t:	Power purchase amount in period t [kWh]
Y_t:	Forecasted private power generation in period t [kWh]
Coefficient	
α:	Coefficient of real power consumption
β_1:	Cost coefficient for maintaining electricity storage [yen/kWh]
β_2:	Cost coefficient for understocked storage [yen/kWh]
β_3:	Cost coefficient for overstocked electricity is storage [yen/kWh]
Evaluation Value	
L_{init}:	Initial storage capacity [kWh]
L_{max}:	Maximum storage capacity [kWh]
L_{loss_t}:	Storage opportunity loss amount in period t [kWh]
TP:	Total amount of power purchase [kWh]
TP_{count}:	Total number of power purchases [number]
TL_{count}:	Total number of storage opportunity loss [number]

Table A2. Weather during surveyed period.

Date	Day and Night	Weather
20 August	Day	Rainy, then sometimes cloudy
	Night	Cloudy, with brief rain
21 August	Day	Cloudy, then rain for a while
	Night	Cloudy
22 August	Day	Cloudy, then sunny
	Night	Cloudy, sometimes sunny
23 August	Day	Cloudy, with brief sun
	Night	Cloudy, with brief sun
24 August	Day	Cloudy
	Night	Cloudy
25 August	Day	Cloudy, then rain for a while
	Night	Rainy, with brief cloud
26 August	Day	Rainy
	Night	Cloudy, sometimes rainy, then sunny for a while
27 August	Day	Cloudy
	Night	Cloudy, then rainy for a while

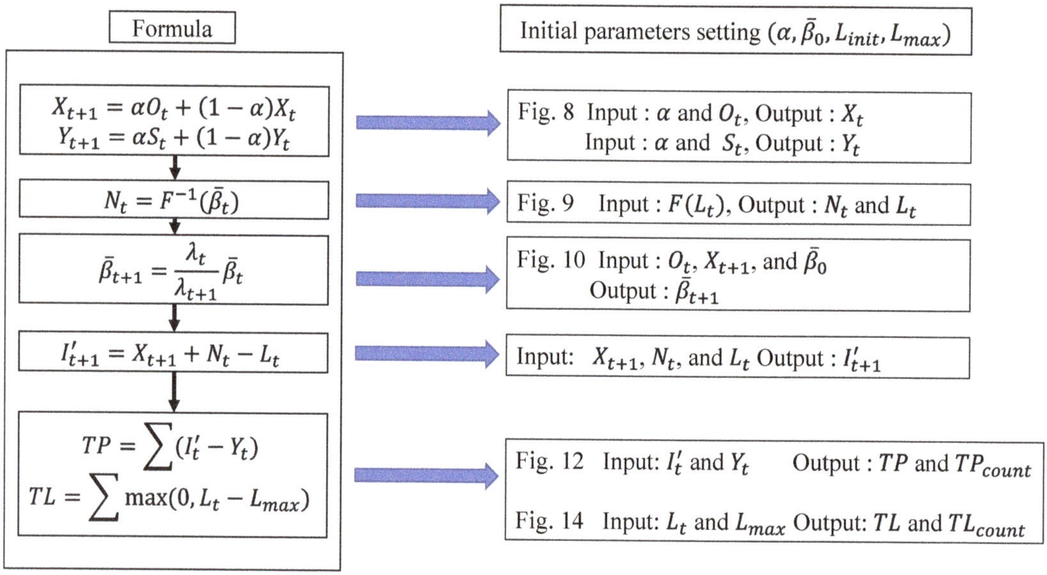

Figure A1. Input data including measurements and output data with the resulted main figures in Section 3.

References

1. UNFCCC. The 26th session of the Conference of the Parties to the United Nations Framework Convention on Climate Change, "Glasgow Climate Pact". Available online: https://unfccc.int/sites/default/files/resource/cop26_auv_2f_cover_decision.pdf (accessed on 9 December 2021).
2. BBC News. Biden: This will be 'Decisive Decade' for Tackling Climate Change. Available online: https://www.bbc.com/news/science-environment-56837927 (accessed on 31 May 2021).
3. Ministry of Economy, Trade and Industry. Government Trends in Green Innovation. Available online: https://www.meti.go.jp/english/press/2020/1111_002.html (accessed on 24 March 2022).
4. Ministry of the Environment. *Annual Environmental White Paper 2020*; Nikkei Printing Inc.: Tokyo, Japan, 2020. (In Japanese)
5. Environmental Business Online. Cabinet Decides on Long-Term Strategy Against Global Warming to Realize a Decarbonized Society. Available online: https://www.kankyo-business.jp/column/022598.php (accessed on 24 March 2022).
6. Agency for Natural Resources and Energy: What is Renewable Energy? 2020. Available online: https://www.enecho.meti.go.jp/category/saving_and_new/saiene/renewable/outline/index.html (accessed on 21 September 2020).
7. Ootake, H. Application of meteorological technology to the field of photovoltaic power generation and energy management. *Commun. Oper. Res. Soc. Jpn.* **2020**, *65*, 5–11. (In Japanese)
8. Agyekum, E.B.; PraveenKumar, S.; Alwan, N.T.; Velkin, V.I.; Shcheklein, S.E. Effect of dual surface cooling of solar photovoltaic panel on the efficiency of the module: Experimental investigation. *Heliyon* **2021**, *7*, e07920. [CrossRef] [PubMed]
9. Agyekum, E.B.; PraveenKumar, S.; Alwan, N.T.; Velkin, V.I.; Shcheklein, S.E.; Yaqoob, S. Experimental investigation of the effect of a combination of active and passive cooling mechanism on the thermal characteristics and efficiency of solar pv module. *Inventions* **2021**, *6*, 63. [CrossRef]
10. Agyekum, E.B.; PraveenKumar, S.; Alwan, N.T.; Velkin, V.I.; Adebayo, T.S. Experimental study on performance enhancement of a photovoltaic module using a combination of phase change material and aluminum fins—Exergy, energy and economic (3e) analysis. *Inventions* **2021**, *6*, 69. [CrossRef]
11. Agyekum, E.B.; PraveenKumar, S.; Eliseev, A.; Velkin, V.I. Design and construction of a novel simple and low-cost test bench point-absorber wave energy converter emulator system. *Inventions* **2021**, *6*, 20. [CrossRef]
12. PraveenKumar, S.; Agyekum, E.B.; Alwan, N.T.; Velkin, V.I.; Yaqoob, S.J.; Adebayo, T.S. Thermal management of solar photovoltaic module to enhance output performance: An experimental passive cooling approach using discontinuous aluminum heat sink. *Int. J. Renew. Energy Res.* **2021**, *11*, 1700–1712. [CrossRef]
13. Matsui, M.; Hujikawa, H.; Ishi, N. *Supply Chain Management Towards Post ERP/SCM*; Asakura Publishing Co., Ltd.: Tokyo, Japan, 2009.
14. Matsui, M. *Management of Production Companies: Profit Maximization and Factory Science*; Asakura Publishing Co., Ltd.: Tokyo, Japan, 2005.

15. Usuki, J.; Kitaoka, M. Classification of cumulative curve from initial inventory and relationship between inventory quantity and lead time. *J. Jpn. Ind. Manag. Assoc.* **2001**, *51*, 558–565. (In Japanese) [CrossRef]
16. Wichmann, M.G.; Johannes, C.; Spengler, T.S. Energy-oriented Lot-Sizing and Scheduling considering energy storages. *Int. J. Prod. Econ.* **2019**, *216*, 204–214. [CrossRef]
17. Moon, J.; Park, J. Smart production scheduling with time-dependent and machine-dependent electricity cost by considering distributed energy resources and energy storage. *Int. J. Prod. Res.* **2014**, *52*, 3922–3939. [CrossRef]
18. Uhlemair, H.; Karschin, I.; Geldermann, J. Optimizing the production and distribution system of bioenergy villages. *Int. J. Prod. Econ.* **2014**, *147*, 62–72. [CrossRef]
19. Trappey, A.J.C.; Trappey, C.V.; Liu, P.H.Y.; Lin, L.; Ou, J.J.R. A hierarchical cost learning model for developing wind energy infrastructures. *Int. J. Prod. Econ.* **2013**, *146*, 386–391. [CrossRef]
20. Pham, A.; Jin, T.; Novoa, C.; Qin, J. A multi-site production and microgrid planning model for net-zero energy operations. *Int. J. Prod. Econ.* **2019**, *218*, 260–274. [CrossRef]
21. Jahanpour, E.; Ko, H.S.; Nof, S.Y. Collaboration protocols for sustainable wind energy distribution networks. *Int. J. Prod. Econ.* **2016**, *182*, 496–507. [CrossRef]
22. Rentizelas, A.A.; Tolis, A.I.; Tatsiopoulos, I.P. Investment planning in electricity production under CO2 price uncertainty. *Int. J. Prod. Econ.* **2012**, *140*, 622–629. [CrossRef]
23. Xydis, G. A techno-economic and spatial analysis for the optimal planning of wind energy in Kythira island, Greece. *Int. J. Prod. Econ.* **2013**, *146*, 440–452. [CrossRef]
24. Takanokura, M.; Matsui, M.; Tang, H. Energy Management with battery system for smart city. In Proceedings of the 33rd Chinese Control Conference, Nanjing, China, 28–30 July 2014.
25. Santana-Vieraa, V.; Jimenezb, J.; Jin, T.; Espiritu, J. Implementing factory demand response via onsite renewable energy: A design-of-experiment approach. *Int. J. Prod. Res.* **2015**, *53*, 7034–7048. [CrossRef]
26. Matsui, M.; Uchiyama, H.; Hujikawa, H. Progressive-curve-based control of inventory fluctuation under on-demand SCM. *J. Jpn. Ind. Manag. Assoc.* **2005**, *56*, 139–145. (In Japanese) [CrossRef]
27. Ecolomy Co., Ltd. Website. Available online: https://www.ecolomy.co.jp/ (accessed on 21 September 2020).
28. Asada, K.; Iwasaki, T.; Aoyama, Y. *An Introduction to Demand Forecasting for Inventory Management*; Toyo Keizai Inc.: Tokyo, Japan, 2008. (In Japanese)
29. Japan Meteorological Agency: Historical Weather Data Search Website. Available online: https://www.data.jma.go.jp/obd/stats/etrn/index.php?prec_no=&block_no=&year=&month=&day=&view=a4 (accessed on 25 November 2020).

energies

MDPI

Review

The Current Status and Future Potential of Biogas Production from Canada's Organic Fraction Municipal Solid Waste

Omid Norouzi and Animesh Dutta *

School of Engineering, University of Guelph, Guelph, ON N1G2W1, Canada; Norouzio@uoguelph.ca
* Correspondence: adutta@uoguelph.ca; Tel.: +1-519-824-4120 (ext. 52441)

Abstract: With the implementation of new policies supporting renewable natural gas production from organic wastes, Canada began replacing traditional disposal methods with highly integrated biogas production strategies. Herein, data from published papers, Canadian Biogas Association, Canada's national statistical agency, and energy companies' websites were gathered to gain insight into the current status of anaerobic digestion plants in recovering energy and resource from organic wastes. The availability of materials prepared for recycling by companies and local waste management organizations and existing infrastructures for municipal solid waste management were examined. Governmental incentives and discouragements in Canada and world anaerobic digestion leaders regarding organic fraction municipal solid waste management were comprehensively reviewed to identify the opportunities for developing large-scale anaerobic digestion in Canada. A range of anaerobic digestion facilities, including water resource recovery facilities, standalone digesters, and on-farm digesters throughout Ontario, were compared in terms of digestion type, digester volume, feedstock (s), and electricity capacity to better understand the current role of biogas plants in this province. Finally, technology perspectives, solutions, and roadmaps were discussed to shape the future in terms of organic fraction municipal solid waste management. The findings suggested that the biogas industry growth in Canada relies on provincial energy and waste management policies, advanced technologies for diverting organic waste from landfills, improving biogas yield using existing pretreatment methods, and educating farmers regarding digester operations.

Keywords: waste management; organic waste; anaerobic digestion; biogas plant; pretreatment; Canada

Citation: Norouzi, O.; Dutta, A. The Current Status and Future Potential of Biogas Production from Canada's Organic Fraction Municipal Solid Waste. *Energies* **2022**, *15*, 475. https://doi.org/10.3390/en15020475

Academic Editor: Dino Musmarra

Received: 12 December 2021
Accepted: 7 January 2022
Published: 10 January 2022

1. Introduction

The Canadian economy currently seems locked into an inefficient system. Production, economics, contracts, regulation, and consumer behaviour all favor the linear model of production and consumption. This model applies to our current food industry and is very wasteful. Between 33% and 50% of food is wasted, and many food production techniques cause widespread environmental degradation [1,2]. The situation will exacerbate within the next 30 years due to population growth and shifting demographics. The current waste management strategies, which mostly rely on thermal conversion by incineration, do not encourage recycling and waste reduction. This is not a strong long-term strategy for society. Nowadays, there is a significant worldwide trend of promulgating regulations and policies banning organic fraction municipal waste (OFMSW) from entering landfills [3]. Due to encouraging or enforcing Canadian policies on biogas plants, anaerobic digestion (AD) has been considered a suitable method for converting organic waste streams into renewable energy and fertilizer. Feed-in tariff (FIT) policies that encourage renewable natural gas (RNG) from AD resulted in the implementation of many biogas plants in Canada [3]. However, to reach the global PARIS Agreement and join the biogas leaders (Germany, the UK, Japan, Italy, Spain, Denmark, Sweden, and China), Canada should focus more on reforming its renewable energy policies and economic incentives [4–6].

Canada generates 35.5 million tonnes of waste containing 20% to 40% OFMSW [7]. This amount of OFMSW can be converted into about 12,000 kWh of renewable energy per year [8]. In Canada, the current waste management infrastructures can only capture 2.6 million tonnes of OFMSW from landfills (Figure 1) [9]. Figure 2 shows the proportion of organic waste diverted from municipal waste in Canada by province. Among the provinces, New Brunswick diverted the most organics (57.57%), followed by Nova Scotia (46.84%), Prince Edward Island (36%), British Columbia (35.33%), Alberta (34.62%), Ontario (34.27%), Manitoba (25.38%), Newfoundland (21.57%), Saskatchewan (16.98%), and Yukon (15.69%) [10]. In these provinces, managing technologies to establish OFMSW includes biological and thermochemical conversion systems such as composting, AD, hydrothermal carbonization (HTC), pyrolysis, gasification, and incineration. Among all others, anaerobic digestion seems to be a more sustainable system and can be integrated with or replace conventional waste management strategies more efficiently. AD is becoming an important technology in the conversion of OFMSW, waste-activated sludge (WAS), agricultural waste, animal manure, and food waste. Further information on life cycle assessment (LCA) and life cycle cost (LCC) of AD of OFMSW can be found in a study by Demichelis et al. [11].

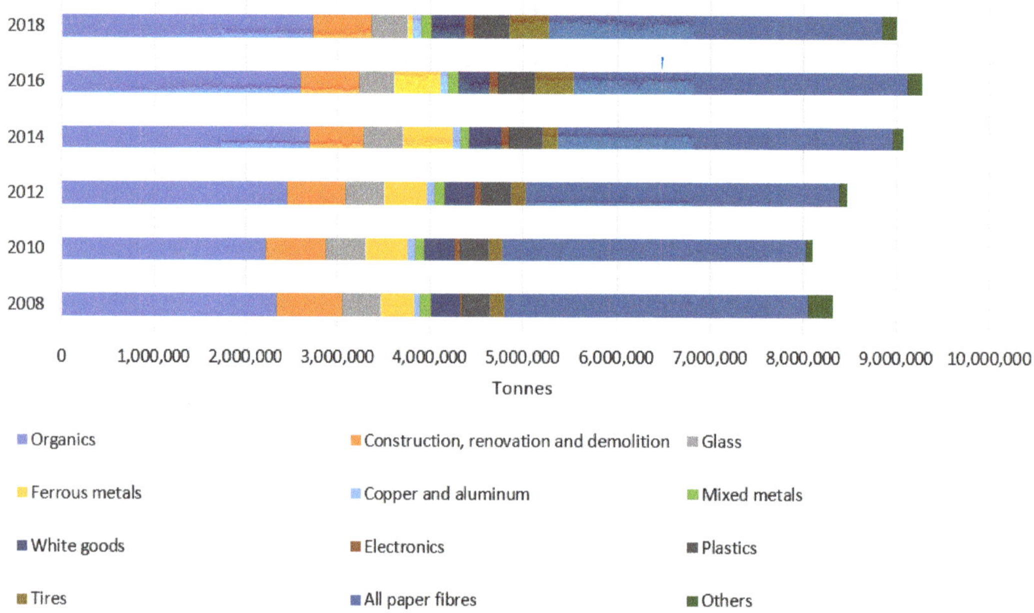

Figure 1. Materials prepared for recycling by companies and local waste management organizations in Canada. The authors originally produced the figure from Canada's national statistical agency [9].

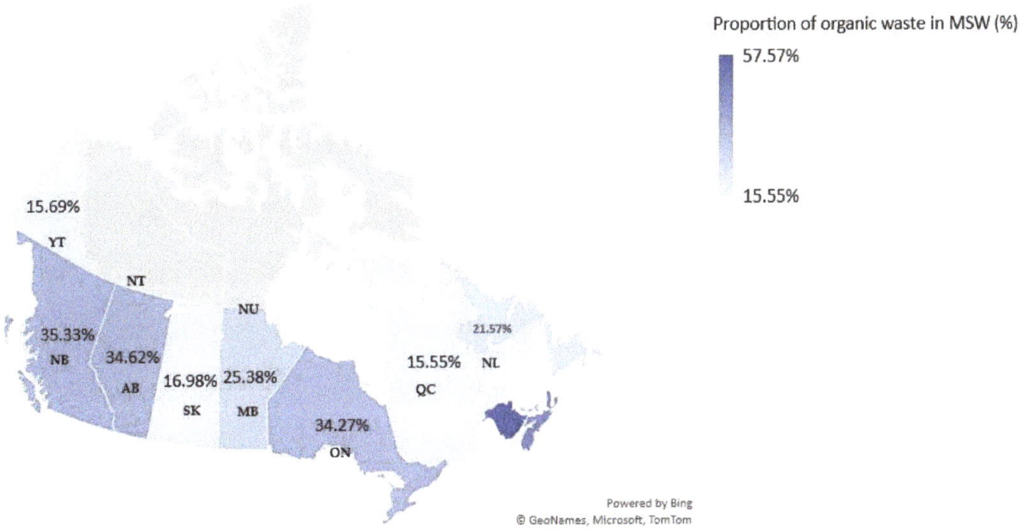

Figure 2. The proportion of organic waste diverted from municipal waste in Canada by province. The authors originally produced the figure from Canada's national statistical agency [9].

AD is a process by which organic waste streams produced in the food and beverage industry, the paper industry, agriculture, WWTPs, and households, are converted into biogas, liquid, and solid fertilizers, owing to the anaerobic bacteria or facultative anaerobic bacteria [12,13]. The largest bioenergy facility to manage OFMSW is the Rialto bioenergy facility implemented in North America. This plant can process 700 tonnes per day of organics from food waste and biosolids from WWTPs to produce renewable electricity to power the facility, export renewable electricity onto the grid, and export RNG to fuel their fleet of NG fueled vehicles. The plant is also able to produce class A fertilizer and soil conditioner [14]. At present, AD technology has been significantly advanced in Europe and turned into a well-established waste management strategy within the continent. With 18,943 biogas plants, 725 biomethane plants, 15.8 billion m^3 of biogas, and 2.4 billion m^3 of biomethane, Europe is considered the leading producer of biomethane [15]. However, compared with Europe, biogas production in Canada is a small but growing industry. Canada should reform its renewable energy policies and economic incentives to enter the league of biogas leaders. There are three reasons for today's fast-growing Canadian RNG marketplace. First, over 480,000 kilometres of natural gas pipelines are already available in Canada. Second, according to Ontario's FID program, the cost of the RNG project is lower than solar, wind, and biogas. Finally, RNG has a great production potential from Canadian organic waste streams [16–19].

To the best of our knowledge, this paper is the first review paper providing a guideline for better selecting OFMSW management methods. This review covers the technological advances in biogas plants, the composition of MSW by province, inventory of publicly owned solid waste assets, existing anaerobic digesters, the current Canadian policies that both encourage and enforce, and suggestions for building a zero organic waste future.

2. Policies/Regulations on AD in Both Recycling and Energy Recovery

Table 1 shows policies/regulations of 12 countries promoting AD for organic waste streams (particularly food waste and OFMSW). Regulations and policies on the treatment of organic waste streams are not yet uniform globally. However, protecting and preserving the environment is the priority in all laws and regulations. Biogas plants are becoming a popular waste management strategy in cities where organic wastes are separated from landfills [20,21]. Due to the European Union (EU) legislation, most countries in the EU have both OFMSW separation infrastructure and biogas plants. With 18,943 biogas plants, 725 biomethane plants, 15.8 billion m^3 of biogas, and 2.4 billion m^3 of biomethane, Europe is considered the leading biomethane producer [15]. To reach the global targets for reducing GHG emissions, the United States enacted new regulations and announced incentives to promote facilities for recovering energy and resources from OFMSW. Currently, the US has 236 facilities in the form of water resource recovery facilities (WRRFs), standalone digesters, and on-farm digesters. To date, the most enacted and potentially important policies are the US EPA Renewable Fuel Standard and California's Low Carbon Fuel Standard (LCFS) [22].

Table 1. Policies/regulations of 12 countries promoting AD for organic wastes. Copyright 2021 by Elsevier [22].

Country	Policy/Regulations	Incentives	AD Applications
Australia	The National Food Waste Strategy identified the role of AD in both recycling and energy recovery.	Support for installing small-scale AD technology.	• Majority operated by the water, agro-waste, or food processing industries, and number of plants accepting post-production FW is increasing. • Co-digestion of organic waste from commercial and industrial sources.
Canada	Policies to support RNG from AD.	Feed-in tariff (FIT) program resulted in 40 AD plants between 2010 and 2017 in Ontario.	• Few plants solely digest FW, or source-separated organics (SSOs). • Majority of plants co-digesting SSO with other feedstocks, most commonly agricultural manures at on-farm facilities, or less commonly with SS.
China	• China's medium- and long-term renewable energy plans of 2006 gave a target of 44 billion m^3 of biogas per year by 2020. • Mandatory garbage sorting for 46 cities.	Forcing the municipalities to resolve urban garbage problems increased number of AD installations.	• Mono FW digestion and co-digestion with other organic feedstocks.
Indonesia	The Ministry of Energy and Mineral Resources introduced favorable tariffs for electricity generated from municipal wastes and biomass.	Limited AD installations, especially for FW, although a number of initiatives, such as Indonesia Domestic Biogas program promoted AD for other feedstocks.	• Source segregated organics digestion in small-scale AD modelled on Indian designs; • Reusability of digestate is one of the main economic barriers.
United States	• The US EPA Renewable Fuel Standard; • California's Low Carbon Fuel Standard (LCFS); • Landfill bans in many states.	• RINs • Renewable electricity production tax credit • Carbon credits • Nutrient credits	• 236 facilities processing FW using AD; • Higher number of WRRFs processing FW, but standalone facilities process higher FW volume; • Tipping fees major source of revenue.
United Kingdom	• Waste Resources Action Program; • The Renewable Heat Incentive (RHI) program ends by 2021, but current UK policy on climate change and its Climate Change Act commitments can bridge the policy by 10–15%.	• The RHI • The Feed-In Tariff for renewable electricity.	• Approximately 8% of FW sent to AD • Gate fees for AD plants receiving FW fall from GBP 35 to GBP 11 per ton.

Table 1. *Cont.*

Country	Policy/Regulations	Incentives	AD Applications
Vietnam	• Little or no separate collection of FW; • No AD schemes have been applied to manage organic FW.	Strategies to 2025 focus on methods to recover energy and materials from MSW in cities.	FWs account for about 60%. The increasing rate of MSW annually is about 12%. However, currently, all 35 MSW treatment plants in Vietnam are using landfilling, incineration, or composting. Some recent studies have indicated that there is a very high potential in producing biogas via AD process from MSW in Vietnam.
Thailand	• AD of FW is not very common due to many operational issues; • Many pilot programs have been initiated, but most were unsuccessful or not sustainable.	Many pilot programs include small AD operations in urban and rural areas.	• Small-scale community/school AD operations suffer from operational and maintenance issues; • Co-digestion of household FW and waste from farm animals.
South Korea	• FW disposal into landfills was banned in 2005; • "Pay-as-you-throw" initiative helped tracking source-separated FW; • Bioenergy strategy has a target to increase biogas production by a factor of 4 by 2030.	The Ministry of Environment has also funded biogas Research on organic wastes to energy with a budget of USD 74 million from 2013 to 2020.	• Source segregated organics digestion in standalone digesters. • Co-digestion with animal manures and sewage sludges.
Singapore	The National Environmental Agency's pilot plan to co-digest FW with SS towards achieving energy neutrality in wastewater treatment.	The co-digestion pilot-scale program will be extended to all sewage treatment plants if it is successful.	• Majority of food waste is processed in incinerators; • Some small-scale community, university, and business AD operations.

China's long-term renewable energy plan, in 2006, aimed at producing 44 billion m^3 of biogas per year by 2020. This plan mandated municipalities to use mono and co-digestion of food waste with other organic feedstocks. They successfully implemented many biogas plants between 2006 and 2020. "Chongqing Black Stone Food Waste Treatment Plant", with a processing capacity of 365 thousand tonnes per year and biogas yield of 28 million m^3 per year, is one of the largest plants in China [5,15]. In the United Kingdom, due to the Waste Resources Action Plan and RHI, and FIT programs dedicated to improving renewable electricity, approximately 8% of food waste streams are sent to AD. Canada's policies for supporting AD-based systems are similar to the UK [23,24]. In Canada, the FIT program supports RNG from AD. This program resulted in the implementation of frothy AD plants between 2010 and 2017 across Canada. Some countries in Asia are also making great progress in shifting from conventional waste management systems to AD-based WWTPs. For example, Japan is one of the pioneers in developing AD facilities. They enacted a new law introducing AD as a promising method to reuse food waste. They installed the first biogas plant in 2000 when incineration was the dominant waste management method for resource recovery from MSW in Asia. In Japan, many food-based companies currently use anaerobic digestion to process food wastes (e.g., soy sauce and shochu by-products). The South Korean Ministry of Environment banned disposal into landfills and invested USD 74 million, from 2013 to 2020, on biogas research, to be able to combine co-digestion to all sewage sludge (SS) treatment plants industrially. The most commonly used feedstock in existing sewage treatment plants is animal manure. They are planning to increase biogas production by a factor of four by 2030.

As seen in Table 1, although the implementation of biogas plants in China, the US, Canada, UK, Japan, Australia, and South Korea was successful, some countries such as Thailand, Singapore, Malaysia, and Vietnam still have had problems commercializing the AD technology. This is technically due to the lack of source segregate collection infrastructure, weak national policies, expertise, and knowledge in the AD process. In

Singapore, the National Environmental Agency recently started investigating the feasibility of running co-digest food waste with SS to obtain energy neutrality in wastewater treatment. They plan to apply this approach to all sewage treatment plants if it is successful [22].

While AD is being implemented commercially in the EU, in Canada, AD projects still suffer a lack of infrastructure for SSO, management, and technical knowledge on the AD process. Herein, the importance of public–private partnerships is highlighted to show its vital role in creating a more stable revenue generation system and overcoming implementation challenges. Recent literature and governmental reports suggest that the following three factors are driving forces for shifting a country from conventional to AD based waste management strategies:

- Governmental incentives and discouragements (e.g., carbon credits, nutrient credits, and tipping fees major);
- Energy expense reductions (renewable electricity production tax credit, RHI, RIN, and FIT);
- Environmental benefits.

Although AD technology has been considered the best solution worldwide to achieve net-zero emissions by 2050, it still needs to be modified from the technical, market, economic, institutional, socio-cultural, and environmental points of view. Table 2 highlights challenges faced by the implementation of AD in various countries and the root cause of the problems in the implementation.

Table 2. Challenges faced by the implementation of AD in various countries. Copyright 2019 by Elsevier [25].

Barriers	Sub-Barriers
Technical	• Infrastructural challenges (e.g., plant size, lack of resource availability, limited number of gas filling stations); • Technical failures and problems and negative images caused by failed biogas plants; • Need for specialized technical staff and expertise (incl. a lack of technical training and knowledge); • Poor collection, improper segregation, a lack of vehicles, and adequate waste transportation; • Insufficient follow-up services; • Specific characteristics of biogas; • Dependency on imported materials.
Economic	• High investments/lack of available capital (low incomes and widespread poverty); • Lack of subsidies and financial support programs (incl. fossil fuel subsidization); • High cost of biogas production, transportation, clean-up, and upgrading; • Unavailability of bank loans (incl. with preferential terms); • Lack of R&D funding.
Market	• Lower prices of fossil fuels; • High price of biogas/ biomethane; • Competition with other fuels; • Easy availability of fuelwood at zero private cost; • Uncertainties related to the injection of biogas into the grid.
Institutional	• Lack of political support/legislation; • Uncertain policy landscape (incl. political instability); • Lack of private sector participation and poor coordination between the public and the private sectors; • High level of bureaucracy (e.g., complex administrative and legal procedures).

Table 2. *Cont.*

Barriers	Sub-Barriers
Socio-cultural	• Lack of public participation and consumer interest; • Desire to maintain the status quo/Resistance to change; • Low level of knowledge; • Lack of information and information sharing; • Lack of literacy rate/Low level of education; • Cultural and religious outlook including stigmatization; • Migration.
Environmental	• Odor complaints; • Noise complaints; • Need for abundant water resources for biogas digesters; • Lack of access to adequate water; • Pollution.

3. Canada's Existing Infrastructures for the Solid Waste Management

Figure 3 shows Canada's existing infrastructures for solid waste management. Seven types of waste management facilities are actively operating in Canada, which are transfer station assets, composting, material recovery facility, anaerobic digestion, engineered landfill, incineration, and energy from waste. As shown in Figure 3, Ontario, Quebec, Alberta, British Columbia, Newfoundland, Manitoba, and Saskatchewan adopted around 1813, 713, 1204, 581, 324, 595, and 1074 waste management facilities, respectively. Currently, the most common approach for municipal waste disposal in Canada is landfilling. Although modern MSW landfills are able to collect and treat leachate and capture greenhouse gasses, it is still not a suitable approach for the disposal of waste. Environmental analysis of incineration, gasification, AD, landfill, and composting showed AD has obvious advantages in the environmental criteria over other methods. Mondello et al. compared the potential environmental impacts of four waste management strategies—namely, landfill, incineration, composting, and AD, to manage organic waste, particularly food waste, using the life cycle assessment (LCA) method [26]. Based on the treatment of a functional unit of 1 tonne of food waste, the energy use in AD is lower than other options. In addition, the main environmental impacts were detected for landfill and incineration options, which is in agreement with the literature. Among studied strategies, AD showed the best environmental performance.

Reducing the amount of organic waste entering landfills is now a high priority worldwide. In Canada, the total amount of MSW diverted to recyclable materials (e.g., organics, plastics, tires, paper, electronics, etc.) increased from 8.3 million tonnes in 2008 to 9.8 million tonnes in 2018. The organic fraction of MSW has a great potential to be converted into renewable energies by AD or other waste-to-energy facilities. Thus, in recent years, Canada's MSW treatment plants are gradually adopting the AD mode of operation in which the OFMSW can be converted into biogas, electricity, and fertilizer. The two most populous provinces, Ontario and Quebec, recycled the most organics and implemented the highest number of AD plants in 2018, with 40 and 9 plants, respectively. In Ontario, such fast-paced green development is due to provincial incentives and discouragements such as the FIT program.

A higher number of waste-to-energy facilities was recorded in New Brunswick. In this province, over seven thousand tonnes of tires and organics are diverted from disposal. These materials are great input for thermochemical technologies such as pyrolysis and gasification. Plastic, in particular, has been the target material of recent programs such as the Federal Action Plan on Zero Plastic Waste. This program aims to reduce the presence of microplastics in oceans and mitigate its negative effect on the ecosystem and human health. British Columbia also recycles over 65 thousand tonnes of plastic, 42 thousand tonnes of tires, and 590 thousand tonnes of organics. With the current incentives and regulations, we can expect an increase in waste-to-energy capacity in the near future in British Colombia [27].

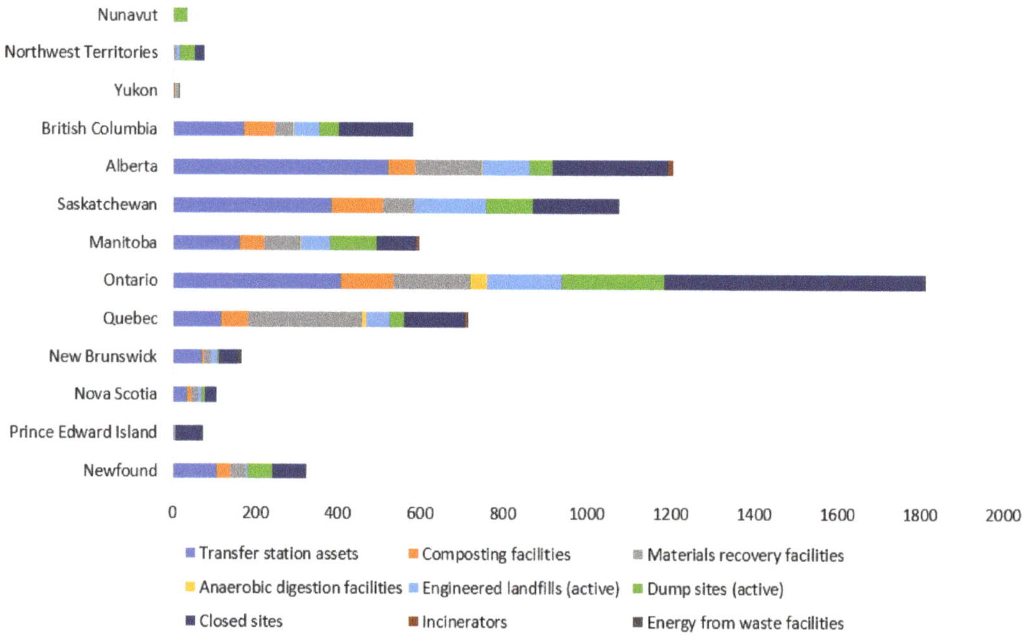

Figure 3. Public-owned MSW management infrastructures in Canada. The authors originally produced the figure from Canada's national statistical agency [28].

4. Biogas Projects in Ontario

Ontario government has been encouraging renewable natural gas (RNG) market in the province through updating its Climate Change Action Plan and founding biogas projects since 2016. Since then, AD has been extensively implemented, from pilot to industrial scale, in the form of WRRFs, standalone digesters, and on-farm digesters [28]. Table 3 shows existing biogas plants projects in Ontario for energy and resource recovery. The information on volume, organic input, and biogas plants' capacity was obtained from the Canadian Biogas Association, Canada's national statistical agency, and manufacturer websites. As can be seen, most biogas plants in Ontario are operating to process agricultural wastes and the manure produced by cattle in farms. Among them, Delft Blue Veal Inc. (Cambridge), Koskamp Family Farms (Stratford), Athlone BioPower (Tavistock), and Donnandale Farms (Stirling) have the highest capacity ($\approx$2000 m^3) and can generate 500 kW of electricity. Some AD plants were installed in different locations for the purpose of training students and farmers. The most famous one is CARES located at the University of Guelph Ridgetown Campus. This biogas plant consists of a 1527 m^3 digester to process dairy and swine manure and crude glycerol to generate 250 kW of electricity. In addition to pilot-scale AD plants, a few cities in Ontario (i.e., Toronto, Bridgeport, and Surrey) are using large-scale AD plants to extract energy and resources from MSW. The biggest project is related to the Dufferin Organics Processing Facility (DOPF) in Toronto. With a high capacity of 55,000 tonnes per year, this facility converts SSO collected annually in the residential and commercial Green Bin Program into electricity. City of Surrey's Organic Waste Biofuel Processing Facility converts the city's MSW into RNG and supplies waste collection and recycling vehicles fuels. County of Oxford has been using Ingersoll WWTP anaerobic digester since 2011. The facility consists of two identical digesters (two-phase AD), with a volume of 1090 m^3 for each digester, and is fed only by waste-activated sludge (WAS).

Table 3. Several biogas plants projects in Ontario for energy and resource recovery. The table was originally produced by the authors from the information shared by the Canadian Biogas Association [29].

Technology	Source	Type	Description	Location
CCI Disco	Source separated organics (SSOs) collected in the residential and commercial Green Bin Program	Single phase	Biogas is used to provide heating for the facility and for the functioning of the anaerobic digestion system and upgraded to renewable natural gas and injected into Ontario's natural gas grid. Digester solids are sent to a composting facility in southern Ontario.	City of Toronto
Dufferin Organics Processing Facility (DOPF.)	Source separated organics (SSOs) collected annually in the residential and commercial Green Bin Program	Single phase	The original DOPF, which had been built to process 25,000 tonnes of organic material annually from the City's Green Bin Program, will be upgraded with new processes and expanded to a capacity of 55,000 tonnes per year.	City of Toronto
Bridgeport Wastewater Treatment Plant	Source-separated organic materials from commercial generators.	Single phase	The anaerobic digestion facility will generate over 10 million kWh of renewable electricity per year–enough to power more than 1000 homes.	City of Bridgeport
Surrey's Organic Waste Biofuel Processing Facility	City's solid waste Stream	Single phase	The system converts organic wastes into renewable natural gas (RNG) for waste collection and recycling vehicles.	City of Surrey
Ingersoll WWTP anaerobic digester	Waste activated sludge (WAS)	Two phase	The anaerobic digestion facility consists of a primary anaerobic digester with an operating volume of 1090 m^3 and a secondary anaerobic digester of identical capacity.	County of Oxford
ZooShare Biogas	Zoo manure and food waste	Single phase	The plant is designed to handle 17,000 tonnes of organic waste and recover 500 kW electricity.	Toronto
Escarpment Renewables	High total solids organic waste	Two phase	The plant is capable of producing 12,000 m^3 of biogas per day.	Grimsby
CARES—University of Guelph Ridgetown Campus	Farmer waste streams, manure, and glycerol obtained from biodiesel plants	Single phase	The AD is connected to a 250 kW MAN engine with an operating volume of 1527 m^3. The plant is built for training students and farmers.	Guelph
StormFisher	Farmer waste streams, manure, organic material, and mixed food scraps	Single phase	StormFisher is built to process 65,000 tonnes of organic wastes into electricity and fertilizer granules.	London
The Gardiner Farms	Farmer waste streams, manure, and organic material	Single phase	The Gardiner Farms produces electricity and thermal energy using two 250 kW CHP units.	Caledon
Greenholm Farms	Recycled digestate solids, organic waste, and the manure produced by cattle	Single phase	The plant is designed with an operating volume of 2077 m^3 and is able to produce 250 kW of energy.	Embro
Escarpment Renewables	Fats, oils, grease, and organic liquids	Single phase	Escarpment Renewables is an industrial AD facility that is permitted to receive 23,000 tonnes of organics annually.	Beamsville
Bayview Flowers Ltd.	Manure, grape pumice, corn silage, pet food	Single phase	The plant is designed with an operating volume of 1200 m^3 and its biogas is sent to a 250 kW Scania generator and a retrofitted boiler.	Jordan
Delft Blue Veal Inc.	Calf manure and discarded organic residuals provided by food processing companies	Single phase	The plant is designed with an operating volume of 1750 m^3 and is able to produce 499 kW of electricity.	Cambridge

Table 3. *Cont.*

Technology	Source	Type	Description	Location
Koskamp Family Farms	Manure and other organic materials	Single phase	The plant is designed with an operating volume of 1500 m^3 and is able to produce 500 kW of electricity.	Stratford
Athlone BioPower	Manure and other organic materials	Two phase	The facility has two anaerobic digesters, a primary and secondary tank, with an operating volume of 2077 m^3 in size for each of them. The plant is able to produce 500 kW of power.	Tavistock
Birchlawn Farms	On-farm materials and outsourced organic waste from food processing plants	Single phase	The plant is designed with an operating volume of 1800 m^3 and is able to produce 440 kW of electricity.	Listowel
Woolwich Bio-en Inc.	Food waste	Single phase	The facility is built to process 70,000 tonnes per year of organic wastes into electricity and fertilizer granules. The CHP's produce 2.852 MW of electricity under a Feed-in Tariff contract with the Ontario Power Authority.	Elmira
Chatsworth/Georgian Bluffs	Biosolids, grease trap waste, source-separated organics, and other organics	Two phase	The anaerobic digester is a two-stage process with a 100 m^3 hydrolysis tank and a 1000 m^3 digester.	Owen Sound
Clovermead Farms	Manure and other organic materials	Single phase	A 1500 m^3 anaerobic digester which supplies fuel for the 250 kW generator, installed by European Power Systems Ltd.	Aylmer
Marl Creek Renewables	Manure, milk, fats, oils, and grease	Single phase	The plant is designed with an operating volume of 4200 m^3 and is able to fuel two 250 kW combined heat and power units.	Elmwood
CCS agriKomp	Manure from the farm's beef herd, silage, crop residues, and FOG	Single phase	The plant is designed with an operating volume of 680 m^3 and is able to supply fuel for a 100 kW engine.	Millbrook
Donnandale Farms	Manure and other organic materials	Single phase	The plant is designed with two anaerobic digesters (1600 m^3 each) and is able to supply fuel for the 500 kW MWM generator.	Stirling

Most AD plants in Canada are single-phase AD, due to this method's simple design and process control. All four hydrolysis, acidogenesis, acetogenesis, and methanogenesis steps are carried out in a single digester. These types of ADs are usually applied for agricultural wastes consisting of materials resistant to ADs such as cellulose, hemicellulose, and lignin. However, as shown in Table 3, two-phase ADs are designed and implemented for easily biodegradable feedstock (e.g., biosolids, grease trap waste, source-separated organics, and other organics). These materials undergo fast-rate hydrolysis and acidogenesis, which, in turn, lead to the inhibition of methanogenesis by volatile fatty acids (VFAs) accumulation. Ingersoll WWTP anaerobic digester, Athlone BioPower, and Georgian Bluffs have two-phase AD digesters. The advantages of two-phase AD over single-phase AD are as follows:

i. Generation of both methane and hydrogen;
ii. Operational conditions and reactions can be easily controlled;
iii. Improving the speed limiting reaction (hydrolysis);
iv. Higher energy capacity.

Although most AD references in Canada are available in Ontario, there are some innovative AD projects in other provinces [30]. For example, a complete system to process high-strength brewery wastewater from a new Molson–Coors brewery was implemented in Longueuil (Montreal), Quebec, Canada. This unit can produce biogas up to 8450 Nm^3/d, including 72% of CH_4, which is equivalent to 3350 kg/day of fuel oil. Biogas generated in this unit is used to heat the inlet wastewater to maximize the anaerobic treatment efficiency and reduce energy costs. In British Columbia, a high-efficiency biogas plant coupled with a water wash biogas upgrading plant has been recently implemented in Metro Vancouvers Lulu Island WWTP. This unit is able to treat up to 800 Nm^3/h of raw biogas to RNG. Upgraded RNG is injected into the FortisBC gas grid [31].

5. Pretreatment Methodologies to Enhance OFMSW Biodegradability

With the development of biogas plants in the past several decades, pretreatment technologies have also gained momentum and have been successfully applied to stabilize and enhance methane production. As shown in Table 4, pretreatment technologies can be categorized into five types: mechanical, thermal, chemical, biological, and additives (hybrid) [32]. Since the OFMSW composition is relatively complex, the hydrolysis step is considered a rate-limiting phase among the four phases of hydrolysis, acidogenesis, acetogenesis, and methanogenesis. The main goal of pretreatment technologies is to promote the hydrolysis step [33–35].

As for mechanical pretreatment, some technologies are commercially available. The OREXTM (A patented press extruder) is capable of recovering over 90% of putrescible organics from mixed waste streams. Organic pulp extracted from the press extrusion process is highly degradable in AD processes. Disc screen process is another pretreatment method widely used in waste processing facilities. The process separates waste into unders containing putrescible organics and overs including coarse recyclables [36]. Thermal pretreatment is the process in which MSW is heated in the range of 100 °C. The process is performed using pyrolysis, hydrothermal carbonization (HTC), hydrothermal liquefaction (HTL), and microwave reactors. Although this method can increase methane yield, remove odor, and improve dewaterability, it is considered an energy-intensive method. High energy demand and capital cost of building and operating thermal systems become obstacles for their practical commercialization [37]. Chemical pretreatment is divided into three types: acidic, alkali, and ozonation, which relies on hydrolysis of hemicellulose, saphonication, and hydroxyl radicals, respectively. Chemical pretreatment is currently being used in full-scale operations in excess sludge reduction in WWTPs [38].

Regarding biological pretreatment, TPAD and MEC are the most widely used methods in promoting the AD process. In the TPAD method, hydrolysis and acidogenesis steps occur in the first tank under thermophilic conditions, and acetogenesis and methanogenesis in the second tank under mesophilic conditions. Two-stage biogas plants can result in increased methane yield, low energy demand, and better solid destruction. TPAD process offers lower VFAs, COD, and suspended solids concentrations in the effluent and higher methane yield [32].

Additives (e.g., zeolite, biochar, bricks, plastic beads, coconut coir, charcoal, $G_{AC,}$ etc.) have gained significant attention due to their ability in facilitating the adsorption of inhibitors, increasing buffering capacity, microbial growth, DIET, H_2S removal, and CO_2 sequestration. For example, zeolite can stabilize the process due to its cation exchange property via ammonia detoxification. Porous materials such as biochar, hydrochar, and GAC are natural molecular sieves and catalysts; thus, they can purify biogas by separating H_2S and CO_2 from CH_4 and N_2 [39].

In the above paragraphs, mechanical, thermal, biological, and physical pretreatment methods were comprehensively introduced and discussed. However, it is worth mentioning that there are only a few references worldwide applying such pretreatment methods. This is mostly due to technical, energy, economic and environmental barriers. Mainardis et al. have recently could develop a reliable and standard protocol based on physicochemical characterization, experimental tests, LCA, and economic analysis to determine the up-scale feasibility of the proposed pretreatment method for AD of sewage sludge. They investigated six different pretreatment technologies: thermal, alkali, combined alkali–thermal, ultrasonication, icing–thawing, and biochar addition. Among the proposed methods, in terms of biomethane potential, low-temperature thermal pretreatment (110%), ultrasonication (53%), and biochar addition (16%) showed the best performance. Figure 4 compares the LCIA results of ultrasonic, thermal, and biochar scenarios with the baseline scenario. In most environmental criteria, proposed pretreatment methods showed better performance. However, the economic analysis showed that the capital costs of these methods could not be recovered in 15 years only if we consider heat recovery for thermal pretreatment and lower price for the biochar addition scenario [11]. We believe that the current protocol is a robust tool to assess a proposed pretreatment technology from the technical, energy, economic and environmental points of view within the Canadian context.

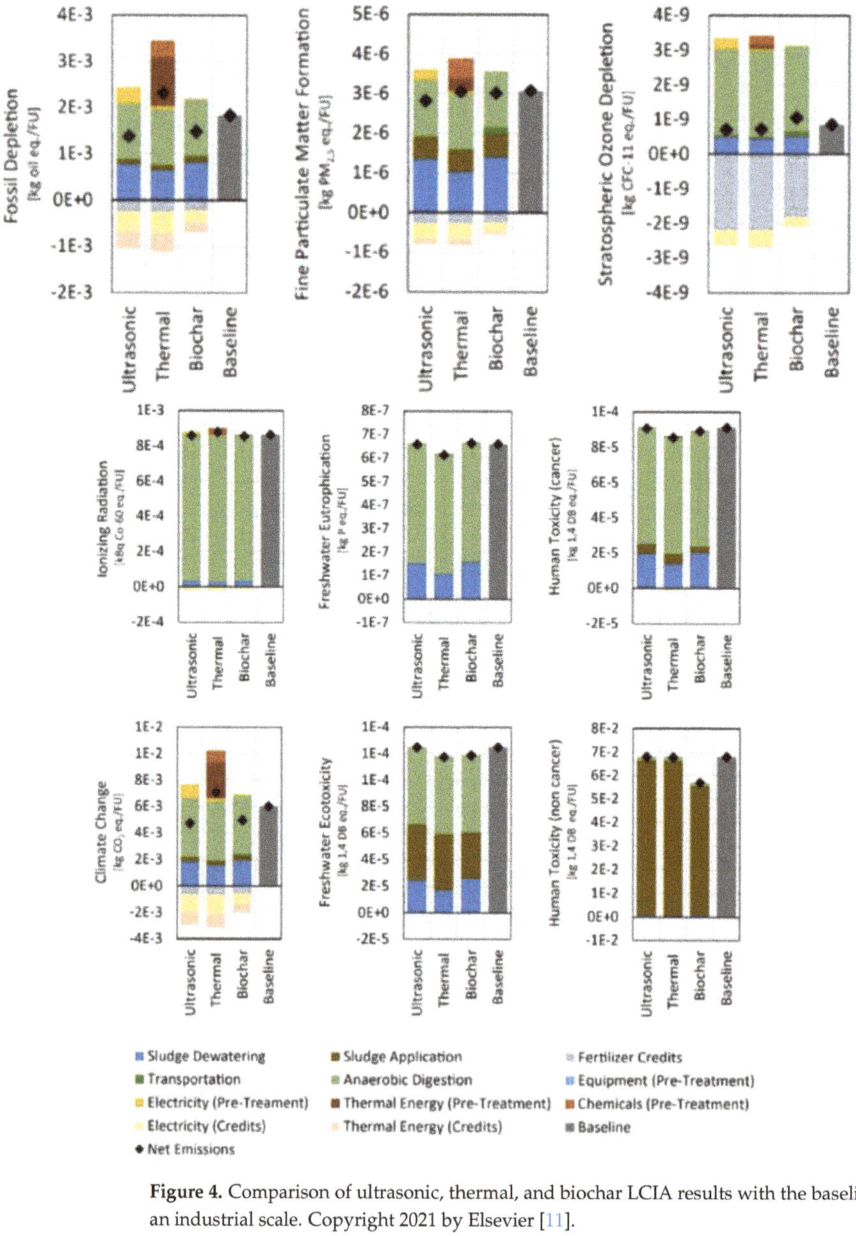

Figure 4. Comparison of ultrasonic, thermal, and biochar LCIA results with the baseline scenario at an industrial scale. Copyright 2021 by Elsevier [11].

Table 4. Summary of available commercial pretreatment AD technologies along with their advantages and disadvantages.

Pre-Treatment	Description	Available Processes	Advantages and Disadvantages	Commercial Technologies	Ref.
Mechanical treatment	Shredding and chopping of raw substrates to enhance the interaction between microorganisms and fragmented organic molecules (e.g., sugar, amino, and fatty acids)	• Press extruder • Disc screen • Bag opener • Wind sifter • High-pressure homogenizer • Sonication • Maceration	• Enhancement of COD solubilization • Improving nutrient availability to microbes • Excessive shredding or chopping of raw substrates may lead to the increased VFAs and inhibition of the AD process • Operational and capital cost is high	• OREX™ • Db-Disc Screen • Db-Wind Sifter • CleanREX™ • BIOREX™	[36]
Thermal treatment	Applying heat to decompose MSW via different approaches	• Hydrothermal • Microwave	• Removing pathogen • Enhancing the dewaterability properties • Polarization of macromolecules • Energy intensive • Treatment at high temperatures (>170 °C) lead to the complex recalcitrant substrates	• Patented Anaergia pyrolysis • CambiTHP™ • Biothelys® • Exelys, • Turbotec • Lysotherm, • Biorefinex	[32,40]
Chemical treatment	Chemical treatment is applied to disrupt the cell walls using strong and concentrated chemicals	• Commercial alkaline materials (e.g., NaOH and $CaCO_3$) • Ozone (O_3) • Peracetic acid • Acetic acid	• Easy operation • Suitable for lignin decomposition • Corrosion • Special materials for reactor construction, • Neutralization before digestion	• Full-scale operations in existing WWTP	[38]
Biological treatment	Promoting microbial growth	• Temperature phased anaerobic digestion (TPAD) • Microbial electrolysis cell (MEC)	• Eco-friendly • Low energy input • Operating at room temperature • Operational and capital cost is low		[32]
Additives	Additives can promote the AD process through adsorption of inhibitors, increasing buffering capacity, and microbial cell immobilization.	• Activated carbon • Biochar • Hydrochar • Conductive materials	• Adsorbing inhibitors such as LCFA, ammonia, limonene, heavy metals, and phenols. • Supporting microbial metabolism • Buffering pH during hydrolysis and acidogenesis steps • Contributing to the circular economic approach • Direct Interspecies Electron Transfer (DIET)		[39,41]

6. Guideline for Better Selection of OFMSW Management Methods

Over the past 10 years, Canada has been investigating the feasibility of various available renewable energy sources such as biomass, wind, and solar. The challenge is to achieve net-zero emissions by 2050 [24,42,43]. As discussed before, many incentives and regulations have been passed to support RNG. These incentives are increasing exponentially in terms of both values and numbers, according to a bench analysis in Canada. Among the solutions for climate change, RNG could be considered a negative carbon fuel. However, other solutions can reduce GHG emissions maybe by 40% in the best-case scenario, and they can not reach net-zero emissions. The cost of large and small scale RNG, biogas, solar, and wind power projects offered by Ontario's feed-in-tariff (FIT) program were estimated by Canadian Gas Association. The average price in 2008 was CAD 8/GJ and has fallen in recent years to CAD 3/GJ due to robust supplies of natural gas. More information on the affordability of natural gas prices can be found here [19]. The lower project cost of RNG and the existing NG pipeline infrastructure are the main reasons for today's fast-growing Canadian RNG marketplace.

Some robust, cutting-edge solutions for OFMSW recovery from MSW in Canada are available. Unlike traditional approaches, these solutions can recover 90% of organics without limitations on in-feed contamination levels. OFMSW is a nutrient-rich feed for AD and can be used in co-digestion plants to increase biogas production. Figure 5 shows the process of waste sorting and recovery in treatment facilities. The OREX™ can separate the OFMSW through a high-pressure extrusion process. The organic fraction is used in the advanced anaerobic digestion process to produce methane-rich biogas and, in turn, electricity, fertilizer, and clean water. In order to convert organic waste into biogas and fertilizer, there are three advanced options—namely, single-phase, two-phase, and high-solid digesters. A single-phase digester is designed for readily biodegradable substrates such as food waste. However, a two-phase digester is ideal for materials needing both primary and secondary fermentation, such as energy crops. The double-ring tank provides two-stage digestion in a decreased footprint.

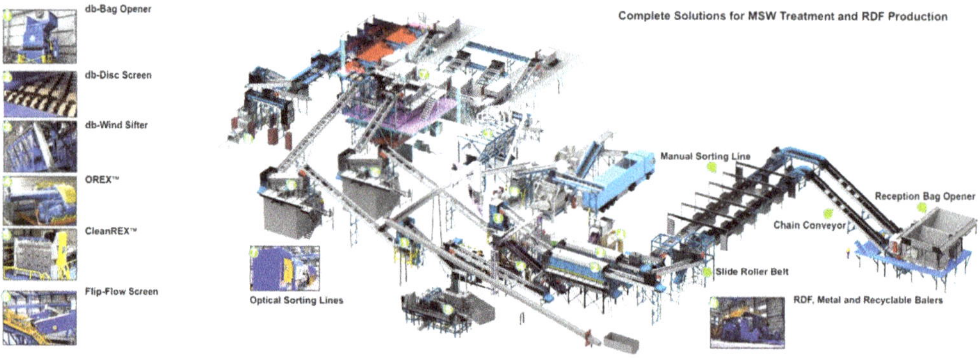

Figure 5. Recovering organics and recyclables from MSW and wet commercial waste. The authors originally produced the figure.

Single- and two-phase digesters can be converted into high-solids digesters using solid Omnivore™ technology to significantly increase capacity and biogas production. The main advantages of a high-solid digester are as follows:

- It transforms municipal WRRFs for co-digestion and works towards energy neutrality;
- It enables reception and co-digestion of high strength waste streams such as fats, oils, and grease, or the organic fraction from municipal waste;
- It reduces foaming potential with high torque mechanical mixing.

Although AD technology is receiving increasing attention due to government incentives and public support, there is still significant room for increasing AD performance via integrating the produced biogas with other renewable energy sources. For example, in WRRFs, anaerobic digesters could be coupled with photovoltaic energy generation, wind turbines, and battery storage, potentially creating a 100% renewable WRRF. Campana et al. developed a model and applied it to a medium-scale Italian municipal WWTP. They also analyzed the costs of installing and operating proposed renewable scenarios [44].

Another important by-product alongside biogas that should be considered in the anaerobic treatment of OFMSW is digestate. Currently, the main focus of Canadian project developers and policymakers is on financial subsidies coming from RNG production. They do not consider the market opportunities of digestate in the agricultural application (i.e., organic fertilizer) and the non-agricultural applications (i.e., soil remediation, biochar production, landfill cover, and landscape restoration). In line with the concept of circular economy, European Commission has specified certain principles to collect revenue from OFMSW digestate. Beggio et al. have recently statistically analyzed the quality of digestate from OFMSW and agro-industrial feedstock. The results suggest that digestate derived from OFMSW could be considered for direct agricultural use as fertilizer. However, it is worth mentioning that the feasibility of using OFMSW digestate should be further investigated by considering hygiene features and ecotoxicological thresholds. This approach is being introduced in other developed countries, including Canada. Policymakers are working on national regulations defining digestate quality to ensure the economic viability and environmental safety of OFMSW digestate use [45]. The most successful reference of using OFMSW digestate as class A fertilizer is the Realto bioenergy facility in California, United States. This facility has the capacity to receive 700 tonnes per day pre-processed SSO and 300 tons per day dewatered WWTP sludge and convert them into 3 MW electricity, 1200 standard, cubic feet per minute RNG, and 26 tonnes per day biochar.

In Canada, to involve stakeholders in AD projects, a national AD guideline document has been developed by the Canadian Biogas Association (CBA) [46]. The guideline includes the following features:

- Best planning, design, and operational practices to assist stakeholders (i.e., project developers, regulators, organizations);
- Recommendations supporting the circular economy concept;
- A clear outline to assist stakeholders in converting food and organic waste streams using AD.

7. Conclusions

Detailed information on the proportion of organic waste diverted from MSW and recent Canadian regulations and incentives enable decision makers to select the best strategies for waste management. The FIT program shifted the conventional waste management strategies to highly integrated biogas production strategies between 2010 and 2017, particularly in Ontario. However, to reach the global Paris Agreement and join leading countries in biogas production, Canada should focus more on the reform of renewable energy policies and economic incentives. The introduction of OFMSW into the cities existing WWTPs provides significant opportunities for Canada's renewable energy market. However, more studies are needed on pretreatment technologies to generate more biogas and accomplish a better rate of organic decomposition to make the process more effective and economically feasible. Most digesters in Ontario work at low volume (1000 to 2000 m^3) and generate 100–500 kW of electricity for being used on-site. The province has only a capacity of around 100 AD facilities that convert organic waste into biofuels, biopower, and bioproducts. Incineration still is the most widely used technology for resource recovery from MSW. Considering how the biogas industry has evolved in Europe, it can be concluded that the biogas industry growth in Canada relies on (1) provincial energy and waste management policies; (2) using advanced technologies for diverting organic waste from landfill; (3) improving biogas yield using existing pretreatment methods; (4) educating farmers regarding digester operations. Future studies in the field should focus on obtaining a 100% renewable energy system and reducing capital investment cost via developing a novel energy system integrating photovoltaic, wind turbine generation, hydrogen, and battery storage.

Author Contributions: Conceptualization, O.N.; methodology, O.N.; software, O.N.; validation, O.N. and A.D.; investigation, O.N.; resources, A.D.; writing—original draft preparation, O.N.; writing—review and editing, A.D. and O.N.; visualization, O.N.; supervision, A.D.; project administration, A.D.; funding acquisition, O.N. and A.D. All authors have read and agreed to the published version of the manuscript.

Funding: This research was funded by the Ministry of Agriculture, Food and Rural Affairs grant number [HQP2019-1592 Project] And Biomass Canada of BioFuelNet Canada Network [Project Number: ASC-16].

Institutional Review Board Statement: Not applicable.

Acknowledgments: The authors would like to express their sincere gratitude to Sasha Rollings-Scattergood, VP Technology at Anaergia, for their helpful advice and support.

Conflicts of Interest: The authors declare no conflict of interest.

Abbreviations

Organic fraction municipal waste	OFMSW	Renewable heat incentive	RHI
Anaerobic digestion	AD	Source-separated organics	SSOs
Feed-in tariff	FIT	California's Low Carbon Fuel Standard	LCFS
Renewable natural gas (NRG)	RNG	Dufferin Organics Processing Facility	DOPF
European Union (EU)	EU	Food waste	FW
Municipal solid waste	MSW	Sewage sludge	SS
Water resource recovery facilities	WRRFs	Microbial electrolysis cell	MEC
Temperature-phased anaerobic digestion	TPAD	Direct interspecies electron transfer	DIET
Hydrothermal carbonization	HTC		

References

1. Parizeau, K.; von Massow, M.; Martin, R. Household-level dynamics of food waste production and related beliefs, attitudes, and behaviours in Guelph, Ontario. *Waste Manag.* **2015**, *35*, 207–217. [CrossRef]
2. Parizeau, K.; von Massow, M.; Martin, R.C. Directly observing household food waste generation using composition audits in a Canadian municipality. *Waste Manag.* **2021**, *135*, 229–233. [CrossRef]
3. Babu, R.; Veramendi, P.M.P.; Rene, E.R. Strategies for resource recovery from the organic fraction of municipal solid waste. *Case Stud. Chem. Environ. Eng.* **2021**, *3*, 100098. [CrossRef]
4. Parvez, A.M.; Lewis, J.D.; Afzal, M.T. Potential of industrial hemp (*Cannabis sativa* L.) for bioenergy production in Canada: Status, challenges and outlook. *Renew. Sustain. Energy Rev.* **2021**, *141*, 110784. [CrossRef]
5. Negri, C.; Ricci, M.; Zilio, M.; D'Imporzano, G.; Qiao, W.; Dong, R.; Adani, F. Anaerobic digestion of food waste for bio-energy production in China and Southeast Asia: A review. *Renew. Sustain. Energy Rev.* **2020**, *133*, 110138. [CrossRef]
6. Zubi, G.; Dufo-López, R.; Carvalho, M.; Pasaoglu, G. The lithium-ion battery: State of the art and future perspectives. *Renew. Sustain. Energy Rev.* **2018**, *89*, 292–308. [CrossRef]
7. Janus, A. More Than Half of All Food Produced in Canada is Lost or Wasted, Report Says. CBC News. Available online: https://www.cbc.ca/news/canada/toronto/food-waste-report-second-harvest-1.49817282019 (accessed on 6 January 2022).
8. Di Maria, F.; Sordi, A.; Micale, C. Energy production from mechanical biological treatment and Composting plants exploiting solid anaerobic digestion batch: An Italian case study. *Energy Convers. Manag.* **2012**, *56*, 112–120. [CrossRef]
9. Canada, S. Table 38-10-0034-01 Materials Diverted, by Type, Inactive. Available online: https://www150.statcan.gc.ca/t1/tbl1/en/tv.action?pid=3810003401 (accessed on 6 January 2022).
10. Canada, S. Table 38-10-0033-01 Materials Diverted, by Source, Inactive. Available online: https://www150.statcan.gc.ca/t1/tbl1/en/tv.action?pid=3810003301 (accessed on 6 January 2022).
11. Mainardis, M.; Buttazzoni, M.; Gievers, F.; Vance, C.; Magnolo, F.; Murphy, F.; Goi, D. Life cycle assessment of sewage sludge pretreatment for biogas production: From laboratory tests to full-scale applicability. *J. Clean. Prod.* **2021**, *322*, 129056. [CrossRef]
12. Norouzisafsari, O.; Di Maria, F.; El-Hoz, M. A short review of comparative energy, economic and environmental assessment of different biogas-based power generation technologies. *Energy Procedia* **2018**, *148*, 846–851. [CrossRef]
13. Di Maria, F.; Sisani, F.; Norouzisafsari, O.; Mersky, R.L. The effectiveness of anaerobic digestion of bio-waste in replacing primary energies: An EU28 case study. *Renew. Sustain. Energy Rev.* **2019**, *108*, 347–354. [CrossRef]
14. Adam Redling Anaergia Begins Construction of North America's Largest WTE Facility. WasteToday 2018. Available online: https://www.wastetodaymagazine.com/article/anaergia-rialto-waste-to-energy/ (accessed on 6 January 2022).
15. Jin, C.; Sun, S.; Yang, D.; Sheng, W.; Ma, Y.; He, W.; Li, G. Anaerobic digestion: An alternative resource treatment option for food waste in China. *Sci. Total Environ.* **2021**, *779*, 146397. [CrossRef]
16. FortisBC Energy Inc. *Renewable Natural Gas Supplier Guide*; FortisBC Energy Inc.: Columbia, Canada, 2018; p. 14.
17. Studies, C. Case Studies Surrey, British Columbia Saint-Hyacinthe, Quebec. 2017. pp. 1–4. Available online: https://biogasassociation.ca/resources/rng_outreach_and_market_development (accessed on 6 January 2022).
18. Audrey. The Renewable Natural Gas Quality Specifications in North America. American Gas Association 2010, pp. 5–7. Available online: https://www.aga.org/research/reports/renewable-natural-gas-rng/ (accessed on 6 January 2022).

19. Canadian Gas Association. *Renewable Natural Gas Technology Roadmap for Canada*; Canadian Gas Association: Ottawa, ON, Canada, 2014; p. 24.
20. Pognani, M.; Barrena, R.; Font, X.; Sánchez, A. A complete mass balance of a complex combined anaerobic/aerobic municipal source-separated waste treatment plant. *Waste Manag.* **2012**, *32*, 799–805. [CrossRef] [PubMed]
21. Alibardi, L.; Cossu, R. Composition variability of the organic fraction of municipal solid waste and effects on hydrogen and methane production potentials. *Waste Manag.* **2015**, *36*, 147–155. [CrossRef] [PubMed]
22. Dalke, R.; Demro, D.; Khalid, Y.; Wu, H.; Urgun-Demirtas, M. Current status of anaerobic digestion of food waste in the United States. *Renew. Sustain. Energy Rev.* **2021**, *151*, 111554. [CrossRef]
23. Zachary, A. Anaerobic digestion can help UK reach renewable energy targets. *Renew. Energy Focus* **2016**, *17*, 21–22. [CrossRef]
24. Ackrill, R.; Abdo, H. On-farm anaerobic digestion uptake barriers and required incentives: A case study of the UK East Midlands region. *J. Clean. Prod.* **2020**, *264*, 121727. [CrossRef]
25. Nevzorova, T.; Kutcherov, V. Barriers to the wider implementation of biogas as a source of energy: A state-of-the-art review. *Energy Strat. Rev.* **2019**, *26*, 100414. [CrossRef]
26. Mondello, G.; Salomone, R.; Ioppolo, G.; Saija, G.; Sparacia, S.; Lucchetti, M.C. Comparative LCA of Alternative Scenarios for Waste Treatment: The Case of Food Waste Production by the Mass-Retail Sector. *Sustainability* **2017**, *9*, 827. [CrossRef]
27. Canada, G. of Zero plastic waste: Canada's actions Canada-wide Strategy on Zero Plastic Waste. Available online: https://www.canada.ca/en/environment-climate-change/services/managing-reducing-waste/reduce-plastic-waste/canada-action.html (accessed on 6 January 2022).
28. Canadian Biogas Association. *RNG Outreach and Market Development*; Canadian Gas Association: Ottawa, ON, Canada, 2017.
29. Canadian Biogas Association. *Biogas Projects in Canada*; Canadian Gas Association: Ottawa, ON, Canada, 2019; pp. 2–3.
30. Canadian Biogas Association. *Current Status and Future Potential of Biogas Production from Canada's Agriculture and Agri-Food Sector*; Canadian Gas Association: Ottawa, ON, Canada, 2018.
31. Biogas World Biogas and Biomethane Projects. 2021. Available online: https://www.biogasworld.com/news/showcase-report-2021-is-out/ (accessed on 6 January 2022).
32. Zhen, G.; Lu, X.; Kato, H.; Zhao, Y.; Li, Y.-Y. Overview of pretreatment strategies for enhancing sewage sludge disintegration and subsequent anaerobic digestion: Current advances, full-scale application and future perspectives. *Renew. Sustain. Energy Rev.* **2017**, *69*, 559–577. [CrossRef]
33. Pagés-Díaz, J.; Pereda-Reyes, I.; Taherzadeh, M.J.; Sárvári-Horváth, I.; Lundin, M. Anaerobic co-digestion of solid slaughterhouse wastes with agro-residues: Synergistic and antagonistic interactions determined in batch digestion assays. *Chem. Eng. J.* **2014**, *245*, 89–98. [CrossRef]
34. Dong, L.; Zhenhong, Y.; Yongming, S. Semi-dry mesophilic anaerobic digestion of water sorted organic fraction of municipal solid waste (WS-OFMSW). *Bioresour. Technol.* **2010**, *101*, 2722–2728. [CrossRef] [PubMed]
35. Ji, C.; Kong, C.-X.; Mei, Z.-L.; Li, J. A Review of the Anaerobic Digestion of Fruit and Vegetable Waste. *Appl. Biochem. Biotechnol.* **2017**, *183*, 906–922. [CrossRef] [PubMed]
36. Anaergia Municipal Solid Waste-Anaergia's Approach for Waste Revalorization. 2021. Available online: https://www.anaergia.com/what-we-do/municipal-solid-waste/materials-recovery (accessed on 6 January 2022).
37. Norouzi, O.; Taghavi, S.; Arku, P.; Jafarian, S.; Signoretto, M.; Dutta, A. What is the best catalyst for biomass pyrolysis? *J. Anal. Appl. Pyrolysis* **2021**, *158*, 105280. [CrossRef]
38. Akbay, H.E.G.; Dizge, N.; Kumbur, H. Enhancing biogas production of anaerobic co-digestion of industrial waste and municipal sewage sludge with mechanical, chemical, thermal, and hybrid pretreatment. *Bioresour. Technol.* **2021**, *340*, 125688. [CrossRef] [PubMed]
39. Chiappero, M.; Norouzi, O.; Hu, M.; Demichelis, F.; Berruti, F.; Di Maria, F.; Mašek, O.; Fiore, S. Review of biochar role as additive in anaerobic digestion processes. *Renew. Sustain. Energy Rev.* **2020**, *131*, 110037. [CrossRef]
40. Ghysels, S.; Acosta, N.; Estrada, A.; Pala, M.; De Vrieze, J.; Ronsse, F.; Rabaey, K. Integrating anaerobic digestion and slow pyrolysis improves the product portfolio of a cocoa waste biorefinery. *Sustain. Energy Fuels* **2020**, *4*, 3712–3725. [CrossRef]
41. Arif, S.; Liaquat, R.; Adil, M. Applications of materials as additives in anaerobic digestion technology. *Renew. Sustain. Energy Rev.* **2018**, *97*, 354–366. [CrossRef]
42. Medrano, J.; Llosa-Tanco, M.; Cechetto, V.; Tanaka, D.A.P.; Gallucci, F. Upgrading biogas with novel composite carbon molecular sieve (CCMS) membranes: Experimental and techno-economic assessment. *Chem. Eng. J.* **2020**, *394*, 124957. [CrossRef]
43. Stephen, J. Renewable Natural Gas (Biomethane) Feedstock Potential in Canada. 2020. Available online: https://www.enbridge.com/~{}/media/Enb/Documents/Media%20Center/RNG-Canadian-Feedstock-Potential-2020%20(1).pdf?la=en (accessed on 6 January 2022).
44. Campana, P.E.; Mainardis, M.; Moretti, A.; Cottes, M. 100% renewable wastewater treatment plants: Techno-economic assessment using a modelling and optimization approach. *Energy Convers. Manag.* **2021**, *239*, 114214. [CrossRef]
45. Beggio, G.; Schievano, A.; Bonato, T.; Hennebert, P.; Pivato, A. Statistical analysis for the quality assessment of digestates from separately collected organic fraction of municipal solid waste (OFMSW) and agro-industrial feedstock. Should input feedstock to anaerobic digestion determine the legal status of digestate? *Waste Manag.* **2019**, *87*, 546–558. [CrossRef]
46. Ellis, D. Canadian Anaerobic Digestion Guideline. 2019; pp. 1–76. Available online: https://biogasassociation.ca/resources/canadian_anaerobic_digestion_guideline (accessed on 6 January 2022).

energies

MDPI

Article

Duty to Address Climate Change Litigation Risks for Australian Energy Companies—Policy and Governance Issues

Prafula Pearce

School of Business and Law, Edith Cowan University, Joondalup, WA 6027, Australia; p.pearce@ecu.edu.au

Abstract: The transition from fossil fuels to renewable energy requires cooperation from all, including corporations, shareholders, and institutional investors. The purpose of this paper is to explore climate change litigation risks for Australian energy companies and investors from a policy and governance perspective. Companies are increasingly reporting their climate policies to satisfy their shareholders and investor demands. In addition, the government and judiciary are making laws and decisions to support the Paris Agreement. This paper explores whether company directors can and, in some cases, should be considering the impact of climate change litigation risks on their business, or else risk breaching their obligation to exercise care and diligence under the Corporation Act 2001 (Cth, Australia). The paper concludes that in addition to reducing climate change litigation risks, Australian energy companies and institutional investment bodies that invest in Australian energy companies can make informed climate risk decisions by aligning their investments with the goal of net-zero or reduced emissions.

Keywords: energy companies; climate change litigation risks; directors duties; policy and governance

Citation: Pearce, P. Duty to Address Climate Change Litigation Risks for Australian Energy Companies—Policy and Governance Issues. *Energies* **2021**, *14*, 7838. https://doi.org/10.3390/en14237838

Academic Editor: Dalia Štreimikienė

Received: 17 October 2021
Accepted: 11 November 2021
Published: 23 November 2021

Publisher's Note: MDPI stays neutral with regard to jurisdictional claims in published maps and institutional affiliations.

1. Introduction

The landmark Dutch court decision *Milieudefensie v Shell* [1] handed down in Hague District Court on 26 May 2021, held that Royal Dutch Shell PLC, a global energy company, had a duty of care to reduce CO_2 emissions in its entire global value chain. The Court ordered Royal Dutch Shell to reduce the group's emissions by 45% through Shell group's corporate policy. This decision has ramifications for energy companies around the world in assessing their climate change litigation risks.

The conceptual measure of global climate change litigation risk can be measured with two main databases that maintain details of climate litigation cases: the Climate Change Laws of the World (CCLW) database maintained by the Grantham Research Institute of Climate Change and the Environment [2]; and the United States Climate Litigation Database maintained by the Sabin Centre for Climate Change Law [3]. The *Global trends in climate change litigation: 2021* report identified 1841 cases of climate change litigation around the world as of 31 May 2021, with the United States having the highest number of cases, totalling 1387, followed by Australia with 115 cases [4]. The forecast is for climate change litigation cases to grow in all countries, especially against corporations. However, the focus of this paper is not on global climate change cases, or the issues involved in the global cases against corporations as listed in Appendix B at the end of this paper. The global listings do, however, provide a conceptual measure of the climate change litigation risks. The purpose of this paper is to explore climate change litigation risks for Australian energy companies and how the directors can prevent the company from being exposed to this risk. Australian companies are increasingly reporting their climate policies to satisfy their shareholders and investor demands. In addition, the Australian government and judiciary are making laws and decisions to support the Paris Agreement.

The starting point for a discussion of global international obligations to climate change is the United Nations Framework Convention on Climate Change (UNFCCC). The UNFCCC has near-universal membership of 197 countries, including Australia, that have ratified the Convention. The UNFCCC has also established the Conference of the Parties (COP), which are the meetings of the participating countries to discuss climate change agenda. The most significant COP has been COP21, the 2015 Paris Agreement [5]. The Paris Agreement has changed from Kyoto's top-down approach of countries complying with legally binding commitments for emission reductions to a bottom-up approach whereby countries make their own commitment by pledging nationally determined contributions. This bottom-up approach recognises that adaptation is a global challenge faced by all. This means a broader responsibility than just through government policies. The responsibility is increasingly put on investors and corporations to make the right choices.

Climate change is also impacting institutional investors, as they are increasingly becoming aware of climate litigation risks, since investments in projects exposed to climate risks often flow into financial risks. Financial institutions may expose themselves to litigation by not addressing the Paris Agreement's objective of making "finance flows consistent with a pathway towards lower greenhouse gas emissions and climate resilient development [6]". The reason for this is that climate litigation can expose financial institutions to stranded assets from abandoned projects if the judge orders the project to halt due to associated climate risks. It is not possible to completely divest from these risks. However, the message from the litigated cases is being responded to through tangible investment decisions. The message of this paper is to make the financial institutions and corporations aware of the litigation risks if they ignore climate risks in investment and business decisions. In addition, they can become exposed to litigation for breaches of the Australian *Corporations Act 2001* (Cth, Australia) and climate legislation, as discussed below.

This review paper is in five parts. Following this brief introduction, Section 2 explores the climate change litigation risks faced by energy and superannuation companies in Australia that invest in nonrenewable energy. Section 3 then examines whether climate change risks may be relevant to an Australian company director's duty of care and diligence under s180(1) of the *Corporations Act 2001* (Cth, Australia) to the extent that they interact with the interests of the company, followed by Section 4, which points to actions that the directors of Australian energy companies can take to minimise the impact of climate change litigation risks on their businesses and prevent the risk of breaching their directors' duties under the *Corporations Act 2001* (Cth, Australia). Part 5 concludes the paper, stating that the role of directors is to manage physical and transitional risks and test company strategy against climate change scenarios, including timely flow of climate-change-related information from management to the board.

2. Climate Change Litigation Risks for Both Energy and Superannuation Companies in Australia

Companies involved in the production of energy resources, by the very nature of their operations, should be considering climate change litigation risk as a potential business risk when their directors and managers make management decisions.

Major companies and financial institutions in Australia are increasingly being challenged in their investment decisions that impact upon climate change, as shown in selected cases listed in the Appendix A of this paper. Some of the litigation issues that have impacted climate change risks for energy and superannuation companies are:

- Shareholders seeking disclosure from the Commonwealth Bank of Australia on internal documents under the *Corporations Act 2001* (Cth, Australia) relating to projects to ascertain whether the projects comply with the goals of the Paris Agreement [7];
- An institution (the Australasian Centre for Corporate Responsibility) challenging whether a company involved in oil and gas made a false and misleading representation about its net-zero emissions plan on the basis that natural gas is a clean fuel [8];
- An institution (Environment Victoria Inc.) challenging whether the Environmental Protection Authority had correctly considered the Climate Change Act when a new license for coal-burning power stations that failed to lower the limits of GHG emissions was granted [9];

- Climate activist Sharma (a teenager) challenging the Australian government's approval of a coal mine on the basis that the Minister owes a duty of care to avoid personal injury to children [10];
- An Australian government bondholder suing the Australian government for failure to disclose climate risks, and hence misleading and deceiving investors by failing to disclose such risks [11];
- A member of an Australian pension fund alleging that the Retail Employees Superannuation Trust (REST) failed to provide information relating to climate change business risks and any plans to address those risks, and hence violated the *Corporations Act 2001* (Cth, Australia) [12];
- The Minister for the Environment and Energy not taking into consideration the physical impacts of climate change on the Great Barrier [13];
- An application to open a new open cut mine was rejected, and one of the grounds for rejection was the impact of the mine on climate change [14].

Australia is a signatory to the Paris Agreement [15], whereby it agrees to contribute to keeping the global average temperature risk to 1.5° to 2° by a reduction in greenhouse gas emissions, namely carbon dioxide and methane from the burning of coal. This causal factor is well known to the judiciary, who have commented that greenhouse emissions "adversely impact upon measure to limit dangerous anthropogenic climate change [16]" That is, the burning of extracted coal, with the subsequent release of greenhouse gases, will have a cumulative effect on climate change globally [17]. As such, Australia has adopted a Carbon Budget Approach in order to highlight trade-offs involved between actions taken now to reduce greenhouse gas emissions and those made necessary later [18]. The carbon budget approach measures carbon sources against carbon sinks. In the case of *Gloucester Resources Limited v Minister for Planning* [14], the Federal Court of Australia defined this approach as being based on the close interplay between planetary temperature warming and the cumulative effect of anthropogenic emissions. The Court argued that this approach is a "scientifically robust approach to estimating the level of greenhouse gas emission reductions required to meet a desired temperature target". It was argued by the appellant that if the respondent refused authority for the mine site to be developed and opened, any savings to the emissions total would be negated globally, as another mine in a (probably lesser-developed) country would open in any event with less stringent supervision, and replace the emissions saved in Australia. The court found that this "market substitution" was a flawed argument to make, as it found there was no certainty of mines in other areas opening up. In fact, the strong position taken by Australia in this scenario could lead those lesser-developed countries to "follow suit". The US courts have also stated that the "market substitution" theory is one that is arbitrary, capricious, and irrational [19].

Investment companies, such as those in superannuation, are equally put at litigation risk for lack of climate change considerations in their investment decisions. Investments in coal-producing mines are no longer seen as a viable option, which could lead to such investments being stranded [20], with courts willing to directly link the burning of fossil fuels to climate change [14]. As a result, private corporations involved in the finance sector are being held to account over expectations of disclosing to shareholders corporate consideration relating to climate change [21]. Where they fail to meet the necessary level of specific disclosure, they face unwanted scrutiny, either through regulatory bodies or through activist third-parties [22].

In *Mark McVeigh v Retail Employees Superannuation Pty Ltd.* [23], a private suit brought against a major superannuation fund, whereby a shareholder sought specific climate change disclosure information, as that information was nonspecific and inadequate in detail. Declaratory relief was sought in the suit because the fund had violated the *Corporations Act 2001* (Cth, Australia) by failing to disclose the information, and further, an injunction was sought against the fund to produce the sought information. It was further alleged that the fund breached the *Superannuation Industry (Supervision) Act* [24], in that a trustee for the fund would have been able to ensure that the investment managers could have produced the required information, and further that such managers could ensure that such climate change specific information could be made easily accessible to beneficiaries of the fund, and thereby comply with recommendation of the task force on Climate-Related Financial Disclosures [12]. Justice Perram commented, during the maximum costs order hearing (paragraph 9) that:

"The case appears to raise a socially significant issue about the role of superannuation trusts and trustees in the current public controversy about climate change. It is legitimate to describe the Applicant's litigation as being of a public interest nature [12]."

The case was settled just before the hearing was due to commence. However, the case points towards a conclusion that "financial institutions, corporates and global investment funds, are likely to face increased scrutiny from their stakeholders with respect to their climate change policies [25]".

Superannuation investments are being adversely impacted by climate change from the transition of economic reliance on fossil fuels and the physical damage arising from the natural disasters that are becoming all the more common [26]. Investing in such industry that cumulatively impacts on climate change could be seen as not acting in the best interests of the investor [27]. Breach of fiduciary duties could then be alleged. One such fund, Unisuper, which operates an AUD 85 million fund with more than 450,000 members, has refused to divest its AUD 170 million investment in coal companies or its AUD 7.8 million investment in other fossil fuel companies, stating that selling the holdings would deprive them of the right to influence the industries in making a change. However, this is becoming the exception and not the rule, with the Australian Centre of Corporate Responsibility finding an increasing number of shareholder proposals for investments in fossil fuel mining companies being rejected by Australian super funds [26].

Regulators in Australia are being guided by the recommendations from the UNFCCC-approved Task Force on Climate-Related Financial Disclosures (TCDF). The TCDF's recommendations direct a company's climate change risk practices to four areas for adequate oversight and operations: governance, strategy, risk management, and metrics and targets [28]. The TCFD framework recommendations have been endorsed by the regulators in Australia, including the Reserve Bank of Australia (RBA), Australian Prudential Regulation Authority (APRA), the Australian Council of Superannuation Investors' (ACSI), and the Australian Securities and Investment Commission (ASIC) [29].

APRA currently supervises AUD 7.7 trillion in assets for Australian depositors, policyholders, and superannuation fund members. APRA has been raising awareness of climate-related risks to the financial sector and is developing prudential practice guidance that will be released at the end of 2021. The APRA guidance has adopted the TCDF recommendations [30].

The ACSI Governance Guidelines (2019) expect companies to disclose their approach to climate-related risks by adopting the TCFD and align their corporate strategy to the Paris Agreement (net zero by 2050), and the council may direct its members to vote against directors that fall short of managing their climate-related risks [31].

The company directors of energy companies should also observe the ASIC Commissioner's direction to have appropriate governance structures in place to manage climate-related risk and comply with reporting requirements established by the TCDF, and thereby provide their shareholders with reliable and useful information on the exposure to material climate-related risks and opportunities [32].This clear shift in focus in forcing compliance with strategies to meet the Paris Agreement's emission levels is beginning to involve the judiciary in enforcing that compliance and reporting obligations against companies that either directly or indirectly have an impact on the emission of greenhouse gases and hence there is a rise in climate change ligitation.

Ultimately, times have changed, and all corporations that directly or indirectly do business with those who emit greenhouse gases have an ever-increasing mandate to report upon and disclose such transactions. In addition to enforcement through Australian regulatory bodies, individuals (and activist groups) are obtaining greater recognition for standing in courts to litigate against companies that fail to consider or adequately disclose climate change risks.

In order to avoid unnecessary litigative risks, companies across the board, especially energy companies, should assess how their interests affect the possibility of changes to climatic conditions.

3. Relevance of Climate Change Risks on Director's Duty of Care and Diligence

As discussed above, Australian regulatory bodies have increasingly begun to focus on potential obligations of directors regarding environmental issues, such as climate change risks, in their management decisions. In Australia, the obligations to consider climate change risks also arise from Section 180 (1) of the *Corporations Act 2001* (Cth, Australia), which requires company directors to exercise their powers with the degree of care and diligence that would reasonably be expected of a director in the same position and responsibility of a corporation in those circumstances [33]. The degree of care refers to the degree of attention and thorough conduct necessary to act for the benefit of the company [34]. Diligence refers to the consistent attention a director pays to their responsibilities and in maintaining the minimum standards expected of that role [35]. An objective reasonable standard is used in determining a breach of the duty, and similar to other risks, climate risks would also be weighed against the magnitude of the risk of harm and the probability of it occurring; the seriousness of the resulting loss, should the harm occur; and the expense, difficulty, and inconvenience of taking alleviating action [36].

In assessing the climate change risks regarding directors' duty of care and diligence, it should be noted that the business judgement rule in s180 (2) of the *Corporations Act 2001* (Cth, Australia) is limited in application where "decisions" are made by a director to act or to not act. Those who fail to make a decision or to turn their mind to the issues presented by climate change will not be afforded the protection of the business judgment rule [37].

Directors' duties and their relevance to climate change risk were considered by the Centre for Policy Development and the Future Business Council, which commissioned the Australian Legal Memorandum of Opinion "Climate Change and Director's Duties" [38]. The report confirmed that climate change risks are capable of representing a risk to the interests of an Australian company and are relevant to a director's duty of care [39]. In addition to the Hutley 2016 and 2019 opinions, a further 2021 Supplementary Memorandum of Opinion was issued as follows:

"In 2016, our focus was the existence of the duty; that is, what directors could and should be doing on climate change to discharge their duty of due care and diligence. That is now uncontroversial. In 2019, we observed that the risk of liability for directors on this front was rising exponentially. In 2021, it appears to us that the focus is increasingly on how the duty is discharged" [40].

The Hutley 2021 report states that directors' disclosures of climate actions should be accurate, as inaccurate statements could lead to misleading conduct commonly known as "greenwashing" [40].

As regards the extent to which climate change risks are of relevance to the care and diligence duty under s180 of the *Corporations Act 2001* (Cth, Australia), the authors of the Hutley report took the view that such risks are relevant to the extent that they intersect with the interests of the company [39]. Particular emphasis was placed on the issue of foreseeability, and the potential ramifications for directors who fail to address foreseeable risks [38]. The general position in Australian law is that a risk is foreseeable so long as it is not far-fetched or fanciful [41], meaning a risk that is unlikely may nonetheless be entirely foreseeable. Furthermore, a plaintiff is not required to prove that a director's conduct in respect to a foreseeable risk resulted in actual loss [33]. Thus, any harm incurred by a company because of failure to mitigate their response to the risk could be deemed foreseeable by a court, as the risks associated with both climate change and attenuating global warming are "significant and well publicised" [39], not "far-fetched or fanciful". Whilst climate change risks have historically been viewed as a future or nonfinancial problem [42], the Hutley supplementary memorandum contends that intersection between such risks and the interests of companies, particularly in sectors such as energy resources and institutional investment, is inevitable. Thus energy company directors should take heed of the Hutley opinion that directors of Australian listed companies can potentially be liable for material harm to their company, should they fail to adequately consider and disclose foreseeable climate change risks [43].

The next section explores how the directors of energy companies may be able to minimise the impact of climate change litigation risks on their businesses, and also avert the risk of breaching their obligations to exercise care and diligence.

4. Direction for Energy Company Directors to Minimise the Impact of Climate Change Litigation Risks

Climate change risks can be classified as transition risks and physical risks [43]. The physical risks brought about by climate change include increased temperatures, change in rainfall patterns, and an increase of frequency and/or intensity of extreme weather events such as heatwaves, drought, storms, flooding, and rising sea levels [44]. A prudent director of an energy company should include these factors in the future planning of the business. The financial implications of physical risks should also be considered, and this may include damage to assets [45]. Transitional risks refer to the regulatory risks and opportunities associated with carbon emissions and related pollutants. Examples include costs associated with retooling to fit within a low-carbon transition. Transitional impact also includes the risks of litigation, exposure to damage claims, operational disruption, and costs of enforcement of disclosure obligations and potential reputational damage [46].

A director of an energy company should also be aware that these climate change risks include advances in scientific discoveries and the attitudes of Australian regulators and investor groups. The courts will take into account all of these matters when deciding if a director of an energy company has appropriately responded to the risk and made proper disclosure of it [39]. The Australian Securities and Investment Commission (ASIC) published a report in 2018 suggesting that directors of listed companies should consider short-term and long-term climate change risk when assessing risks to a company [47].

The courts will, of course, balance the climate risk posed with potential benefits to the corporation [38], but the corporation needs to heed the commercial consequences that will flow from any breach of duty, as any adverse scrutiny could lead to reputational damage [48] and possible corporate failure.

Though no court in Australia has yet specifically ruled on whether a director's duty under s180 of the *Corporations Act 2001* (Cth, Australia) includes the appropriation of climate change risk, it has been reported by Kenneth Hayne QC that the duty to act in the best interest of the corporation incorporates climate change considerations [49]. The 2019 Hutley legal opinion that views climate change as being a "foreseeable risk to the interest of a company [39]", has been endorsed by the Australian Securities and Investment Commission (ASIC) as being legally sound and accurate as it pertains to prevailing laws in Australia [50]. It has also been supported by the ASX Corporate Governance Council, the Australian Prudential Regulation Authority, the Australian Accounting Standards Board, and the Auditing and Assurance Standards Board.

The Hutley 2021 opinion provides the following practical steps for directors:

- Develop a net-zero strategy that is integrated with a company's operational strategy;
- Document and test the assumptions underpinning the strategy, including any offsets;
- Explain which emissions are included in the strategy and express the scope and timing of the commitment; and

- Disclose promptly the circumstances that affect the strategy from being accomplished [40].

The number of recent climate change cases in the courts points to a conclusion that climate change is a serious factor for consideration for a director of an energy company in demonstrating the exercise of the duty of care and diligence in the management of the corporation. Company directors may risk breaching their duty of care and diligence if they fail to consider the impact of climate change risks on their businesses. The director has a duty to consider the gravity (seriousness) of risk, probability of outcome, cost of prevention, and other obligations when deciding whether to incorporate climate change risks into their decision making. This means that the director must consider climate change risks at least to the degree of determining whether they are a relevant factor in the decision they are making. Failure to contemplate climate change risks, and the reporting thereof, could place the company at the risk of facing litigation.

5. Conclusions

The pathway to avoid the risk of Australian energy company directors breaching duties relating to climate change litigation risks is to follow the Hutley 2021 recommendations, and in particular to include more detailed disclosures in line with TCDF recommendations. Directors should consider who may be relying on climate-related disclosures. They should have systems in place to regularly check whether appropriate disclosures pertaining to climate change have been made and determine the accuracy and completeness of climate-related disclosures.

In addition, the boards of energy companies should regularly consider climate change matters and risks, including the management of physical and transitional risks, and test company strategy against climate change scenarios, including timely flow of material; i.e., climate-change-related information from management to the board.

In addition to reducing climate change litigation risks, corporations and institutional investment bodies can provide support to their government by making informed climate risk decisions and by aligning their investments with the goal of net-zero or reduced emissions. However, this is not an easy task for directors of energy companies to reduce litigation risk exposure by setting short, medium, and long-term emission-reduction targets that align to the Paris Agreement and the objective of net-zero emissions by 2050. Energy company directors have to continue to attract investments from investors and also play a major part in global reduction of emissions by pledging major emission reductions in support of the Paris agreement, as many large companies have already done [51].

Funding: This research received no external funding.

Institutional Review Board Statement: Not applicable.

Informed Consent Statement: Not applicable.

Data Availability Statement: Not applicable.

Acknowledgments: The author would like to acknowledge the assignment research of the students of Environmental Law in 2 Semester 2020 at the School of Business and Law, Edith Cowan University, in particular Clifford Warner and Alicia Nowak.

Conflicts of Interest: The authors declare no conflict of interest.

Appendix A

Table A1. Selected climate change litigation in Australia.

Case Name	Filing Date	Court	Status as of October 2021	Summary of Finding Relevant to Climate Change
Abrahams v Commonwealth Bank of Australia	2021	Federal Court of Australia	Pending	Shareholders of Commonwealth Bank of Australia brought an action to obtain documents to verify whether the bank carried out an environmental assessment of its involvement in projects in line with the Paris agreement.
Australasian Centre for Corporate Responsibility v Santos	2021	Federal Court of Australia	Pending	The Environmental Defenders Office challenged the company Santos' claims of net-zero emissions target by 2040.
Complaint to Ad Standards on HSBC's Great Barrier Reef ad	2021	Ad Standards	Pending	An action challenging HSBC's greenwashing practice on its promotion of Great Barrier Reef protection.
Environment Victoria vs. the EPA et al.	2021	Victorian Supreme Court	Pending	Environment Victoria is challenging the EPA for failing to protect the community by granting license to operate coal mines.
Sharma and others v Minister for the Environment	2020	Federal Court of Australia	Appeal Pending	Climate activist Sharma (a teenager) is challenging the Australian Government for approving a coal mine on the basis that the Minister owes a duty of care to avoid personal injury to children.
O'Donnell v Commonwealth	2020	Federal Court of Australia	Pending	Action brought by a holder of Austrtalian bonds against the Commonwealth Government for failure to disclose climate risks.
Youth Verdict v Waratah Coal	2020	Queensland Land Court	Pending	Youth Verdict Limited application for objection to develop a thermal coal mine in Galilee Basin.
KEPCO Bylong Australia v Independent Planning Commission and Bylong Valley Protection Alliance	2019	New South Wales Court of Appeal	Decided	The Independent Planning Commission has discretion to reject the planning application of a coal mine on the basis of climate change impacts.
McVeigh v Retail Employees Superannuation Trust	2018	Federal Court of Australia	Decided	A pension fund member sued a superannuation fund, who settled the case and agreed to take into considration climate risks and reach net zero by 2050.
Australian Conservation Foundation Incorporated v Minister for the Environment and Energy	2017	Federal Court of Australia	Decided	The appeal related to the Minister for the Environment and Energy not taking into consideration the physical impacts of climate change on the Great Barrier Reef.
Gloucester Resources Limited v Minister for Planning	2017	New South Wales Land and Environment Court	Decided	An application to open a new open cut mine was rejected, and one of the grounds for rejection was the impact of the mine on climate change.

SOURCE: For further information, see the Climate Change Litigation Database, Columbia University, at http://climatecasechart.com/climate-change-litigation/non-us-jurisdiction/australia/ (accessed on 12 October 2021).

Appendix B

Table A2. Selected non-US climate change litigation cases against corporations.

Country	Name of Case	Issue
Australia	Mullaley Gas and Pipeline Accord Inc v Santos NSW (Eastern) Pty Ltd.	Challenge to the development consent of the Narrabri Gas Project by the Independent Planning Commission.
UK	ASA Ruling on Ryanair Ltd. t/a Ryanair Ltd.	Challenge to the accuracy of advertisement relating to emissions from an airline.
Germany	Barbara Metz et al. v Wintershall Dea AG	Petitioners challenging car companies' emissions by producing internal combustion engines.
Germany	Deutsche Umwelthilfe (DUH) v Mercedes-Benz AG	Petitioners challenging car companies' emissions by producing internal combustion engines.
Germany	Deutsche Umwelthilfe (DUH) v BMW	Petitioners challenging car companies' emissions by producing internal combustion engines.
Australia	Complaint to Ad Standards on HSBC's Great Barrier Reef ad	An action challenging HSBC's greenwashing practice on its promotion of Great Barrier Reef protection.
Australia	Abrahams v Commonwealth Bank of Australia (2021)	Shareholders of Commonwealth Bank of Australia brought an action to obtain documents to verify whether the bank carried out an environmental assessment of its involvement in projects in line with the Paris Agreement.
Brazil	Instituto Preservar et al. v Copelmi Mineração Ltd.a. and IBAMA	Whether precautionary measures should be taken to develop open-pit coal mining.
Australia	Australasian Centre for Corporate Responsibility v Santos	The Environmental Defenders Office challenged the company Santos' claims of net-zero emissions target by 2040.
New Zealand	Complaint by Lawyers for Climate to the Advertising Standards Board	Greenwashing advertisement challenged.
South Africa	South Durban Community Environmental Alliance v Minister of Environment and Others	South Africa's authorization of oil exploration challenged on the grounds of climate violations.
Brazil	Ministério Público Federal v de Rezende	Deforestation in the Amazon challenged.
UK	R v Bramwell et al. ("The Shell Six case")	Environmental protestors charged for criminal damage.
UK	Attorney General v Crosland	Unregistered barrister convicted of criminal contempt of court for breaching a court embargo of publication on a court judgement with an environmental issue.
Belgium	ClientEarth v Belgian National Bank	NGO seeks to stop the bank from directing its capital to programmes that affect climate.
France	Envol Vert et al. v Casino	NGO is suing a French supermarket for seeking supplies of cattle from areas that impact the environment.
France	Friends of the Earth et al. v Prefect of of Bouches-du-Rhône and Total	Challenging a permit to operate a biorefinery on the basis of environmental impact.
Australia	Conservation Council of Western Australia v Hatton and Woodside	Challenges to approval of gas projects without full climate impact assessment.
Argentina	Carballo et al. v MSU S.A., UGEN S.A., & General Electric	Whether environmental impact assessment was flawed.
Argentina	OAAA v Araucaria Energy SA.	Whether environmental impact assessment was flawed.
Argentina	Hahn et al. v Araucaria Energy Sociedad Anonima	Whether environmental impact assessment was flawed.
Argentina	Hahn et al. v APR Energy S.R.L	Whether environmental impact assessment was flawed.
Argentina	FOMEO v MSU S.A., Rio Energy S.A., & General Electric	Whether environmental impact assessment was flawed

Table A2. *Cont.*

Country	Name of Case	Issue
Australia	Friends of Leadbeater's Possum Inc v VicForests (No 3)	Conservation group seeking protection of the environment.
Australia	EH v Queensland Police Service; GS v Queensland Police Service	Climate protesters' convictions for protesting against coal mine.
Australia	Australasian Centre for Corporate Responsibility (ACCR) v Commonwealth Bank of Australia	Shareholders sought climate disclosures.
Canada	Trans Mountain Pipeline ULC v Mivasair	Pipeline protesters sought to assert necessity defense
Poland	Development YES—Open-Pit Mines NO v Group PZU S.A.	NGO claimed that Polish NCP did not observe OECD National Guidelines relating to environmental protection.
Australia	Youth Verdict v Waratah Coal	Youth Verdict Limited application for objection to develop a thermal coal mine in Galilee Basin.
Germany	Germanwatch vs. Volkswagen	Whether Volkswagen violated climate obligations.
Norway	Norwegian Climate Network et al. v Statoil	Claim that oil sands must not be exploited for climate stability.
Netherlands	BankTrack and Friends of the Earth Netherlands v ING Bank	Complaint against the bank for not committing to OECD guidelines on climate change.
Japan	Market Forces v SMBC, MUFG and Mizuho	Complaint against funding of coal mines in Vietnam by Japanese banks.
Brazil	Federal Environmental Agency (ibama) v Siderúrgica São Luiz Ltd.	Action against steel company to prevent deforestation by not sourcing coal.
Poland	Greenpeace Poland v PGE Giek	Greenpeace sues to stop fossil fuel investment.
New Zealand	Smith v Fronterra Co-Operative Group Limited & Ors	Maori heritage spokesperson claims company owes duty to cease contributing to climate change.
Japan	Citizens' Committee on the Kobe Coal-Fired Power Plant v Kobe Steel. Ltd., et al.	Construction of coal-fired plant was challenged.
Switzerland	Credit Suisse Protesters v Credit Suisse	Activists protesting for Credit Suisse to heed climate change.
UK	ClientEarth v BP.	Whether BP misled the public about its presentation of low-carbon activities.
France	Friends of the Earth et al. v Total.	Whether company owes duty of vigilence to assess threats of oil projects to human rights.
Argentina	Mapuche Confederation of Neuquén v YPF.	Dispute about dumping dangerous waste and harming the environment.
	R v Roberts	Protestors dispute about anti-fracking.
UK	R v Basto	Climate activists challenged for protesting near airport.
Poland	ClientEarth v Polska. Grupa Energetyczna.	Seeking Europe's largest power plant operator to reduce emissions.
France	The Take Down Macron campaign Cases	Activists challenged for protesting against France's failure to meet climate targets.
France	Notre Affaire a Tous and Others v Total.	Whether climate change risks were adequately reported.
Netherlands	Milieudefensie et al. v Shell	Whether a multinational corporation is obliged to curtail carbon dioxide emissions.
Poland	ClientEarth v ENEA	Breach of fiduciary duties for decision to construct coal plant.
Australia	McVeigh v Retail Employees Superannuation Trust	A pension fund member sued a superannuation fund, who settled the case and agreed to take into considration climate risks and reach net zero by 2050.

Table A2. *Cont.*

Country	Name of Case	Issue
Brazil	Public Prosecutor's Office v H Carlos Schneider S/A Comércio e Indústria & Others.	Dispute about draining and clearing mangrove forest.
Brazil	Public Prosecutor's Office v Oliveira & Others.	Challenged low-tech sugar refining's impact on emissions.
Brazil	Sao Paulo Public Prosecutor's Office v United Airlines.	Seeking airlines to use regional airports to offset emissions.
Germany	Lliuya v RWE AG.	Whether the German company that produces energy contributes to climate change that caused damage to property.
UK	The Kingsnorth Six Trial	Greenpeace activists attempted to shut down coal-power station.
UK	Heathrow Airport Ltd. & Ors v Garman & Ors.	Protestors campaign near Heathrow Airport challenged.
UK	Grainger plc and others v Nicholson.	Employees claimed that climate change belief is not merely an opinion.
UK	Deutsche Bank AG v Total Global Steel Ltd.	Breach of contract by using "surrendered" Certified Emissions Reductions (CERs).
UK	CF Partners (UK) LLP v Barclays Bank PLC.	Misuse of confidential information relating to carbon credits.
Philippines	In re Greenpeace Southeast Asia et al.	Fossil fuel companies challenged for climate change impacts.
New Zealand	Royal Forest and Bird Protection Society of New Zealand Incorporated v Buller Coal Ltd.	Dispute about whether regard should be given to effects of climate change arising from use of coal.
Canada	Weaver v Corcoran.	Dispute over articles about global warming.
Canada	Chicago Climate Exchange, Inc v Montreal Green Exchange.	Challenge of a trademark application, as the name was similar
Australia	Australian Competition and Consumer Commission v V8 Supercars Australia Pty Ltd.	Claims made about offsetting carbon emissions from its V8 car racing series challenged.
Australia	Australian Competition and Consumer Commission v Prime Carbon Pty Ltd.	Challenge to representations about carbon sequestration program.
Australia	Australian Competition and Consumer Commission v Goodyear Tyres.	Challenge to misleading consumers about the environmental benefits.
Australia	Australian Competition and Consumer Commission v GM Holden Ltd.	Green claims challenged in advertising Saab vehicles.
Australia	Australian Competition and Consumer Commission v Global Green Plan Ltd.	ACCC pursues court order for Green Plan to puchase Renewable Energy Certificates.
Australia	Australian Competition and Consumer Commission v De Longhi Australia Pty Ltd.	De Longhi's environmental claims challenged by ACCC.

Extracted from: http://climatecasechart.com/climate-change-litigation/non-us-case-category/corporations/ (accessed on 9 November 2021).

References

1. District Court of The Hague. *Milieudefensie and others v. Royal Dutch Shell*; 26 May 2021 ECLI:NL:RBDHA:2021:5337; District Court of The Hague: The Hague, The Netherlands.
2. Available online: https://climate-laws.org/ (accessed on 9 November 2021).
3. Available online: http://climatecasechart.com/climate-change-litigation/ (accessed on 9 November 2021).
4. Grantham Research Institute on Climate Change and the Environment and the Centre for Climate Change Economics and Policy. *Global Trends in Climate Change Litigation: 2021 Snapshot*. Policy Report July 2021. Available online: <https://www.lse.ac.uk/

granthaminstitute/wp-content/uploads/2021/07/Global-trends-in-climate-change-litigation_2021-snapshot.pdf10 (accessed on 9 November 2021).
5. Available online: https://unfccc.int/process-and-meetings/the-paris-agreement/the-paris-agreement/key-aspects-of-the-paris-agreement (accessed on 31 March 2021).
6. International Institution for Sustainable Development. *Making Finance Flows Consistent with the Paris Agreement*. Commentary 11 February 2020. Available online: https://sdg.iisd.org/commentary/policy-briefs/making-finance-flows-consistent-with-the-paris-agreement/ (accessed on 31 March 2021).
7. Equity Generation Lawyers. *Abrahams v Commonwealth Bank of Australia 2021*. Available online: https://equitygenerationlawyers.com/abrahams-v-commonwealth-bank-of-australia-2021/ (accessed on 26 August 2021).
8. Federal Court of Australia, New South Wales Registry. *Australasian Centre for Corporate Responsibility v. Santos NDS858/2021*. Available online: https://www.comcourts.gov.au/file/Federal/P/NSD858/2021/actions (accessed on 9 November 2021).
9. Environment Victoria. Why We're Taking the EPA to Court. Available online: https://environmentvictoria.org.au/2021/09/23/why-were-taking-the-epa-to-court/ (accessed on 23 September 2021).
10. Federal Court of Australia. *Sharma v. Minister for the Environment [2021] FCA 560*. Available online: http://climatecasechart.com/climate-change-litigation/non-us-case/raj-seppings-v-ley/ (accessed on 16 July 2021).
11. Federal Court of Australia. *Kathleen O'Donnell v Commonwealth of Australia & Ors*; 22 July 2020; VID482/2020. Available online: https://www.comcourts.gov.au/file/Federal/P/VID482/2020/actions (accessed on 23 September 2021).
12. Federal Court of Australia. *McVeigh v. Retail Employees Superannuation Trust [2019] FCA 14*. Available online: http://climatecasechart.com/climate-change-litigation/wp-content/uploads/sites/16/non-us-case-documents/2019/20190117_NSD13332018_judgment-1.pdf (accessed on 20 October 2021).
13. Full Court of the Federal Court of Australia. *Australian Conservation Foundation Incorporated v Minister for the Environment and Energy [2017] FCAFC 134*. Available online: http://climatecasechart.com/climate-change-litigation/non-us-case/australian-conservation-foundation-incorporated-v-minister-for-the-environment-and-energy/ (accessed on 20 October 2021).
14. Land and Environment Court of New South Wales. *Gloucester Resources Limited v. Minister for Planning [2019] NSWLEC 7*. pp. 440–539. Available online: http://climatecasechart.com/climate-change-litigation/non-us-case/gloucester-resources-limited-v-minister-for-planning/ (accessed on 20 October 2021).
15. United Nations. *The Paris Agreement art 2*. Available online: https://unfccc.int/sites/default/files/resource/parisagreement_publication.pdf (accessed on 22 April 2019).
16. Land and Environment Court of New South Wales. *Gloucester Resources Limited v. Minister for Planning [2019] NSWLEC 7 Preston CJ at 422*. Available online: http://climatecasechart.com/climate-change-litigation/non-us-case/gloucester-resources-limited-v-minister-for-planning/ (accessed on 20 October 2021).
17. Lesley, H. The Rocky Hill Decision: A watershed for climate change action. *J. Energy Nat. Resour. Law* **2019**, *37*, 341–351.
18. Climate Change Authority. *Targets and Progress Review*; Final report; 9 February 2014. Available online: https://www.climatechangeauthority.gov.au/sites/default/files/2020-06/Target-Progress-Review/Targets%20and%20Progress%20Review%20Final%20Report.pdf (accessed on 20 October 2021).
19. *WildEarth Guardians v US Bureau of Land Management*. 870 F 3d 1222 (10th Cir, 2017), 1234. Available online: https://casetext.com/case/wildearth-guardians-v-us-bureau-of-land-mgmt-2 (accessed on 20 October 2021).
20. Potter, B. Landmark decision blocks Gloucester Resources coal mine in NSW. *Australian Financial Review*. Available online: https://www.afr.com/politics/landmark-decision-blocks-gloucester-resources-coal-mine-in-nsw-20190208-h1b0bo (accessed on 8 February 2019).
21. Carrick, D. Climate Change Takes Front Seat in NSW Court. *ABC Law Report*. Available online: https://www.abc.net.au/radionational/programs/lawreport/10799006 (accessed on 12 February 2020).
22. Land and Environment Court of New South Wales. *Strarford Coal Pty Ltd. v Minister for Planning [2019] NSWLEC 8*. Available online: http://www.austlii.edu.au/cgi-bin/viewdoc/au/cases/nsw/NSWLEC//2019/8.html (accessed on 12 February 2020).
23. Federal Court of Australia. *McVeigh v. Retail Employees Superannuation Trust*. [2019] FCA 14. Available online: http://climatecasechart.com/climate-change-litigation/wp-content/uploads/sites/16/non-us-case-documents/2018/20180921_NSD13332018_complaint-1.pdf (accessed on 20 October 2021).
24. Commonwealth Government of Australia. *Superannuation Industry (Supervision) Act 1993 (Cth)*.
25. Chance, C. Climate Change Test Case Settles: $57BN Australian Super Fund Responds to Pressure on Climate Change Policy. Available online: https://www.cliffordchance.com/content/dam/cliffordchance/briefings/2020/11/climate-change-test-case-settle-client-briefing.pdf (accessed on 20 February 2020).
26. Grieve, C. Super Giants Funnel Billions into Fossil Fuels, Vote down Climate Push. Available online: https://www.smh.com.au/business/banking-and-finance/super-giants-funnel-billions-into-fossil-fuels-vote-down-climate-push-20200211-p53zt1.html (accessed on 13 February 2020).
27. Yoo, T. Can You Sue Your Super Fund for Climate Change? *Yahoo Finance AU*. Available online: https://au.finance.yahoo.com/news/sue-super-fund-climate-change-232121951.html (accessed on 13 February 2020).
28. *Task Force on Climate-related Financial Disclosures, 2019 Status Report: Task Force on Climate-Related Financial Disclosures 2*. Available online: https://www.fsb.org/2018/09/task-force-on-climate-related-financial-disclosures-status-report/ (accessed on 10 October 2020).

29. Debelle., G. *Climate Change and the Economy*; Speech, Reserve Bank of Australia: Sydney, Australia, 14 October 2021.
30. APRA. APRA Releases Guidance on Managing the Financial Risks of Climate Change. Available online: https://www.apra.gov.au/news-and-publications/apra-releases-guidance-on-managing-financial-risks-of-climate-change (accessed on 22 April 2021).
31. Australian Council of Superannuation Investors. ACSI Launches New Climate Change Policy. Available online: https://acsi.org.au/media-releases/acsi-launches-new-climate-change-policy/ (accessed on 25 April 2021).
32. Armour, C.; Commissioner, Australian Securities; Investment Commission. Managing Climate Risk for Directors. *Australian Institute of Company Director Magazine 2021 Back Editions February*. Available online: https://aicd.companydirectors.com.au/membership/company-director-magazine/2021-back-editions/february/managing-climate-risk-for-directors (accessed on 1 February 2021).
33. New South Wales Supreme Court. *Australian Securities and Investment Commission v Rich [2003] NSWSC 85.*
34. New South Wales Law Reports. *Daniels v Anderson (1995) 37 NSWLR 438.*
35. High Court of Australia. *Australian Securities and Investment Commission v Hellicar [2012] HCA 12.*
36. New South Wales Supreme Court. *Australian Securities and Investment Commission v Vines [2005] NSWSC 738.*
37. Troiano, R. Climate change: Corporate liability, disclosure requirements and shareholders' remedies. *Co. Secur. Law J.* **2018**, *26*, 418–426.
38. Hutley, N. SC and Sebastian Harford-Davis. Climate Change and Directors Duties; Memorandum of Opinion, The Centre for Policy Development, 7 October 2016. ('Hutley 2016').
39. Hutley, N. SC and Sebastian Hartford Davis. Climate Change and Director's Duties; Supplementary Memorandum of Opinion; The Centre for Policy Development, 26 March 2019. ('Hutley 2019').
40. Hutley, N. SC and Sebastian Hartford Davis, 'Climate Change and Director's Duties; Supplementary Memorandum of Opinion, The Centre for Policy Development, 23 April 2021. 18 ('Hutley 2021').
41. Australian Commonwealth Law Reports. *Wyong Shire Council v Shirt*, 146 CLR 40. 1980; 47–48.
42. di Lernia, C. Climate Risk Disclosure: Tracking the Uptake of the Taskforce on Climate-related Financial Disclosures (TCFD) Recommendations in the Australian Market. *Co. Secur. Law J.* **2020**, *37*, 470–471.
43. Foerster, A.; Peel, J. US fossil fuel companies facing legal action for misleading disclosure of climate risks: Could it happen in Australia? *Aust. Environ. Rev.* **2017**, *32*, 56–57.
44. Australian Government, Department of the Environment and Heritage, Australian Greenhouse Office. Climate Change Impacts & Risk Management A Guide for Business and Government. Available online: https://sciencepolicy.colorado.edu/students/envs_5120/australia_RCC_2006.pdf (accessed on 10 October 2020).
45. Task Force on Climate-Related Financial Disclosures. Recommendations of the Task Force on Climate-related Financial Disclosures; 14 December 2016. Available online: https://assets.bbhub.io/company/sites/60/2020/10/16_1221_TCFD_Report_Letter.pdf (accessed on 7 October 2021).
46. Governance Institute of Australia. Climate Change Risk Disclosure: A Practical Guide to Reporting against ASX Corporate Governance Council's Corporate Governance Principles and Recommendations, (Report, February 2020) 9. Available online: https://www.governanceinstitute.com.au/advocacy/thought-leadership/climate-change-risk-disclosure/ (accessed on 25 August 2021).
47. Australian Securities and Investment Commission. Climate Risk Disclosure by Australia's Listed Companies; Report, 593, 3 September 2018. Available online: https://asic.gov.au/regulatory-resources/find-a-document/reports/rep-593-climate-risk-disclosure-by-australia-s-listed-companies/ (accessed on 25 August 2021).
48. Baxt, R. Duties and Responsibilities of Directors' and Officers. *The Australian Institute of Company Directors*, 21st ed, 2016; 72.
49. Kenneth Hayne QC, 'What Kenneth Hayne says about climate change'. *The Financial Review*. Available online: https://www.afr.com/politics/federal/what-kenneth-hayne-says-about-climate-change-20191206-p53hiw (accessed on 19 December 2019).
50. Price, J. *ASIC Commissioner, Climate Change*; Speech, Centre for Policy Development: Sydney, Australia, 18 June 2018.
51. Available online: https://www.wri.org/insights/6-signs-progress-adoption-paris-agreement (accessed on 10 September 2021).

energies

MDPI

Article

The Impact of Atmospheric Precipitation on Wastewater Volume Flowing into the Wastewater Treatment Plant in Nowy Targ (Poland) in Terms of Treatment Costs

Piotr Bugajski [1,*], Elwira Nowobilska-Majewska [1] and Michał Majewski [2]

[1] Department of Sanitary Engineering and Water Management, Faculty of Environmental Engineering and Land Surveying, University of Agriculture in Krakow, 30-059 Kraków, Poland; elwiranowmaj@gmail.com

[2] Miejski Zakład Wodociągów i Kanalizacji w Nowym Targu, ul. Długa 21, 34-400 Nowy Targ, Poland; michal.majewski86@o2.pl

* Correspondence: piotr.bugajski@urk.edu.pl; Tel.: +48-12-662-4039

Abstract: This study determined the influence of precipitation occurring in the sewerage catchment basin in Nowy Targ (Poland) on the amount of wastewater inflow to the wastewater treatment plant, and determined the costs resulting from the treatment of accidental (rain) water entering the analyzed sewerage system. The research was conducted from 2016 to 2019, for which daily precipitation and average daily wastewater inflows in the so-called dry, normal, and very wet periods were analyzed. The research period was divided into six characteristic intervals in terms of precipitation. It was found that, on days with different precipitation intensity, the amount of accidental water as a proportion of the total amount of wastewater flowing into the plant ranges from 9.6% to 34.1%. The annual costs incurred by the operator resulting from the environmental fee are 1625.8 EUR/year. Alternatively, the costs resulting from financial expenditures for wastewater treatment processes amount to 337,651 EUR/year. The results of the research provide important information for sewage network operators to take effective actions to eliminate illegal connections of roof gutters and/or yard inlets to the sanitary collectors, and to replace the combined sewage system in Nowy Targ with a distributed sewerage system. This would reduce the costs of wastewater treatment and the irregularity of wastewater inflow.

Keywords: accidental water; precipitation; sewer system; wastewater treatment costs

Citation: Bugajski, P.; Nowobilska-Majewska, E.; Majewski, M. The Impact of Atmospheric Precipitation on Wastewater Volume Flowing into the Wastewater Treatment Plant in Nowy Targ (Poland) in Terms of Treatment Costs. *Energies* **2021**, *14*, 3806. https://doi.org/10.3390/en14133806

Academic Editor: José Carlos Magalhães Pires

Received: 18 May 2021
Accepted: 21 June 2021
Published: 24 June 2021

1. Introduction

In recent years, numerous initiatives have been undertaken in Europe related to increasing rainfall water retention to mitigate the effects of the progressive phenomenon of drought [1,2]. According to the measures taken, rainwater should be managed at the site of precipitation (rain or snow) and not discharged to a watercourse in the short term [3–5]. Many combined sewerage systems discharge rainwater—known as accidental water—into rivers within a short period of time after a precipitation event; this action has only negative ecological and economic consequences [6]. The negative ecological effects of rainwater discharge into sewers are the problems of maintaining wastewater treatment processes at a sufficiently high level, which arise due to the high variability of the quantity and quality of wastewater. High variability in both the amount of inflowing wastewater and the concentrations of pollutants in the water, in addition to sudden changes in other parameters e.g., wastewater temperature and pH, make it difficult for operators to optimize treatment processes [7,8]. As indicated by reports described in the literature, rainwater illegally introduced into sewerage systems, which is referred to as accidental water, negatively affects the operation of wastewater treatment plants, in which the dimensions and capacities of individual technological facilities are not adapted to periodically increasing and decreasing wastewater flow rates [9–11]. Accidental water mixed with municipal wastewater also has a major impact on reducing pollutant concentrations in wastewater by diluting it with rainwater, thereby depleting it of the organic matter necessary for the growth of activated sludge microorganisms [12–14]. The negative impact of accidental water on the functioning of wastewater treatment plant processes consequently leads to a periodic decrease in the efficiency of their operation, and thus the occurrence of a threat of pollution of receiving waters with insufficiently treated wastewater [15,16]. Moreover, in terms of ecology, an important but negative effect of introducing rainwater into the sewerage system is the lack of capacity to recharge groundwater with rainwater in the territory of the sewerage basin where the rainfall occurred, thus lowering the groundwater level [17]. As already mentioned, apart from the ecological risks resulting from the introduction of rainwater into the sewerage system, there are also equally important economic effects of such an action. Each cubic meter of treated wastewater in a wastewater treatment plant translates into real costs related to electricity [18], chemical consumption, wear and depreciation of equipment, or environmental fees for discharging treated wastewater into the environment [19,20]. These costs are directly borne by the sewage system operator, but are indirectly borne by all users (residents, industrial plants) who use the sewage network.

The results of the research described in this publication contribute to the discussion regarding the need to reduce the inflow of rainwater to sewer systems as a negative phenomenon that limits rainwater retention and causes excessive operating costs.

The aim of the research was to determine the influence of precipitation height in the sewerage system of Nowy Targ (Poland) on the amount of wastewater flowing into the collective wastewater treatment plant, and to determine the costs associated with its treatment and discharge into the environment.

2. Materials and Methods

2.1. Characteristics of the Sewage System and Wastewater Treatment Plant

Nowy Targ is a town located in the southern part of the Małopolska Province, with a current population of 31,850. The analyzed sewage system is made of PVC and stoneware material, with diameters ranging from 200 to 400 mm. It is 86.9 km long and has approx. 4800 connected houses. The sewage network operates using a gravity system and is intended for the collection of domestic wastewater from residential buildings and for the collection of wastewater from tannery (furrier) shops. Sixty furrier shops are legally connected to the sewage network. Many furrier shops are also connected to the network illegally [21]. All sewage, i.e., domestic and industrial wastewater, flows to the collective wastewater treatment plant with the designed capacity of 21,000 m^3/d and PE = 116,000 inhabitants assumed in the project [21,22]. The wastewater treatment plant in Nowy Targ, which was modernized in 2016, is located at 49°29′ N, 20°3′ E, as presented in Figure 1. The wastewater treatment plant in Nowy Targ is a mechanical and biological type, where the mechanical part includes a dense screen, a horizontal sand trap, and a basic settling tank (two-chamber horizontal). In the biological part of the wastewater treatment plant there are 3 sequential SBR biological reactors, each 70 m long, 23 m wide, and 4.5 m deep. Biological reactors operate on an 8 h cycle and there are 5 phases in each cycle. In the technological process, the wastewater is raised in front of the grates to a height of 7.5 m so that the entire flow of wastewater in subsequent devices is gravitational. Wastewater from cesspools, which are installed next to buildings not connected to the sewage system, is also delivered to the wastewater treatment plant by means of slurry tanks. Sewage sludge, both from basic settling tanks and from biological reactors, is collected in a gravity thickener, and then biogas is produced from them in separate fermentation chambers, from which electricity and heat are generated. The treated wastewater inflows through a sewage collector with a diameter of DN = 1000 mm, a distance of 197 km to the Dunajec River.

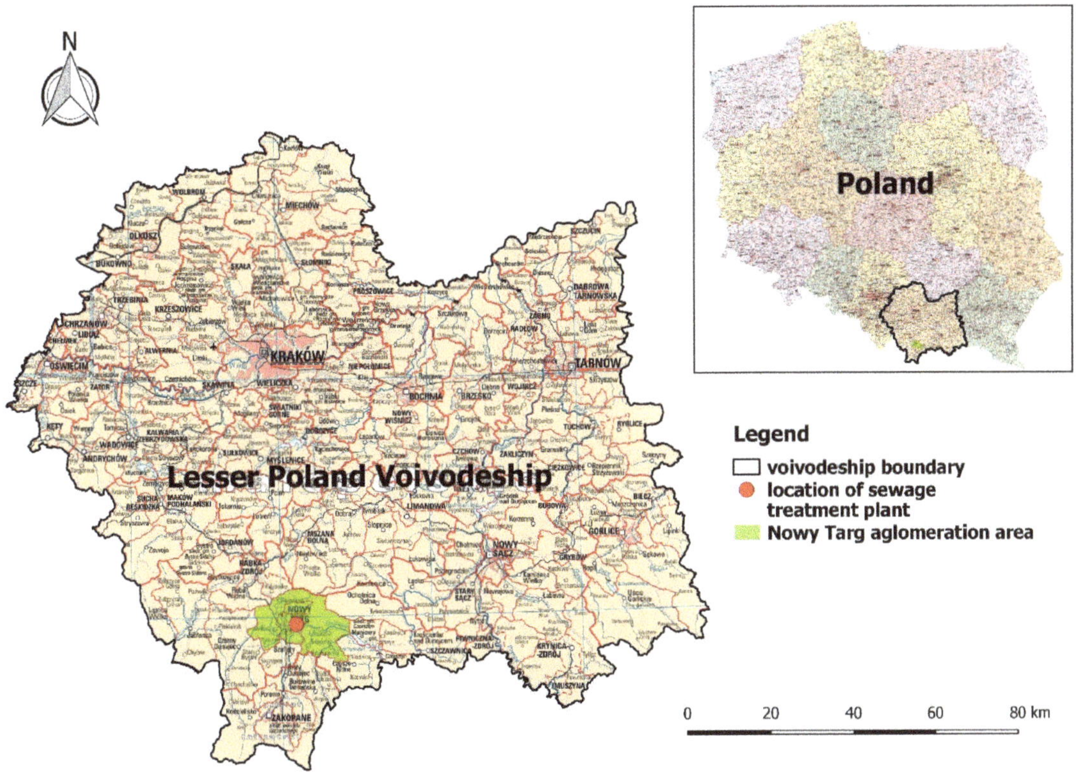

Figure 1. The location of the rural agglomeration of Nowy Targ against the background of the Lesser Voivodeship and Poland [source: own study].

2.2. Analytical and Statistical Methods

Surveys of daily wastewater outflow to the wastewater treatment plant and daily precipitation were conducted from 2016 to 2019, a period of 4 years. There was no increase in the number of sewer connections within the sewerage system during the analyzed time period, therefore the period of four years, i.e., 1461 days, was taken as homogeneous and authoritative. Wastewater outflow intensity was measured by means of an ultrasonic level probe (Waterpilot FMX 167) placed in an open channel over a triangular overflow behind the process line. The amount of precipitation on individual days was determined on the basis of indications from a Hellmann rain gauge installed in the analyzed sewage basin. According to the guidelines indicated by Kaczor [13], daily wastewater inflows were assigned to dry weather or wet weather. According to Kaczor [13], a given day of the year was classified as dry weather if, during its duration and in the five preceding days, no precipitation occurred, or it occurred but its daily amount did not exceed 1 mm. The average daily wastewater outflow during rainless (dry) weather was used to determine the average (averaged) amount of the so-called proper wastewater, i.e., not containing accidental water. As proposed by Chmielowski [23] in the case of wet weather, i.e., in days with precipitation, 6 characteristic groups were generated:

A—0.1 to 1.0 mm/d,
B—1.1 to 5.0 mm/d,
C—5.1 to 10.0 mm/d,
D—10.1 to 15.0 mm/d,
E—15.1 to 20.0 mm/d,

F—over 20.0 mm/d.

The calculation of the accidental water quantity Q_{dp} and its share Uwp in the wastewater mixture flowing into the wastewater treatment plant was based on the guidelines of Kaczor [13].

The first parameter Q_{dp} defines the volume of accidental (rainfall) water in the total volume of wastewater. Therefore, in the wet weather period the volume of accidental water in a given day was calculated according to Formula (1).

$$Q_{dp} = Q_{dm} - Q_{ds} \tag{1}$$

where:

Q_{dp}—daily inflow of accidental water to the sewer, [m^3/d];
Q_{dm}—daily inflow of the mixture of wastewater and accidental water during wet weather, (m^3/d);
Q_{ds}—average daily urban wastewater inflow (excluding accidental water) during dry weather, (m^3/d).

The second parameter, Uwp, represents the proportion of accidental water, which indicates, as a percentage, how much accidental water is contained in the total volume of the mixture of municipal wastewater together with accidental water [24]. The Uwp parameter was calculated according to Formula (2).

$$Uwp = \frac{Q_{dp}}{Q_{dm}} \cdot 100 \tag{2}$$

where:

Uwp—daily contribution of accidental water to sewerage flowing out of sewer, (%);
Qdp—daily inflow of accidental water to the sewer, (m^3/d);
Qdm—daily inflow into the sewerage system of the mixture of wastewater and accidental water during wet weather, (m^3/d).

Because the total retention time of wastewater in the sewer system is approximately 24 h, the daily precipitation amount was assigned to the amount of wastewater flowing on the following day.

3. Results and Discussion

3.1. Analysis of Atmospheric Precipitation in the Sewage Catchment Area

To verify the influence of precipitation on the amount of wastewater inflow to the sewerage system in Nowy Targ, years with different annual totals of precipitation were considered. Like Cebulska et al. [25], we assumed that the average annual precipitation in the area of Nowy Targ is 825 mm (825 dm^3/m^2) according to the classification proposed by Kaczorowska [26]; thus, the years 2016 and 2019 were average years. In 2016, total precipitation was 8.2% higher than normal, and in 2019, total precipitation was 4.7% higher. In contrast, 2018 was a very dry year, because the annual rainfall total was 64.5% of normal precipitation for this region. In contrast, 2017 was a wet year because the total precipitation was 23.2% higher than the multi-year average precipitation. Characteristic monthly totals and annual totals of precipitation in the sewerage system of Nowy Targ are presented in Table 1.

3.2. Characteristics of the Inflow of Wastewater

In the study period of 2016–2019, a high degree of irregularity in the amount of wastewater inflow to the wastewater treatment plant in Nowy Targ was found. In the four-year period, irregular average daily wastewater inflow to the wastewater treatment plant was found in individual years and in individual months. The main reason for the high variability (irregularity) of the amount of wastewater inflow to the wastewater treatment plant is the inflow of accidental water in the combined sewerage systems, in whole or in a significant proportion [24–28]. Moreover, in sewerage systems designed exclusively for the disposal of municipal sewage (distribution sewers), the reason for the appearance of an increased volume of wastewater during periods of intense precipitation is the illegal connection to sewers of outlets of roof gutters by residents [29]; outlets of drains used for land-property drainage [24]; or, as noted by Kaczor [30], manholes or ventilation holes of wastewater wells through which water flowing on the street surface enters the sewerage network.

Table 1. Monthly and annual precipitation amounts in the analyzed sewage system in Nowy Targ.

Month	2016	2017	2018	2019
		[mm]		
January	44.1	19.9	7.0	96.6
February	82.2	35.9	8.8	35.2
March	28.6	42.4	17.3	39.4
April	54.3	106.5	20.8	68.1
May	93.6	51.6	47.0	154.9
June	24.2	90.7	110.5	41.9
July	132.9	85.9	96.7	107.4
August	86.9	106.0	53.2	83.0
September	64.8	227.8	43.6	61.6
October	127.1	124.9	46.1	41.4
November	68.1	61.0	13.3	55.6
December	85.9	64.0	68.1	79.1
Total	892.7	1016.6	532.4	864.2

The average daily wastewater inflow at the WWTP in the 4 year research period was $Q_{a.d.}$ = 13,719 m^3/d with the variability expressed by the irregularity coefficient Cv = 23%, which indicates that the variability of the wastewater inflow was at an average level [31]. In average precipitation years, i.e., 2016 and 2019, the average daily wastewater inflow was 13,772.2 m^3/d and 14,050.8 m^3/d, respectively. In the wet year of 2017, the average daily wastewater inflow was 14,702.3 m^3/d, and in the very dry year of 1018, the average daily wastewater inflow was 12,351.5 m^3/d. Extreme wastewater inflows also occurred in each year. The lowest recorded mean daily wastewater inflow during the research period was 8.472 m^3/d, whereas the maximum was 41.754 m^3/d. Characteristic values of wastewater inflow to the wastewater treatment plant in Nowy Targ in particular years are presented in Figure 2.

The histograms shown in Figure 3A–D were developed to illustrate the variability of the mean daily inflows to the WWTP. The histograms include seven class intervals with a class span of 2000 m^3/d. In 2018, the year with the lowest annual precipitation, the average daily wastewater inflow appeared most frequently between 10,000 and 12,000 m^3/d (47.1% of cases) and between 12,000 and 14,000 m^3/d (34.8% of cases). The total wastewater inflow to the wastewater treatment plant in 2018 was 4,508,291 m^3/year. In 2017, a wet year, 10.1% of incidents were recorded in the wastewater inflow range of 10,000 to 12,000 m^3/d and 41.6% of incidents were recorded in the range of 12,000 to 14,000 m^3/d· In 2017, wastewater inflows in the range of 14,000 to 16,000 m^3/d were found to occur 34.8% of the time. Compared to the dry year (2018), the average daily wastewater inflow in the wet year (2017) occurred in the range of 14,000 to 16,000 m^3/d, representing an increase of 15.9%. In 2017, the total wastewater inflow to the WWTP was 5,453,719 m^3/year, which was more than 945,000 m^3 higher compared to the dry year (2018). In years with average rainfall, i.e., 2016 and 2019, the wastewater inflows to the wastewater treatment plant in the range of 12,000 to 14,000 m^3/d represented 40.4% and 36.4% of the cases, respectively. In these years, the inflows in the range of 10,000 to 12,000 m^3/d, i.e., 25.7% of cases in 2016 and 22.5% of cases in 2019, were recorded at similar levels. In addition, in the next range of 14,000 to 16,000 m^3/d, the number of cases of inflows was recorded at a similar level; in 2016, this represented 17.8% of cases, and in 2019 it was 15.1% of cases. The total inflow to the wastewater treatment plant in 2016 was 5,040,628 m^3/year and in 2019 it was 4,116,000 m^3/year.

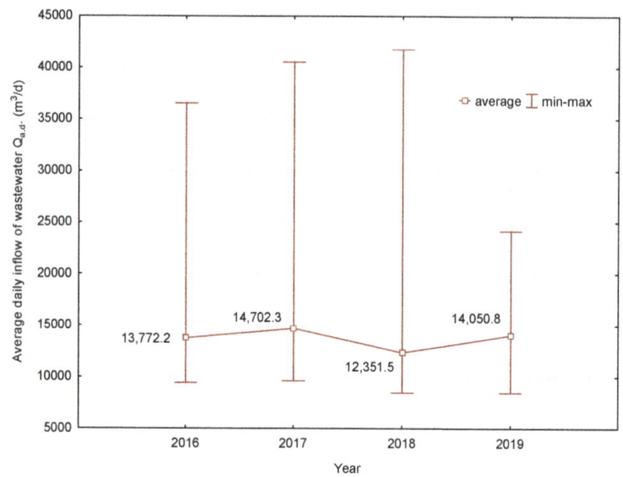

Figure 2. Daily average and extreme daily inflows of municipal wastewater and accidental water to the wastewater treatment plant in Nowy Targ.

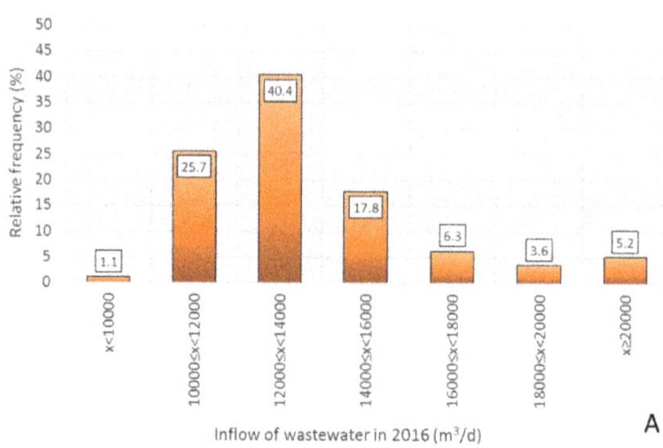

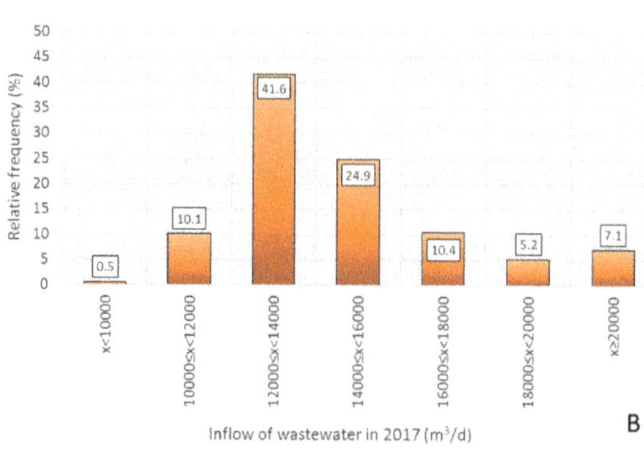

Figure 3. *Cont.*

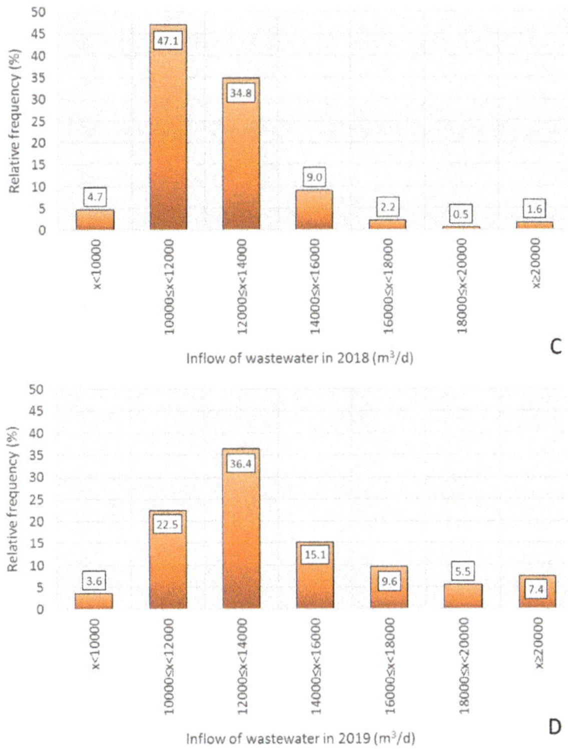

Figure 3. The histogram of the frequency of dependence of characteristic daily wastewater inflows in individual surveyed years. (**A**)-2016 year, (**B**)-2017 year, (**C**)-2018 year, (**D**)-2019 year.

As can be seen from the analysis, the volume of inflowing wastewater in particular years depends on the amount of precipitation in a given calendar year; this indicates the inflow of water from precipitation into the sewage network, which is defined as accidental water [32]. The next step in the analysis is to determine the volume of accidental water entering the wastewater treatment plant (Q_{dp}) and its contribution to the total volume of wastewater (Uwp). The calculation of the volume of accidental water and its contribution to the wastewater mixture was based on guidelines developed by Kaczor [13].

The analysis concerning the determination of the quantity of accidental water and the share of accidental water in the total quantity of wastewater inflow to the wastewater treatment plant was performed for the total of the four analyzed years, 2016–2019, assuming that, regardless of the type of year in terms of precipitation, the quantity of municipal sewage generated in dry weather is at the same level. Based on the described research methodology, during the four-year research period, 316 daily wastewater inflows were generated from 1461 daily wastewater inflows, which occurred during the rainless so-called dry weather. On the basis of recommendations proposed by Kaczor [13], it was assumed that the inflow of wastewater on a given day in dry weather was equivalent to a day during which there was no precipitation, and precipitation did not occur or was less than 1 mm in the 5 days preceding the measurement of wastewater inflow. In the research period the average wastewater inflow during dry weather (rainless) was Q_{ds} = 12,046 m^3/d (Table 1). Variability of wastewater inflow in dry weather, as described by the coefficient of variation, was Cv = 13%, which indicates a low variability of wastewater inflow in dry weather, according to the scale proposed by Wawrzynek [31]. The dry weather (rainless) wastewater inflow was 42.6% lower than the assumed design capacity of 21,000 m^3/d of the wastewater treatment plant in Nowy Targ [33]. Currently, there is no danger of exceeding the design capacity of the WWTP of 21,000 m^3/d, because the maximum daily dry weather inflow was 20,196 m^3/d, which is an inflow of 96% of the design inflow (Table 2).

Table 2. Statistical characteristics of wastewater inflow $Q_{a.d.}$ in dry weather.

Parameters	Statistics					
	Average m³/d	**Median m³/d**	**Min. m³/d**	**Max. m³/d**	**Standard Deviation m³/d**	**Coefficient of Variation %**
$Q_{a.d.}$	12,046	11,824	8910	20,196	1609.8	13

3.3. Analysis of the Quantity of Accidental Water in the Sewerage System

Based on the average daily wastewater inflow at dry weather (rainless), which is 12,046 m^3/d, the average daily rainwater inflow was calculated using Formula (1) for the period of days on which precipitation occurred. The calculations of accidental water volumes were undertaken for particular groups—A, B, C, D, E and F—with appropriately assigned precipitation ranges. In group A, i.e., on days in which precipitation ranged from 0.1 to 1.0 mm/d, the average daily inflow of accidental water to the WWTP was 1276.2 m^3/d. On days with precipitation from 1.1 to 5.0 mm/d (group B), the average inflow of accidental water to the WWTP was 2202.7 m^3/d. On days with precipitation from 5.1 to 10.0 mm/d (group C), the amount of rainwater inflow was 3010.0 m^3/d. On days with precipitation between 10.1 and 15.0 mm/d (group D), the amount of rainwater entering the sewer system was 3451.4 m^3/d. In the next interval (group E), i.e., during the days with daily precipitation from 15.1 to 20.0 mm/d, the amount of inflow rainwater was 4838.3 m^3/d. In the last of the analyzed intervals (group E), in the days with precipitation above 20 mm/d, the inflow of accidental water to the WWTP increased to 6239.6 m^3/d. Average daily inflows of urban wastewater and inflow of rainwater on days with characteristic precipitation are presented in Figure 4.

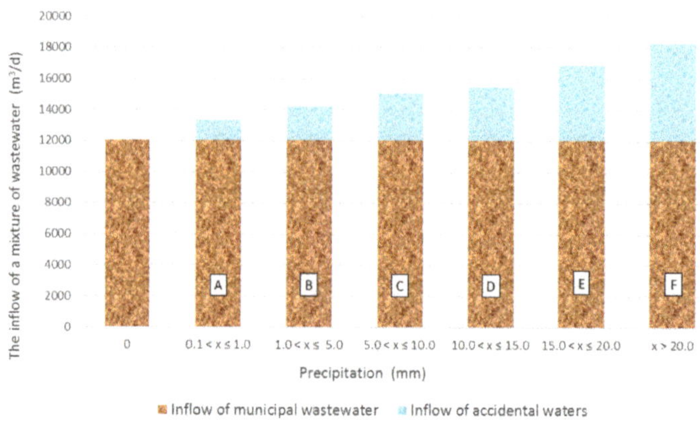

Figure 4. Volume of municipal wastewater and volume of accidental water (rainwater) flowing into the wastewater treatment plant against the background of precipitation in particular groups.

Based on Formula (2), the percentage share of accidental water (Uwp) in the total mixture of wastewater flowing into the wastewater treatment plant was calculated for days with precipitation. In the days with precipitation ranging from 0.1 to 1.0 mm/d, the percentage of accidental water in the total amount of wastewater was 9.6% on average. On the days with precipitation from 1.1 to 5.0 mm/d, the percentage of accidental water in the total amount of wastewater was 15.5%. On days with precipitation between 5.1 and 10.0 mm/d, accidental water contributed 20.0% to the total volume of wastewater. In the interval with precipitation from 10.1 to 15.0 mm/d, accidental water in the total mixture of inflowing wastewater accounted for 22.3%. In the days with precipitation from 15.1 to 20.0 mm/d in the sewerage catchment area, the proportion of accidental water in the total amount of wastewater was 28.7%. On days with precipitation above 20.1 mm/d, the percentage of accidental water in the total amount of wastewater flowing to the WWTP was 34.1%. The share of accidental water (rainfall) in the total amount of wastewater flowing to the wastewater treatment plant depending on the amount of precipitation in the basin area of Nowy Targ is presented in Figure 5.

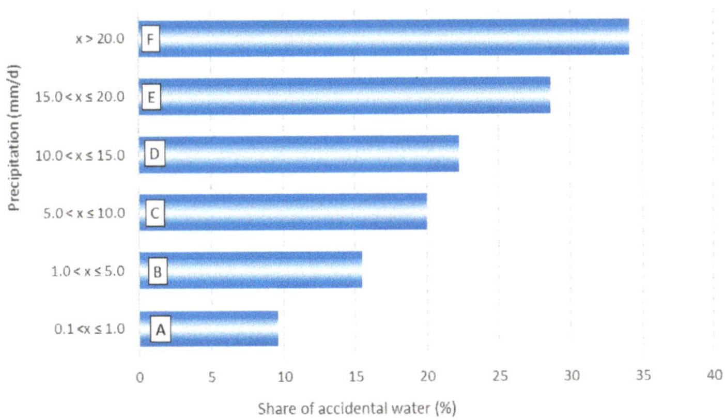

Figure 5. Percentage share of accidental water in the total amount of wastewater flowing into the wastewater treatment plant in Nowy Targ against the background of characteristic precipitation events.

3.4. Cost of Treatment and Discharge of Accidental Water

(*-current exchange rate of 1PLN = 4.5 €)

As indicated by Pajares et al. [34], Matej-Łukowicz and Wojciechowska [35], and Królikowska and Królikowski [36], the basic costs incurred by a wastewater treatment plant operator are the costs of the environmental fee for discharged wastewater and the costs of wastewater treatment processes. Thus, if accidental (rain) water flows into the treatment plant in addition to municipal wastewater, the costs increase relatively.

The analysis of treatment costs related to excessive inflow of accidental water was developed in terms of the environmental fee related to the discharge of treated wastewater into the receiving water body, and in terms of the costs resulting from the wastewater treatment processes at the treatment plant.

The cost of the environmental fee for the discharge of treated wastewater into the environment (watercourse) was based on the guidelines contained in the Ordinance [37]. Assuming that only municipal wastewater flowed into the WWTP during the research period (2016–2019) totaled 1461 m^3, the environmental fee resulting from the indicated tariff would be PLN 338,961 (€75,324.7) i.e., 84,740 PLN/year (18,831.1 €/year). However, due to the inflow of accidental (rain) water to the sewerage system and the related increase in the volume of wastewater discharged to the receiving body (River Dunajec), the total costs in the analyzed multi-year period amounted to PLN 368,225 (€ 81,827.8), i.e., 92,056 PLN/year (20,456.9 €/year). Therefore, during the period of 4 years, as a result of discharging an excessive amount of wastewater to the receiving body (the Dunajec river), the costs were higher by PLN 29,264 (€6503.1) i.e., in each year by PLN 7316 (€1625.8) on average.

The second financial aspect analyzed is the cost of wastewater treatment, in which accidental water has a significant share. According to Boruszko et al. [19], the unit cost of treating 1 m^3 of wastewater of a collective treatment plant is 4.01 PLN/m^3. A similar unit cost for wastewater treatment of 4.24 PLN/m^3 is given by Przybyła et al. [38] and

0.83 EUR/m^3 is given by Pajares et al. [34]. Thus, this analysis of the total cost of treating the mixture of municipal wastewater and accidental water at the wastewater treatment plant in Nowy Targ assumed a unit cost of treatment of 4 PLN/m^3. In the analyzed multi-year period, the total inflow of wastewater to the WWTP was $Q_{2016–2019}$ = 19,118,638 m^3. Based on the aforementioned analysis, of this total amount of mixed wastewater, accidental water represents $Q_{dp2016–2019}$ = 1,519,432 m^3. Based on the assumed unit cost of treating 1 m^3 of wastewater of 4 PLN/m^3, it was concluded that the cost of treating accidental water at the analyzed treatment plant was PLN 6,077,728 (€1,350,606), which, in terms of cost, is 1,519,432 PLN/year (337,651.5 €/year). These represent the real costs, because the environmental fee and costs of wastewater treatment processes are borne by the wastewater treatment plant operators, i.e., by the residents using the sewerage system in Nowy Targ.

4. Conclusions

As a result of the analysis, it is concluded that the inflow of accidental water from precipitation to the sewage system in Nowy Targ is a significant problem in terms of the variable hydraulic load of the wastewater treatment plant and the high financial costs associated with treatment. During the period of precipitation, rainwater constitutes 9% to 34% of the wastewater flowing into the wastewater treatment plant. In this sewage system, which should be regarded as a medium-sized system, the annual costs of treating the additional amount of wastewater (accidental water) averaged around 340,000 EUR/year. Therefore, it is recommended to take measures to limit the inflow of accidental water by modernizing (eliminating) sections of the sewage network, by upgrading from a combined system to a distribution system. Because the main source of accidental water inflow to the sewage system is rainwater discharged from roof gutters, it is recommended to inspect the sewage network to detect illegally connected roof gutters to the sewage system and eliminate this phenomenon. The developed research methodology and analysis of the costs of accidental water treatment in the sewage system in Nowy Targ (Poland) is universal and can be applied to any other sewage system. The basic ecological message of the conducted research is to raise awareness about how much rainwater, which can be a source of water for people and the economy, is wasted.

Author Contributions: Conceptualization, P.B.; methodology, P.B.; and E.N.-M.; software, P.B.; validation, P.B.; E.N.-M. and M.M.; formal analysis, P.B.; investigation, E.N.-M.; resources, M.M.; data curation, M.M.; writing—original draft preparation, P.B.; writing—review and editing, P.B.; and E.N.-M.; visualization, P.B.; supervision, P.B.; project administration, P.B.; and M.M., funding acquisition, P.B. All authors have read and agreed to the published version of the manuscript.

Funding: This research received no external funding.

Institutional Review Board Statement: Not applicable.

Informed Consent Statement: Not applicable.

Data Availability Statement: Not applicable.

Conflicts of Interest: The authors declare no conflict of interest.

References

1. Hynčica, M.; Huth, R. Long-term changes in precipitation phase in Europe in cold half year. *Atmos. Res.* **2019**, *227*, 79–88. [CrossRef]
2. Młyński, D.; Cebulska, M.; Wałęga, A. Trends, Variability, and Seasonality of Maximum Annual Daily Precipitation in the Upper Vistula Basin, Poland. *Atmosphere* **2018**, *9*, 313. [CrossRef]
3. Bugajski, P.; Kaczor, G.; Chmielowski, K. Variable dynamics of sewage supply to wastewater treatment plant depending on the amount of precipitation water inflowing to sewerage network. *J. Water Land Dev.* **2017**, *33*, 57–63. [CrossRef]
4. Domínguez, I.; Ward, S.; Mendoza, J.-G.; Rincón, C.-I.; Oviedo-Ocaña, E.-R. End-User Cost-Benefit Prioritization for Selecting Rainwater Harvesting and Greywater Reuse in Social Housing. *Water* **2017**, *9*, 516. [CrossRef]
5. Mańkowska-Wróbel, L. Basic problems of rainwater and meltwater management in urbanized areas. *Pragmata Oikonomias* **2014**, *8*, 209–2020. (In Polish)

6. Marszelewski, W.; Piasecki, A. Analysis of the development of wastewater infrastructure in Poland in ecological and economical aspects. *Eur. Policies Financ. Mark.* **2014**, *11*, 127–137.
7. Kaczor, G.; Bugajski, P. Impact of Snowmelt Inflow on Temperature of Sewage Discharged to Treatment Plants. *Pol. J. Environ. Stud.* **2012**, *21*, 381–386.
8. Correa de Oliveira, D.-B.; Soares, W.-A.; Ribeiro de Holanda, M.-A. Effects of rainwater intrusion on an activated sludge sewer treatment system. *Rev. Ambient. Água* **2020**, *15*, 2497.
9. Kaczor, G.; Bergel, T.; Bugajski, P. Impact of extraneous waters on the proportion of sewage pollution indicesregarding its biological treatment. *Infra. Ecol. Rural Areas* **2015**, *IV/3*, 1251–1260.
10. Wąsik, E.; Bugajski, P.; Chmielowski, K.; Cupak, A. The influence of precipitation in Sądecki basin on the variability of quantitative wastewater inflowing into the treatment plant Wielopole. *Infra. Ecol. Rural Areas* **2016**, *2/II*, 543–555. (In Polish)
11. Kaczor, G.; Chmielowski, K.; Bugajski, P. The Effect of Total Annual Precipitation on the Volume of Accidental Water Entering Sanitary Sewage System. *Rocz. Ochr. Sr.* **2017**, *19*, 668–681. (In Polish)
12. McMahan Erin, K. Impacts of Rainfall Events on Wastewater Treatment Processes. Graduate Theses and Dissertations, University of South Florida. 2006. Available online: http://scholarcommons.usf.edu/etd/3846 (accessed on 18 May 2021).
13. Kaczor, G. Effect of infiltration and inflow waters on the performance of small sewer systems. *Sci. Noteb. Agric. Univ. Krakow* **2012**, *495*, 229. (In Polish)
14. Bugajski, P.; Chmielowski, K.; Kaczor, G. Influence of rainwater inflow on the quality of sewage in a small sewage system. *Acta Sci. Pol. Form. Circumiectus* **2016**, *15*, 1–9. (In Polish)
15. Kowalik, T.; Bogdał, A.; Borek, Ł.; Kogut, A. The effect of treated sewage outflow from a modernized sewage treatment plant on water quality of the Breń river. *J. Ecol. Eng.* **2015**, *16*, 96–102. [CrossRef]
16. Bugajski, P.; Nowobilska–Majewska, E. A Weibull Analysis of the Reliability of a Wastewater Treatment Plant in Nowy Targ, Poland. *Rocz. Ochr. Śr.* **2019**, *21*, 825–840.
17. Pawęska, K.; Duda, P. Impact of precipitation on the balance of wastewater treated in municipal wastewater treatment plant. *Ecol. Eng.* **2018**, *19*, 49–56. (In Polish) [CrossRef]
18. Cottes, M.; Mainardis, M.; Goi, D.; Simeoni, P. Demand-Response Application in Wastewater Treatment Plants Using Compressed Air Storage System: A Modelling Approach. *Energies* **2020**, *13*, 4780. [CrossRef]
19. Boruszko, D.; Miłaszewski, R.; Piotrowski, P. Evaluation of Economic Effectiveness of the Municipal Wastewater Treatment Plant in the Community of Sokoły. *Rocz. Ochr. Śr.* **2013**, *15*, 1086–1097. (In Polish)
20. Remiszewska-Skwarek, A.; Fudala-Książek, S.; Łuczkiewicz, A. The Influence of Industrial Wastewater on the Energy Consumption and the Efficiency of Technological Processes in Municipal Wastewater Treatment Plant. *Rocz. Ochr. Śr.* **2016**, *18*, 110–121. (In Polish)
21. Nowobilska–Majewska, E.; Bugajski, P. The Impact of Selected Parameters on the Condition of Activated Sludge in a Biologic Reactor in the Treatment Plant in Nowy Targ, Poland. *Water* **2020**, *12*, 2657. [CrossRef]
22. Nowobilska-Majewska, E.; Bugajski, P. The determination of limit of tannery wastewater flowing to the wastewater treatment plant in Nowy Targ (Poland) in terms of the impact of chromium concentration on treated wastewater quality. *Desalination Water Treat.* **2021**, 1–10, in press. [CrossRef]
23. Chmielowski, K. Impact of atmospheric precipitation on the variability of wastewater discharge from a selected sewage system in Jaworzno. *Acta Sci. Pol. Formatio Circumiectus* **2019**, *18*, 39–49. [CrossRef]
24. Pecher, R. Fremdwasseranfall im Kanalnetz—Ein wasserwirtschaftliches Problem? *Korresp. Abwasser* **1998**, *45*, 2250–2258. (In German)
25. Cebulska, M.; Szczepanek, R.; Twardosz, R. *Spatial Distribution of Precipitation in the Upper Vistula Basin. Average Annual Rainfall (1952–1981)*; W I S P K, I G i G P U J, Politechnika Krakowska im. T.; Kościuszki Wydział Inżynierii Środowiska: Kraków, Poland, 2013. (In Polish)
26. Kaczorowska, Z. Opady w Polsce w przekroju wieloletnim. *Przegląd Geogr.* **1962**, *33*, 72. (In Polish)
27. Młyński, D.; Chmielowski, K.; Młyńska, A. Analysis of hydraulic load of a wastewater treatment plant in Jasło. *J. Water Land Dev.* **2016**, *28*, 61–67. [CrossRef]
28. Cieślak, O.; Pawełek, J. The inflow of foreign water to the sanitary sewage system on the example of the municipality of Mézos in France. *Instal* **2014**, *7–8*, 90–95. (In Polish)
29. Butler, D.; Davies, J.W. *Urban Drainage*, 3rd ed.; Spon Press an Imprint; Taylor & Francis: London, UK; New York, NY, USA, 2011.
30. Kaczor, G. Holes in the sewage canals' hatches as one of the cause for the accidental water infiltration to the separate sewer system. *Infra. Ecol. Rural Areas* **2009**, *9*, 163–165. (In Polish)
31. Wawrzynek, J. Methods of description and statistical inference. In *Wydawnictwo Akademii Ekonomicznej im*; Oskara Langego we Wrocławiu: Wrocław, Poland, 2007; Volume 37. (In Polish)
32. Müller, T.; Schütze, M.; Bárdossy, A. Temporal asymmetry in precipitation time series and its influence on flow simulations in combined sewer systems. *Adv. Water Resour.* **2017**, *107*, 56–64. [CrossRef]
33. Nowobilska–Majewska, E.; Bugajski, P. The Analysis of the Amount of Pollutants in Wastewater after Mechanical Treatment in the Aspect of their Susceptibility to Biodegradation in the Treatment Plant in Nowy Targ. *J. Ecol. Eng.* **2019**, *20*, 135–143. [CrossRef]
34. Pajares, E.-M.; Valero, L.-G.; Sánchez, I. M-R. Cost of Urban Wastewater Treatment and Ecotaxes: Evidence from Municipalities in Southern Europe. *Water* **2019**, *11*, 423. [CrossRef]

35. Matej-Łukowicz, K.; Wojciechowska, E. Fees for the discharge of stormwater. *Res. Pap. Wrocław Univ. Econ.* **2015**, *411*, 104–114.
36. Królikowska, J.; Królikowski, A. Rainwater drainage fees—Needs and opportunities. *Rocz. Ochr. Sr.* **2013**, *15*, 1143–1152. (In Polish)
37. Regulation of the Council of Ministers of December 22, 2017 on Unit Rates of Charges for Water Services. (Dz. U. z dnia 30 Grudnia 2017, poz. 2502)—Law Act. Available online: http://isap.sejm.gov.pl/isap.nsf/download.xsp/WDU20170002502/O/D20172502.pdf (accessed on 18 May 2021).
38. Przybyła, C.; Bykowski, J.; Filipiak, J. Effectiveness of municipal wastewater treatment plants. *Rocz. Ochr. Śr.* **2009**, *11*, 231–239. (In Polish)

energies

Article

Transition to Zero Energy and Low Carbon Emission in Residential Buildings Located in Tropical and Temperate Climates

Modeste Kameni Nematchoua [1,*], José A. Orosa [2], Paola Ricciardi [3], Esther Obonyo [4], Eric Jean Roy Sambatra [5] and Sigrid Reiter [1]

1 Local Environment Management & Analysis (LEMA), Department of Architecture, Geology, Environment and Constructions, Allée de la Découverte 9, Quartier Polytech 1, BE-4000 Liège, Belgium; sigrid.reiter@uliege.be
2 Department of N.S. and M.E. ETSNyM, University of A Coruña, Paseo de Ronda 51, 15011 A Coruña, Spain; jose.antonio.orosa@udc.es
3 Department of Civil Engineering and Architecture, University of Pavia, Via Ferrata 1, 27100 Pavia, Italy; paola.ricciardi@unipv.it
4 School of Engineering Design and Architectural Engineering, College of Engineering, Pennsylvania State University, University Park, PA 16802, USA; eao4@psu.edu
5 Department of Industrial Engineering, Higher Institute of Technology Antsiranana, Antsiranana 201, Madagascar; ericsambatra@gmail.com
* Correspondence: mkameni@uliege.be

Abstract: Different methods to achieve zero-energy and low carbon on the scale of a building are shown by most of the research works. Despite this, the recommendations generally offered by researchers do not always correspond to the realities found during the construction of new buildings in a determined region. Therefore, a standard may not be valid in all climate regions of the world. Being aware of this fact, a study was carried out to analyse the design of new buildings respecting the "zero-energy and low carbon emission" concept in tropical climatic regions when they are compared with a base case of temperate regions. To reach this objective, the comparison between real and simulated data from the different buildings studied was developed. The results showed that the renovation of existing residential buildings allows for reducing up to 35% of energy demand and a great quantity of CO_2 emissions in both climate types. Despite this, the investment rate linked to the construction of zero-energy buildings in tropical zones is 12 times lower than in temperate zones and the payback was double. In particular, this effect can be related to the efficiency of photovoltaic panels, which is estimated to be, at least, 34% higher in tropical zones than temperate zones. Finally, this study highlights the interest and methodology to implement zero-energy buildings in tropical regions.

Keywords: zero-energy; low carbon; residential buildings; tropical; temperate climate

Citation: Nematchoua, M.K.; Orosa, J.A.; Ricciardi, P.; Obonyo, E.; Sambatra, E.J.R.; Reiter, S. Transition to Zero Energy and Low Carbon Emission in Residential Buildings Located in Tropical and Temperate Climates. *Energies* **2021**, *14*, 4253. https://doi.org/10.3390/en14144253

Academic Editors: Chi-Ming Lai and Jae-Weon Jeong

Received: 18 April 2021
Accepted: 9 July 2021
Published: 14 July 2021

1. Introduction

Between 2000 and 2020, the average concentration of carbon dioxide (CO_2) emitted increased by approximately 2–3% each year [1]. In 2018, China (29.7%), the United States (13.9%), India (6.9%), the EU28 (9.1%), Russia, (4.6%) and Japan (3.2%)—the world's largest CO_2 emitters—together accounted for 51% of the population, 80% of total fossil fuel consumption, and 67.5% of total fossil CO_2 emissions in the world [2]. Nevertheless, efforts to reduce the carbon rate have been observed by the most polluting countries with activities like reducing building energy consumption. In this sense, a net-zero-energy building (ZEB) means that the total amount of energy used, calculated on an annual basis, is roughly equal to the amount of renewable energy created on the site [3].

Nowadays, in European Union countries (Belgium, France, Italy), primary energy demand in the building sector represents almost 40% of total energy [4]. Moreover, for the past ten years, it has been noticed that the building sector emitted up to 36% of greenhouse gas, and produced 38% of wastes [4]. In particular, in Brussels, construction and operation represent 98% of the water flow, 75% of energy demand, 65% of greenhouse gas emissions and about 33% of waste generated each year [5]. In consequence, this city has set itself the ambition of reducing energy consumption up to 30% by 2030 and up to 40% of carbon emissions in the construction sector [5]. Another example of a European city is Paris, where the building sector (residential and tertiary) accounts for 80% of the region's energy consumption, and 20% of the emission of greenhouse gas. In consequence, the city of Paris aims to reduce energy consumption by 1/3 by 2030, and by 50% by 2050. To reach this objective, a new climate plan provides for massive renovations of social and tertiary housing in this city [6]. Another example of non-European cities is Washington, where the carbon emission intensity decreased by 4.9% in 2019 [7]. Finally, in Sub-Saharan Africa (Senegal, Cameroon, and Madagascar), the building sector consumes more than 25% of total primary energy, which is generated by fossil fuels and emitting thousands of tons of carbon into the atmosphere each year. This CO_2 concentration is almost negligible compared to those produced in developed countries.

In general, it is interesting to highlight that, although most of these countries have a lot of potential resources (solar, hydro, wind), they are exploited at less than 5%. This situation remains very serious due to how strongly carbon emissions have a negative impact on global warming and, in consequence, it is important to take action to protect the environment. In this sense, the energy renovation of old buildings and the design of new buildings, mainly powered by renewable energy, can contribute enormously to the reduction of greenhouse gas emissions. What is more, stand-alone buildings can be the start of the solution but there are few guides to reach this objective.

Different recent research works aim to be examples of standalone buildings. Some of them evaluated the possibility to reach zero-energy and low carbon in buildings, yet the regulations and objectives set by several international organizations are not enough to be applied on a large scale [8].

In this sense, it is interesting to highlight that, although zero-energy building is more favourable in tropical and hot zones due to more favourable climatic factors, it was observed that countries located in temperate zones are more involved in the implementation of this new technology. In particular, most of the zero-energy buildings use the power grid for energy storage, but some are grid independent [9].

The development of zero-energy buildings has become possible not only thanks to advances in new energy technologies and construction techniques like solar panels, heat pumps, and low-emissivity triple glazing, but also thanks to researchers, who collect precise data on the energy performance of traditional and experimental buildings and provide parameters for advanced computer models to predict the efficiency of engineering designs [10]. In this sense, some recent research works were carried out in this field in the temperate region. For instance, after a strong study on one residential building located in Boston city in 2006, Szejnwald et al. [11] showed that renovation is the best technique for allowing the reduction of the significant part of energy demand. In 2017, the results of studies conducted by Yi et al. [12] explained that although the studies detail step-by-step the different processes of implementing zero-energy in new constructions, it is often difficult in practice to make this objective, because of the variety of outdoor climates according to the seasons. The results of studies conducted by Szalay and Zold [13] aiming to design zero-energy buildings showed that it is easier to achieve the ZEB by grouping buildings according to their geometrical shape and age. The research conducted by Zhou et al. [14] in an office building in China showed that the choice of HVAC systems has a significant impact on the implementation of zero-energy in buildings in temperate zones. This study recommended adopting energy efficiency techniques according to the local climate of each region. With the aim of increasing the energy performance of buildings over their entire life cycle, in order to increase the possibilities of achieving zero-energy building, Srinivasan et al. [15] found a new method aimed at optimizing the production of energy generated by renewable sources. The research carried out by Robert and Kummert [16] based on the "Morphing" method made it possible to analyse, on the basis of hourly data for the last 50 years and a general circulation model, the impacts of global warming on the implementation of NZEB in the temperate zone. The results showed that global warming significantly impacted zero-energy buildings. The study carried out by Pikas et al. [17] aimed to estimate the optimal cost of the energy performance of the residential building with almost zero energy consumption. The results explained that very few methods allow estimating the optimal cost of the energy used in these residences.

Despite these previous works about temperature regions, there is a lack of information about tropical cities and the possibilities to improve the energy consumption in their buildings. In this sense, tropical cities are some of the most favourable regions for the implementation of ZEB but the rate of buildings respecting the zero-energy concept is very low in this region of the world. In this sense, the study carried out in 2019 by Nematchoua and Sigrid [18] aimed to assess the different possibilities for the implementation of residential buildings with zero-energy and low carbon emissions in sub-Saharan Africa. The results showed that the most efficient solution to consider a zero-energy building in the city of Antananarivo requires an additional expenditure estimated at 40% of the initial cost of the building. This will result in savings of $475 per year starting in 2030, and also about a 99% reduction in carbon emissions.

In consequence, the objective of this study is to assess and analyse the cost and energy use with the implementation of residential buildings with zero-energy and low carbon emissions in tropical regions (sub-Saharan Africa), taking as reference this same analysis in temperate zones. In particular, paying special attention to the energy production generated by photovoltaic panels in hot zones and then the impact of heating on energy demand in temperate zones. The final objective is to show to architects, engineers, urban planners, investors, politicians, even non-specialists in energy and the environment, a simple methodology to analyse the energetic and economic interest into building new zero-energy and low carbon buildings in different climate regions.

To reach this objective, the present work is constituted of several parts: an analysis of some keywords in the second part; a methodology, results, and discussion, and finally the conclusions.

2. Keywords Analysis

The systematic structure of this manuscript is shown in Figure 1.

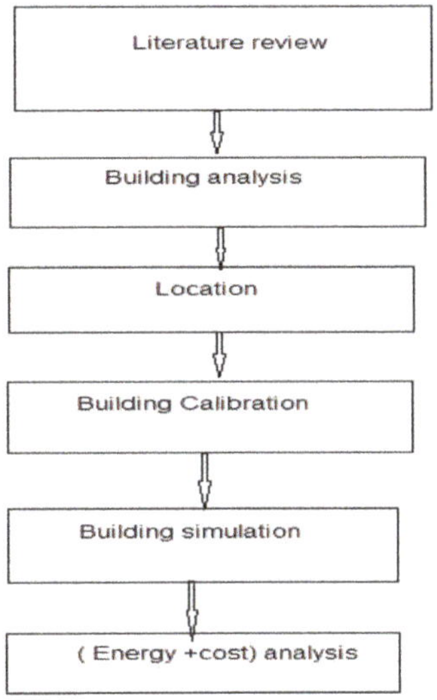

Figure 1. Research conceptual framework.

2.1. Buildings and Zero-Energy Concept

Before carrying out different research works and case studies about zero-energy and low-carbon buildings, it is important to do a previous analysis about the different concepts of existing low environmental impact housing in order to make more understandable some practical cases which will be presented in this research.

2.1.1. Nearly Zero-Energy Building

A nearly zero-energy building is economical in primary energy. The nearly zero quantity required for heating, cooling, lighting, domestic hot water, and ventilation is thus covered by solar heat or by renewable energy sources as well as by reasonable technical installations. In this sense, in the European Union, a building energy performance requirement was set between 2014 and 2021 as shown in Table 1.

Table 1. Evolution of the energy performance requirements of buildings from 2014 to 2021 for the residential and commercial buildings in some European Union countries [19].

Type of Building Construction	Building Energy Performance (2014)	Building Energy Performance (2017)	Nearly Zero-Energy Building (2021)
Overall level of thermal insulation (kWh/m^{2}year)	≤35	≤35	≤35
Overall energy performance of the building (residential building) (kWh/m^{2}year)	≤80	≤65	≤45
Overall energy performance of the building (commercial building) (kWh/m^{2}year)	≤80	€ (65–90)	€ (45–90)
Primary energy demand (kWh/m^{2}year)	≤130	≤115	≤85

2.1.2. Passive Building

The objective of a passive house is to consume very little energy while ensuring high thermal comfort in summer and winter as well as good air quality. Once again, in the European Union, some criteria must be respected to obtain the label "passive building", as shown in Table 2.

Table 2. Criteria for a passive building [20].

Passive Building	Criteria
Heating energy (kWh/m^{2}year)	≤15
Primary energy (kWh/m^{2}year)	≤120
Air tightness (vol/h)	≤0.6
Annual overheating hours (>25 °C)	≤5%

It is interesting to highlight that heating requirements of less than 15 kWh per m^2 and year are four to six times less than a new conventional building and up to ten times less than an existing building over ten years old.

2.2. Zero-Energy Building

It is a building that produces all the energy that it needs. It is interesting to note that a zero-energy building does not necessarily meet the criteria for a passive building. We must distinguish a building called energetically sufficient from a building called energetically autonomous.

(i) An energy-efficient building is capable of producing, over a year, an amount of energy proportional to the amount of energy it consumes. However, it will not necessarily consume the energy it needs when it produces it. Example: consider a building for which we install photovoltaic panels. During sunny periods (in summer), the panels will generate electricity at their optimum efficiency [21], and the house will supply energy to the grid. During the rest of the time or during a dead period such as in winter, the panels will not always supply enough electricity. Consequently, the dwelling cannot do without the electrical network. However, the total annual balance is zero since the excess production in summer compensates for the lack in winter.

(ii) An energetically autonomous building must not be connected to the electrical distribution network. When the panels cannot produce as much electricity as needed, the batteries used to store the excess electricity produced during sunny periods provide the electricity. To achieve energy independence, the building has above-average levels of insulation.

The principle of the zero-energy building, therefore, differs from that of the passive house, since it consists of compensating for total consumption, whatever it is, and not in optimizing the conditions favouring the energy sobriety of the house.

2.3. Positive-Energy Building

A building is said to be positive energy when it produces more energy than it consumes. A positive-energy building can be a passive building with enough renewable energy sources, or a building that does not meet the criteria to be a passive building, but still has a surplus of overall energy production.

3. Methodology

As it was shown before, the present paper aims to show a methodology for designing new buildings respecting the zero-energy concept, adding, as original consideration, the climatic regions where the buildings are placed and taking as reference the temperate regions. To reach this objective, three temperate countries and three tropical countries, with their typical building constructions, were selected. In consequence, weather data and simulations were employed to define the energy consumption and economical investment needed in each case study and its comparison will let define the more interesting region to implement the zero-energy concept. All these items will be described in the next sections.

3.1. Location

This study was carried out in three developed countries (Belgium, France, and the United States) located in a temperate climate, and three developing countries (Cameroon, Senegal, and Antananarivo) located in Sub-Sahara Africa (tropical climate).

The study took place in the capital of each country, as these study locations were chosen based on their environmental (climatic), energy, and social differences. The three cities selected in this study, and placed in Sub-Saharan Africa, are of very high solar potential (which is favourable to the supply of PV) and also the wind speed, which is favourable to the installation of wind turbines (speed between 5 m/s and 10 m/s).

The countries located in temperate zones such as Belgium, France, and the United States, have climatic conditions less favourable to zero-energy objectives because of the low sunshine in winters. However, important reforms are being carried out in these countries with the objective of achieving zero-energy and low carbon in 2050. This study aims to support this approach.

3.1.1. Studied Locations

Sub-Saharan Africa is made up of 48 countries with varying climates. It is one of the regions of the world that is very rich in biodiversity, and, above all, one of the most vulnerable to climate change. More than 80% of the population from this region are under 35 years old. Three of the cities in this study are located in this region: the city of Douala located in Cameroon, the city of Antananarivo in Madagascar, and Dakar in Senegal.

(1) Located between 4°03′ N and 9°4′ E, the city of Douala is the economic capital of Cameroon. Douala covers 923 km^2 and is strongly dominated by the tropical climate essentially made up of two seasons: the dry season from November to April and the rainy season from April to November.

(2) Dakar is the main city of Senegal. It is a coastal city located on the edge of the Atlantic Ocean. This city is dominated by the tropical climate and crossed by the monsoon coming from the southwest which is a humid wind bringing rain, and also a dry wind (the Harmattan). Dakar is one of the largest metropolises in Africa with a very high growth rate.

(3) Located 1435 m above the sea, the city of Antananarivo, the political capital of Madagascar, is spread over 350 km of surface. The city of Antananarivo is strongly dominated by the tropical climate of such an altitude. It has notably cool, mild winters and very rainy summers. Figure 2 gives the geographical location of these different cities studied, while Table 3 shows some climatic characteristics.

(4) Brussels is the capital of Belgium, the French Community of Belgium, the Flemish Community, and the seat of several European Union institutions. The climate of Brussels is temperate and influenced by the Gulf Stream. The proximity to the sea has a strong influence on this climate. The climate of Brussels is generally characterized by mild and rainy winters and relatively cool and humid summers.

(5) Washington DC is one of the world's largest cities, located in the mid-Atlantic region of the East Coast of the United States, between Virginia and Maryland. The nation's capital is around 40 miles south of Baltimore, 30 miles west of Annapolis and the Chesapeake Bay, and 108 miles north of Richmond. Founded in 1791, Washington DC is the place where the seat of the American congress is located [22]. Major factors determining Washington's climate include the large semi-permanent high-pressure and low-pressure systems of the North Pacific Ocean, the continental air masses of North America, and the Olympic and Cascade Mountains [22]. In spring and summer, a high-pressure anticyclone system dominates the North Pacific Ocean, causing air to spiral out in a clockwise fashion [22].

(6) Paris is the capital city of France, and the largest city in France. The area is 105 km^2. Paris is also the centre of the French economy, politics, traffic, and culture. The climate of Paris is said to be a warm temperate. The rainfall in Paris is significant, with precipitation even during the driest month. The average annual temperature in Paris is 11.3 °C. Figure 3 shows location of some cities studied in this research.

Figure 2. Location of three studied cities.

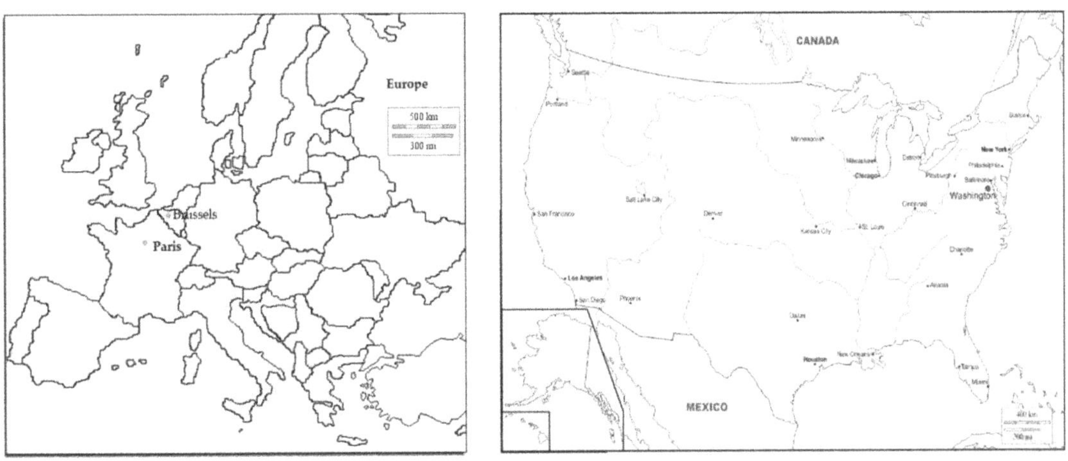

Figure 3. Location of Brussels and Paris (**on right**) and Washington DC (**on left**).

3.1.2. Climatic Data

In this research, all the climate data for simulations are downloaded from American Meteonorm software (Version 7.3.3, Meteotest AG (Bern, Switzerland)), based on the geographical coordinates of each city. The software provided the possibility to download data in hours, days, or months according to our requirements. Hourly data of temperature, relative humidity, airspeed, solar radiation, and precipitation for the last 30 years were collected for all the climate regions.

3.2. Buildings

This research was carried out about residential buildings. The type of building and construction materials varied according to the countries and this is why there were two categories of residential buildings (one more adapted in a temperate climate and another in a tropical climate). In this sense, the different buildings have different shapes with different structure materials and different areas, which can accommodate several people as shown in Table 4.

- The first residential building designed for this project, and located in the tropical region, was a simple one-floor family house consisting of four bedrooms, a shower room, and a kitchen. The building materials mainly consisted of Earth bricks and essentially glazed windows (glass thickness: 5.5 cm).
- The second residential building designed for this project, and located in the temperate region, was a familiar residence composed of two bedrooms, a living room, kitchen, and restroom. The different building materials are detailed in Table 4.

Table 3. Information regarding the six selected representative countries.

No	Country	City	Climate	Location	Temp.(°C)		RH (%)	
					Min	Max.	Min.	Max.
1	Belgium	Brussels	temperate	50.85° N, 4.35° E	−1	30	30	90
2	Cameroon	Douala	tropical	4.05° N, 9.76° E	18	37	30	90
3	France	Paris	temperate	48.85° N, 2.35° E	−10	31	30	90
4	Madagascar	Antananarivo	tropical	18.87° S, 47.50° E	10	35	55	90
5	Senegal	Dakar	tropical	14.72° N, 17.46° W	15	35	30	100
6	United States	Washington	temperate	38.90° N, 77.03° W	−2	33	30	98

From this Table 4, it can be observed that most of the materials chosen in the case of renovated residence buildings have low thermal conductivity and embodied carbon almost equal to zero. The goal is to select the materials that are more sustainable (low conductivity and embodied carbon). The lower the thermal conductivity of a material, the more this material has a huge capacity to limit heat transfer.

3.3. Simulation Tools

In this study, version (5.5.2) of the Design Builder software (Stroud, UK) was employed. This software is highly renowned in this field and has served as the basis for thousands of works of scientific research. Design Builder software, the same as TRNSYS, DOE, Pleiades, Helios, etc. software, is very well known in the field of simulation, optimization, modelling, BIM (Building Information Modelling), LCC (Life Cycle Cost), LCA (Life Cycle Assessment), etc.

This software is coupled with the Energy Plus (version 8.2, funded by the U.S. Department of Energy's (DOE) Building Technologies Of-fice (BTO)) tool to assess the consumption and energy demand of a building [23] and offers the most building materials with their physical thermal property. The modelling of the building selected is shown in Figure 4.

Table 4. Thermal characteristics of building constructions.

Region	Building Category	Building Clement	Layer	Component	Thickness (m)	Thermal Conductivity (W/mK)	Density (kg/m^3)	Specific Heat Capacity (J/kg K)	Embodied Carbon (kg CO_2/kg)	U-Value (W/(m^{2}K))
Tropical region	Residence in the initial state	Exterior wall	Layer1	Plaster	0.025	0.500	1300	1000	0.12	2.750
			Layer2	Concrete block	0.120	1.630	2300	1000	0.08	
			Layer3	Plaster	0.050	0.500	1300	1000	0.12	
		Partition wall	Layer	Concrete block	0.120	1.630	2300	1000	0.08	5.850
		Roof	Layer1	Clay Tile	0.025	1.000	2000.0	800	0.46	0.160
			Layer2	Stone Wool	0.242	0.040	30.0	840	1.05	
			Layer3	Roofing felt	0.005	0.190	960.0	837	-	
	Residence after renovation	Exterior wall	Layer	Wood	0.153	0.040	110	1800	0.00	0.250
		Partition wall	Layer	Wood	0.070	0.040	110	1800	0.00	0.250
		Roof	Layer1	Clay Tile	0.025	1.000	2000.0	800	0.46	0.160
			Layer2	Stone Wool	0.242	0.040	30.0	840	1.05	
			Layer3	Roofing felt	0.005	0.190	960.0	837	-	
Temperate region	Residence in the initial state	Exterior wall	Layer1	Plaster	0.030	0.500	1300	1000	0.12	0.330
			Layer2	Concrete	0.140	0.510	1400	1000	0.08	
			Layer3	Extruded polystyrene	0.120	0.034	35	1400	2.88	
			Layer4	Facing brick	0.090	0.620	1700	800	0.22	
		Partition wall	Layer1	Plaster	0.030	0.500	1300	1000	0.12	1.570
			Layer2	Brick	0.140	0.720	1920	840	0.22	
			Layer3	Plaster	0.030	0.500	1300	1000	0.12	
		Roof	Layer1	Roof tiles	0.030	0.550	1900	837	0.05	0.180
			Layer2	Wooden lathing	0.038	0.130	2800	896	0.45	
			Layer3	Air gap	0.025	-	-	-	-	
			Layer4	Wood	0.042	0.120	510	1380	0.45	
			Layer5	Rock wool	0.400	0.100	500	1000	0.98	
			Layer6	composite wood	0.018	0.040	160	1888	0.19	

Table 4. *Cont.*

Region	Building Category	Building Clement	Layer	Component	Thickness (m)	Thermal Conductivity (W/mK)	Density (kg/m^3)	Specific Heat Capacity (J/kg K)	Embodied Carbon (kg CO_2/kg)	U-Value (W/(m^{2}K))
	Residence after renovation	Exterior wall	Layer1	Plaster	0.030	0.500	1300	1000	0.12	0.350
			Layer2	Concrete	0.140	0.510	1400	1000	0.08	
			Layer3	Wood frame	0.030	0.120	510	1380	0.45	
			Layer4	ISOCELL cellulose	0.020	0.100	400	1360	-	
			Layer5	Composite wood panel	0.020	0.250	900	1000	0.12	
			Layer6	Wooden cladding	0.022	0.130	160	1800	0.05	-
		Partition wall	Layer1	Plaster	0.030	0.50	1300	1000	0.12	1.570
			Layer2	Brick	0.140	0.720	1920	840	0.22	
			Layer3	Plaster	0.030	0.500	1300	1000	0.12	
		Roof	Layer1	Roof tiles	0.030	0.550	1900	837	0.05	0.180
			Layer2	Wooden lathing	0.038	0.130	2800	896	0.45	
			Layer3	Air gap	0.025	-	-	-	-	
			Layer4	Wood	0.042	0.120	510	1380	0.45	
			Layer5	Hemp wool	0.400	0.037	800	1000	0.00	
			Layer6	Composite wood	0.018	0.130	1000	1000	0.01	

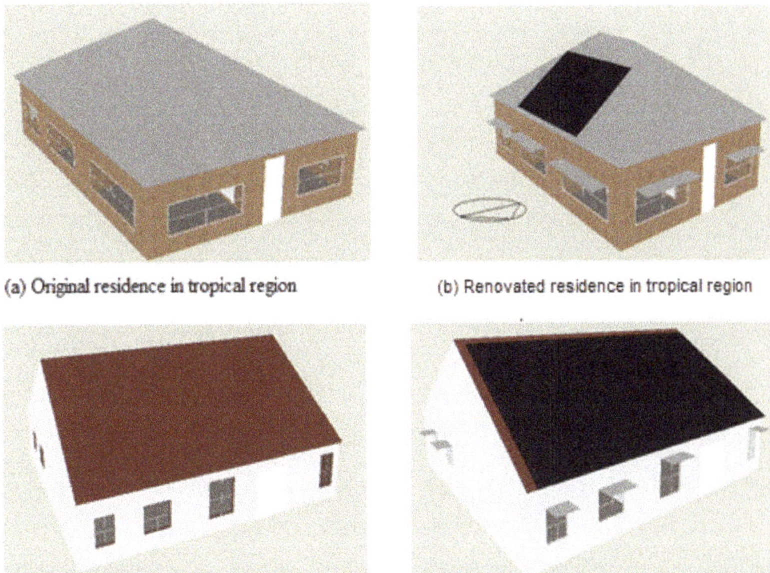

(a) Original residence in tropical region

(b) Renovated residence in tropical region

(c) Original residence in temperate region

(d) Renovated residence in temperate region

Figure 4. Case of (**a**,**b**) renovated residence building in the tropical region and (**c**,**d**) temperate climate.

In this sense, it is interesting to highlight that the design builder software automatically gives the embodied carbon of each type of material used during the construction of the building, as well as the quantity of carbon produced during the operational phase of the building. As a consequence, the total carbon emission rate during the life cycle assessment of the building can be freely obtained.

In the case of a building located in the tropical climate, the efficiency of the solar panels is very high due to the high concentration of solar radiation throughout the year, and it is not needed to add another source of energy such as a wind turbine to achieve the "Zero-Energy Building" objective. In contrast, in the case of a building located in a temperate climate, it was necessary to associate the photovoltaic panels (PV) and wind turbines to reach the "Zero-Energy Building" to maintain the same area of PV installed in a building located in the tropical region. Indeed, in the temperate region, the sunshine is high only in summer, which is why the PV has a low yield in this region. The input data of the simulation software is shown in Table 5.

Table 5. Input data for the simulations.

Parameters (Renovated Residence = After Applying Renovation, Passive Strategy, and Green Energy)	Building Located in the Tropical Region		Building Located in the Temperate Region	
	Initial State	**After Renovation**	**Initial State**	**After Renovation**
Height (m)	3.5	3.5	3.5	3.5
Area (m^2)	342.0	342.0	342.0	342.0
Activity template	Domestic House	Domestic House	TM_3-Bed Living Kitchen	TM_3-Bed Living Kitchen
Occupancy density (people/m^2)	0.0230	0.0230	0.0303	0.0303

Table 5. *Cont.*

Parameters (Renovated Residence = After Applying Renovation, Passive Strategy, and Green Energy)	Building Located in the Tropical Region		Building Located in the Temperate Region	
	Initial State	After Renovation	Initial State	After Renovation
Activity (met)	0.9	0.9	0.9	0.9
Clothing (Clo)	0.5–1.0	0.5–1.0	0.5–1.0	0.5–1.0
DHW: consumption rate (l/m^2-day)	0.53	0.53	0.72	0.52
Fresh air (l/s-person)	10	10	10	10
Lighting: target luminance (lux)	100	100	125	100
Computer: power density(W/m^2)	0.200	0.180	0.200	0.105
Other equipment: power density (W/m^2)	3.58	3.58	3.58	3.58
Occupancy schedule	24/7	24/7	24/7	24/7
Construction template	Project construction template	Project-construction template	Project construction template	Project construction template
Air tightness (vol/h)	0.5	0.5	0.5	0.5
Glazing template	Project glazing template	Project glazing template	Project glazing template	Project glazing template
Glazing type	Preferred height 1.5 m, 30% glazed Thickness: 5.5 cm	Preferred height 1.5 m, 30% glazed Thickness: 5.5 cm	Preferred height 1.5 m, 30% glazed Thickness: 5.5 cm	Preferred height 1.5 m, 30% glazed Thickness: 5.5 cm
Local shading	No	1.0 m overhang	No	1.0 m overhang
Lighting template	Incandescent	LED	Incandescent	LED
Lighting control	No	Yes	No	Yes
Lighting schedule	24/7	Mon.–Sun. 6 p.m.–7 a.m.	24/7	Mon.–Sun. 6 p.m.–7 a.m.
HVAC template	Fan coil unit (4-pipe) Air Cooler chiller	Fan coil unit (4-pipe) with district cooling	Oil heating	Fan coil unit (4-pipe) with district heating+ cooling
HVAC Schedule	12/7	6/7	24/7	24/7
Heating: Fuel	No	No	Natural gas, COP = 0.9	Natural gas, COP = 0.9
Cooling: Fuel	Electricity from grid	Electricity from green energy	Electricity from grid	Electricity from green energy
Other ventilation	Natural ventilation (NV)	Natural ventilation (NV)	Natural ventilation (NV)	Natural ventilation (NV)

3.4. Description of the Photovoltaic Panel

In this study, photovoltaic panels with different areas, according to the cities, placed on the roof of a building located in the tropical and temperate climates, were introduced. The different cells were made of polycrystalline, with a base load of direct current by an inverter. Its optimal inclination was fixed at 37°, oriented toward the south in the case of Paris, Brussels, and Washington and 45° oriented toward the north in Douala, Dakar, and Antananarivo cities [24].

3.5. Model Validation

Validation is applied by comparing simulation and measured data. We analysed two parameters: monthly energy consumption and hourly air temperature. In this study, two formulas mentioned in ASHRAE guideline 14 [25] were applied: coefficient of variation of square root error (RMSE) and mean bias error (MBE). The different RMSE and MBE values were evaluated, taking into account the two Equations (1) and (2) [25].

$$\text{RMSE}(\%) = \frac{1}{M}\sqrt{\frac{\sum_{i=1}^{N}(Mi - Si)^2}{N}} \tag{1}$$

$$\text{MBE}(\%) = \frac{\sum_{i=1}^{N}(Mi - Si)}{\sum_{i=1}^{N} Mi} \tag{2}$$

where Si and Mi were simulated data and measured data over a given interval I, respectively, the total number of data implemented, M is the total number of measured data.

In Guideline 14 of ASHRAE [26], it is recommended that a simulation model can be considered calibrated if the following conditions are fulfilled:

- Hourly MBE between ±10% and hourly RMSE smaller than 30%.
- Monthly MBE between ±5% and monthly RMSE of less than 15%.

The present research work is a continuation of a previous study [23] where a comparison between measurement and simulation data is shown and a small difference and negligible error between these two categories of data are concluded.

The various results of the calculations carried out show that the MBE value found is 0.6%, and that of RMSE is 0.2%. However, the values required in the ASHRAE-14 directive are between (−10%; +10%), counting as (MBE), and (−30%; +30%), counting as (RMSE). By observing these results, we deduce that the hourly values of MBE and RMSE found in this research are within the range requested by ASHRAE. On the basis of the previous results, we conclude that this simulation model is calibrated with different hourly data. In addition, a comparison between the monthly energy values was made. The value of the MBE obtained is −4.6%, while that of the RMSE obtained after the calculation is 0.7%. In accordance with ASHRAE, the various acceptable limits are between (−5%; +5%) representing the (MBE) and (−15%; +15%) representing the (RMSE). By comparing these results, it can easily be deduced that the MBE and RMSE values for the various monthly data are within the range recommended by ASHRAE (2002). Thus, the new simulation model can be considered as calibrated with the different monthly data and, therefore, the set of these two results shows that the new model can be validated.

3.6. Experiment

Before analysing the developed experiment, it is of interest to highlight that the building studied in the case of countries located in the tropical region is a copy of a residential building designed in Antananarivo city.

As it was shown before, it is of interest to comment that an experiment in this building in 2017 [27] was carried out. In this experiment, the new adaptive approach recommended in the ASHRAE-55 standard was applied, consisting of distributing questionnaires and simultaneously taking physical measurements of air temperature, relative humidity, and wind speed, between other variables. Finally, the measurement data and response of occupants allowed us to evaluate the comfort rate of residential buildings, as was described in reference [27].

4. Results and Discussion

4.1. Case of Residence Building Located in the Tropical Region

In this section, the simulation results of buildings placed in the tropical regions are showed and the sum-up is in Tables 6 and 7. In particular, Table 6 shows a comparison of monthly air temperature (°C) before and after the building revision.

Table 6. Variation of indoor air temperature in tropical regions.

Month		January	February	March	April	May	June	July	August	September	October	November	December
Original Residence building	Antananarivo	25.21	24.83	25.47	24.64	23.45	21.74	20.75	21.10	23.27	24.51	25.56	25.12
	Douala	26.66	26.59	26.36	26.13	26.17	25.80	25.35	25.04	25.10	25.62	26.05	26.54
	Dakar	24.66	24.84	25.18	25.19	26.00	26.64	27.04	26.98	27.20	27.20	26.91	25.90
Residence after renovation	Antananarivo	25.25	25.02	25.33	24.48	23.27	21.75	20.75	21.03	23.08	24.56	25.44	25.21
	Douala	26.19	26.24	26.10	25.88	25.89	25.49	25.14	24.96	25.08	25.41	25.52	26.04
	Dakar	24.78	24.96	25.21	25.30	25.82	26.46	26.92	26.91	27.05	27.02	26.71	25.75

Table 7. Annual operational carbon and energy demand in the residential building of tropical regions.

Cities		Antananarivo	Douala	Dakar	Total
Residence in the initial state	Energy demand (kWh)	3149.31	10,276.65	21,593.81	
	Carbon emission rate (kg CO_2)	1722.95	7366.85	13,085.86	
Renovated residence (applying: renovation + passive strategies + PV)	Surface of installed photovoltaic panels (m^2)	16	34	58	
	Cooling energy (kWh)	0.00	0.00	0.00	
	Energy demand (kWh)	2564.80	6944.28	13,874.95	0
	Green energy generated (kWh)	−2565.00	−6944.30	−13,875.01	
	Carbon emission rate (kg CO_2)	−3149.30	−140.51	3278.49	

4.1.1. Analysis of Indoor Conditions and Energy Consumption in Tropical Regions

First of all, it must be explained that indoor conditions are dependent on outdoor conditions, which were obtained from the Meteonorm tool. In particular, in Figure 5, it can be observed that Antananarivo has a wet tropical climate with a range of air temperature, relative humidity, and Predicted Mean Vote (PMV) [28] of 20.99 °C to 26.04 °C, 45% to 75%, and −2.07 to 0, respectively. At the same time, in Douala, the air temperature was between 25.45 °C and 27.37 °C and the relative humidity varied from 68.81% to 73.65%. From these results, it was concluded that, despite the fact that they have the same building construction, it is less comfortable in Douala than Antananarivo, which is located on several mountains. Indeed, climate conditions seem to be more favourable in the cities located in an altitude than in the coastal cities [25], which are dominated by heat [27].

In Antananarivo:

$$PMV = 0.106Top - 4.11, (R = 0.43); Rh = 1.208Top + 24.19, (R = 0.42) \tag{3}$$

In Dakar:

$$PMV = 0.286Top - 7.99, (R = 0.57); Rh = 3.993Top - 43.78, (R = 0.73) \tag{4}$$

In Douala:

$$PMV = 0.980Top - 26.41, (R = 0.85); Rh = -2.199Top + 128.8, (R = 0.88) \tag{5}$$

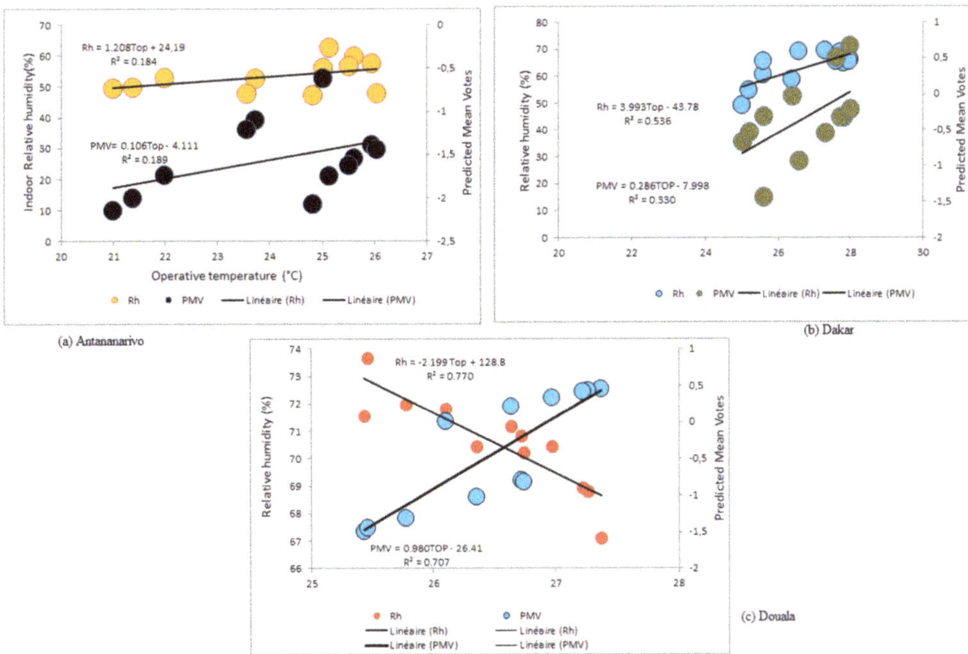

Figure 5. Variation of relative humidity and Predicted Mean Votes function of operative temperature in: (**a**) Douala; (**b**) Antananarivo; (**c**) Dakar.

The average indoor air temperature in the residence building placed in three Sub-Sahara Africa cities is given in Table 6.

In Table 6, the average indoor air temperature before and after the revision process can be observed. From this table, it can be concluded that the temperature decreases in the renovated residence when it is compared with the existing one. As a consequence, the cooling energy demand showed in Figure 5 and Table 7 was equal to zero after a renovation. This result shows that the choice of the most adapted materials to the local climate can significantly reduce the cooling energy demand. From this table, it was observed that renovation allows reducing the energy consumption between 9.7% and 35.5% in these three cities. Furthermore, the energy demand is 63% higher in the coastal region than in high altitude regions, which confirms the previous study carried out by Nematchoua et al. [29–32]. In particular, in Table 7, it can be observed that, after the revision process, the zero-energy objective is respected (the sum of the energy demand and the green energy generated by some photovoltaic panels (PV) is equal to zero).

At the time of analysing the energy savings, it is of interest to pay special attention to PV behaviour. In this sense, it is important to note that the size of the PV varies according to the region. Figure 5 shows that the PV is more efficient in Antananarivo than in Douala and Dakar cities. For these two cities, the solution to PV's inability, to meet monthly energy, is to store the excess electricity produced during sunny periods. At the same time, it was obtained that the energy generated by photovoltaic panels is the most efficient in the dry season. Indeed, it decreases from 8% to 23% in the three cities between rainy (May–September) and dry seasons (Figure 6a–c).

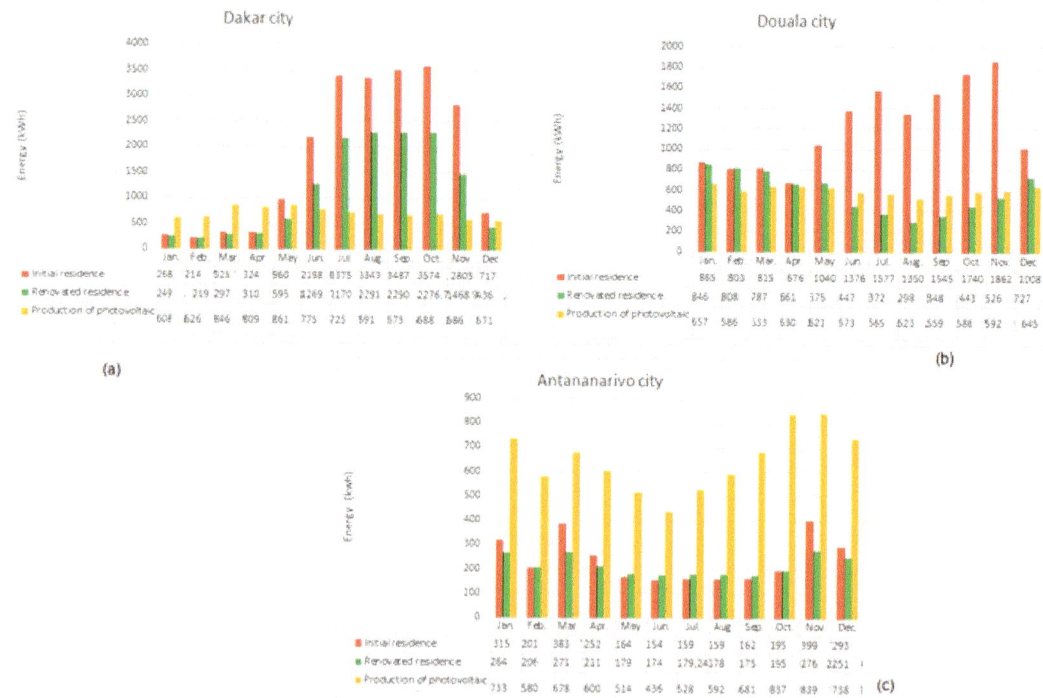

Figure 6. Monthly energy demand and production by photovoltaic panels in Dakar city (**a**); Douala (**b**); and Antananarivo (**c**).

This energy saving has some implications. In this sense, the carbon emission rate is negative in Antananarivo and Douala, which means that these residences remove more CO_2 than it emits into the atmosphere. Furthermore, it was observed that this rate is very low in Dakar (9.58 kg CO_2/m^2) compared to the initial concentration of carbon set to 38.26 kg CO_2/m^2. This reduction of carbon, estimated to be 74.96%, may be due to the implementation of sustainable materials and the application of passive strategy techniques [29].

Finally, from these results, it can be concluded that the renovation of the building requires an additional cost, but it improves indoor air quality [33]. Due to its related effect, in the next subsection, the impact of renovation by analysing investment cost will in turn be analysed.

Similar to air temperature, relative humidity has a significant effect on energy use in the buildings. Indeed, as explained by Aktas et al. [34], improving thermal comfort while reducing energy consumption in a building requires careful management of humidity levels, as well as careful selection of building materials.

4.1.2. Cost Analysis

At the time of evaluating the economic investment, it must be considered that the standards show an estimated investment of a low energy building from 35% to 55% higher than a conventional residential building. In this particular case study, the analysis of the cost is based on many assumptions, as detailed in Tables 8 and 9. In this sense, an increase between 40% and 70% was observed when compared to the basic price to reach the standard building with zero-energy and a payback period of 6.3 years. Furthermore, it is of interest to highlight that the transition between traditional structural work, respecting the energy performance of the building, and structural work respecting the low energy standard, leads to an increase in costs, which can vary from 35 to 55%.

Table 8. The average cost parameter applied at the building level in Sub-Saharan Africa [35].

Currency and Exchange Rate	
Currency Symbol	**$**
Applicable construction labour Hours and local cost index	
Regional material cost index	0.06
Hourly labour rate worker	$0.23–3.01
Hourly labour rate craftsman	$0.21–1.9
Discount factor (capital cost) and inflation	
Discount rate (cost of capital)	18.0%
General inflation rate	8.0%
Energy inflation rate	6.2%
Water inflation rate	5.2%
EOL as % of capex	1.5%

Table 9. Evaluation of residential building cost.

	Parameters	Values
Initial state of the building	Residence area	342 m^2
	Annual heating requirements	0
	building life cycle	50 years
	Price of the closed building structural work	30,000 € including VAT
	Electricity price	0.15 € per kWh
	Electricity consumption	2564.8 kWh per year
	Electricity cost	19,236 € for 50 years
Residence building+ PV installed	The price of the closed building structural work	32,400 € including VAT
	PV cost	2400 €
	Saving cost	16,836 € for 50 years
	Payback	6.23 years

4.2. Case of Residence Building Located in the Temperate Region

4.2.1. Analysis of Indoor Conditions and Energy Consumption

As a base case, to be compared with the results obtained in the tropical region, the building placed in the temperate climate region was analysed. In this sense, three cities located in the countries of France, Belgium, and the United States were chosen. The building placed in this new climatic region shows some differences in morphology and occupation in accordance with its typical design. In particular, the occupation rate is 0.0303, which means a person for 33 m^2 of area. Finally, it is important to notice that, at the time of the revision process, the PV area varied depended on micro-climate.

The main results were summed up in Tables 10 and 11 shown. In the revised building, there is a more comfortable indoor air temperature. In particular, Figure 7 shows the comparison of heating energy demand in the temperate region before and after the revision process. After renovating this residential building, it was noticed that the heating energy decreases to 31.4% (from 89.45 kWh/m^2 to 61.37 kWh/m^2) in Paris; to 33.3% (between 93.3 kWh/m^2 and 62.2 kWh/m^2) in Washington; and to 30.4% (between 95.7 kWh/m^2 and

66.6 kWh/m^2) in Brussels. Finally, from Table 11, it can be deduced that, in the three cities, green energy generated is equal to the energy demand. Furthermore, the results showed that the heating energy represents between 60% and 91% of total energy consumption in a residential building in the temperate region, which confirms the results found by Olonscheck et al. [36]. In this sense, in order to produce green electricity equal to the building energy demand, it is necessary to combine two different renewable technologies: gas cogeneration and photovoltaic panels. The required PV area is 58 m^2 in Brussels, 45 m^2 in Paris, and 42 m^2 in Washington DC with this combined cogeneration solution, but would extend to 194 m^2 in Brussels, 151 in Paris, and 141 m^2 in Washington DC without the cogeneration. This combined solution produces nearly zero annual operational carbon in the three cities, so an economical study will be shown in the next section.

Table 10. Comparison of indoor air temperature (in °C) in one residence building located in three cities.

Month		Jan.	Feb.	Mar	Apr	May	Jun.	Jul.	Aug	Sep.	Oct.	Nov	Dec
Original residence	Brussels	16.42	16.56	17.42	18.09	16.94	18.65	21.56	21.19	17.65	18.57	17.41	16.85
	Paris	16.74	16.81	17.53	18.30	18.20	20.33	22.53	23.12	19.50	18.69	17.53	16.87
	Washington	15.90	16.66	17.83	18.60	20.72	23.80	25.02	24.32	22.23	20.08	17.78	16.60
Renovated residence	Brussels	15.87	15.93	16.61	17.36	15.75	17.63	20.55	20.00	16.53	17.94	16.78	16.24
	Paris	16.11	16.05	16.74	17.62	17.02	19.24	21.64	22.16	18.12	18.03	16.76	16.03
	Washington	16.13	16.26	16.89	17.45	14.16	16.87	21.10	21.19	19.19	17.70	16.23	16.18

Table 11. Annual operational carbon and electricity demand in the residential buildings located in temperate climates.

	Cities	Brussels	Paris	Washington
Initial building	Cooling electricity (kWh/m^2)	1.02	2.46	2.48
	Heating and domestic hot water (Gas) (kWh/m^2)	37.71	34.96	33.20
	Electricity demand (kWh/m^2)	25.46	24.90	22.06
	Carbon emission rate (kg CO_2/m^2)	18.64	18.39	16.89
Revised building	Surface of installed photovoltaic panels (m^2) for a building area of 342 m^2	58.00	45.00	42.00
	Green electricity produced by gas cogeneration (kWh/m^2)	40.96	38.99	37.03
	Heating and domestic hot water (Gas cogeneration) (kWh/m^2)	37.71	34.96	33.20
	Cooling electricity (kWh/m^2)	0	0	0
	Electricity demand (kWh/m^2)	20.36	18.77	14.54
	Green electricity generated (kWh/m^2)	−17.11	−14.74	−10.71
	Carbon emission rate (kg CO_2/m^2)	−2.60	−0.11	2.17

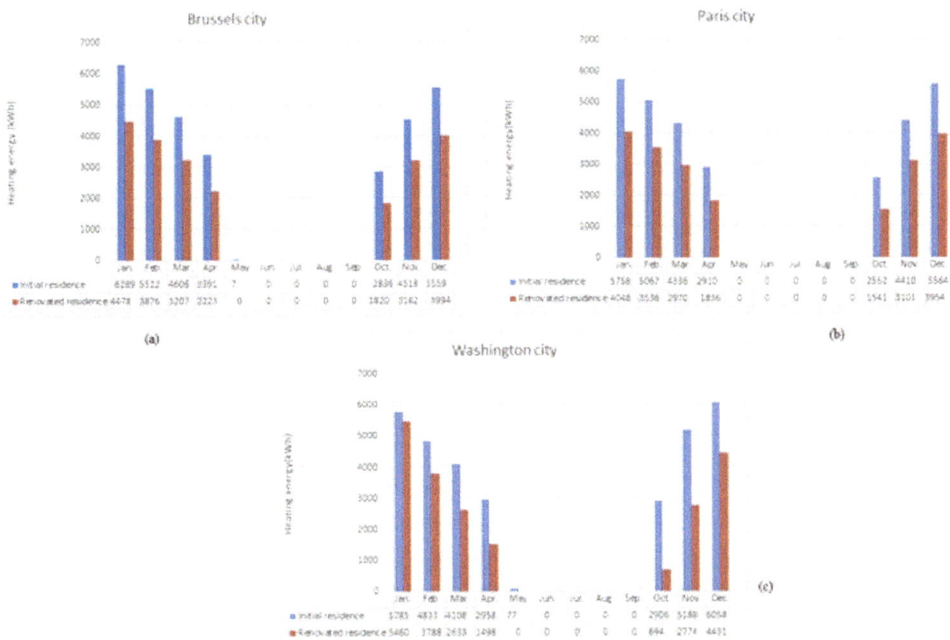

Figure 7. Heating energy in Brussels (**a**), Paris (**b**), and Washington DC (**c**).

4.2.2. Cost Analysis

Some details regarding the evaluation of cost are given in Table 12. From this table, it was concluded that this house meets the criteria of the passive house and a payback period of 3.8 years to amortize the initial economical investment in the revised building.

Table 12. Evaluation of cost.

Buildings	Parameters	Values
Initial state of the building	Initial investment	350,000 € including VAT
	Area	342 m^2
	Annual heating requirements (case of Brussels)	13.81 kWh/m^2
	Electricity price	0.25 €/kWh
Residence building + PV installed	The price of the closed building structural work	385,000 €
	Price of energy demand (cooling+ heating+ electricity+ domestic hot water)	9163.5 € per year
	Payback	3.8 years

Finally, from these results, it can be concluded that the investment cost and energy consumption are more important in a residential building located in temperate rather than in tropical climates. In particular, the energy consumption is around 78.2% higher in temperate than tropical climates and, at the same time, the investment cost is 12 times higher in temperate than tropical climates.

5. Conclusions

With the aim to define the energetic and economic interest of the transition to zero-energy in residential buildings in tropical climates, the energy performance of two standard residential buildings selected in tropical and temperate regions was improved towards zero-energy and low carbon emissions. The results showed that the zero-energy building concept is the most likely in the tropical region in favour of its geographical position. In particular, the zero-energy buildings are reached with an average indoor air temperature between 20.8 °C and 25.6 °C in the continental tropical region; from 24.7 °C to 27.2 °C in the coastal tropical regions; and from 16.4 °C to 25.1 °C in the temperate regions. In consequence, it was obtained that the renovation allowed a reduction of the energy consumption up to 36% in the tropical region with a payback of 6.3 years, whereas heating energy decreases from 30 to 34% in the three cities studied. These results are related to the location of the city and the seasonal efficiency of the PV, between other parameters.

Although a similar energy saving was obtained in temperate and tropical regions, and the investment cost is higher in the temperate region than in the tropical one, it was obtained that the payback is nearly half.

Finally, the results of this study can be applied in other countries with a similar climate. Furthermore, this study can serve as a guide for all those who wish to invest in sustainable construction. Another future, more detailed study will allow us to understand in depth the impact of the choice of construction materials in green buildings.

Author Contributions: Conceptualization, M.K.N.; methodology, M.K.N., J.A.O. and P.R.; formal analysis, M.K.N., E.O. and E.J.R.S.; investigation, M.K.N., E.O. and P.R.; data curation, M.K.N. and P.R.; writing—original draft preparation, M.K.N., S.R., J.A.O. and E.J.R.S., writing—review and editing, M.K.N., J.A.O., E.O., P.R., S.R. and E.J.R.S. All authors have read and agreed to the published version of the manuscript.

Funding: The work presented in this paper has not received external funding.

Institutional Review Board Statement: Not applicable.

Informed Consent Statement: Not applicable.

Data Availability Statement: Data available under request of reader.

Conflicts of Interest: The authors declare no conflict of interest.

References

1. Liu, Z.; Davis, S.; Feng, K.; Hubacek, K.; Liang, S.; Anadon, L.; Chen, B.; Liu, J.; Yan, J.; Guan, D. Targeted opportunities to address the climate—Trade dilemma in China. *Nat. Clim. Chang.* **2015**, *1*, 5. [CrossRef]
2. Crippa, M.; Oreggioni, G.; Guizzardi, D.; Muntean, M.; Schaaf, E.; Lo Vullo, E.; Solazzo, E.; Monforti-Ferrario, F.; Olivier, J.G.J.; Vignati, E. *Fossil CO_2 and GHG Emissions of Allworld Countries—2019 Report*; Publications Office of the European Union: Luxembourg, 2019.
3. Torcellini, P.; Pless, S.; Deru, M.; Crawley, D. Zero energy buildings: A critical look at the definition. In Proceedings of the 2006 ACEEE Summer Study on Energy Efficiency in Buildings, Pacific Grove, CA, USA, 14–18 August 2006.
4. Nematchoua, M.K.; Yvon, A.; Roy, S.E.J.; Ralijaona, C.G.; Mamiharijaona, R.; Razafinjaka, J.N.; Tefy, R. A review on energy consumption in the residential and commercial buildings located in tropical regions of Indian Ocean: A case of Madagascar island. *J. Energy Storage* **2019**, *24*, 100748. [CrossRef]
5. Feuille de Route des Acteurs de la Construction à Bruxelles: Vers une Économiecirculaire. Available online: https://www.circulareconomy.brussels/wp-content/uploads/2019/06/BE_beCircular_feuille-de-route-CD_def_FR1.pdf (accessed on 15 November 2020).
6. Agence Parisienne du Climat. Batiments. Available online: https://www.apc-paris.com/plan-climat/batiments (accessed on 15 November 2020).
7. U.S. Energy Information Administration. U.S. Energy-Related Carbon Dioxide Emissions. 2019. Available online: https://www.eia.gov/environment/emissions/carbon/ (accessed on 13 July 2021).
8. Nematchoua, M.K. From existing neighbourhoods to net-zero energy and nearly zero carbon nęighbourhoods in the tropical regions. *Sol. Energy* **2020**, *211*, 244–257. [CrossRef]
9. Sameti, M.; Haghighat, F. Integration of distributed energy storage into net-zero energy district systems: Optimum design and operation. *Energy* **2018**, *153*, 575–591. [CrossRef]

10. Shaw-Williams, D.; Susilawati, C.; Walker, G.; Varendorff, J. Towards net-zero energy neighbourhoods utilising high rates of residential photovoltaics with battery storage: A techno-economic analysis. *Int. J. Sustain. Energy* **2020**, *39*, 190–206. [CrossRef]
11. Brown, H.S.; Vergragt, P.J. Bounded socio-technical experiments as agents of systemic change: The case of a zero-energy residential building. *Technol. Forecast. Soc. Chang.* **2008**, *75*, 107–130. [CrossRef]
12. Yi, H.; Srinivasan, R.S.; Braham, W.W.; Tilley, D.R. An ecological understanding of net-zero energy building: Evaluation of sustainability based on emergy theory. *J. Clean. Prod.* **2017**, *143*, 654–671. [CrossRef]
13. Szalay, Z.; Zöld, A. Definition of nearly zero-energy building requirements based on a large building sample. *Energy Policy* **2014**, *74*, 510–521. [CrossRef]
14. Zhou, Z.; Feng, L.; Zhang, S.; Wang, C.; Chen, G.; Du, T.; Li, Y.; Zuo, J. The operational performance of "net zero energy building": A study in China. *Appl. Energy* **2016**, *177*, 716–728. [CrossRef]
15. Srinivasan, R.; Braham, W.W.; Campbell, D.E.; Curcija, C.D. Re (De) fining Net Zero Energy: Renewable Energy balance in environmental building design. *Build. Environ.* **2012**, *47*, 300–315. [CrossRef]
16. Robert, A.; Kummert, M. Designing net-zero energy buildings for the future climate, not for the past. *Build. Environ.* **2012**, *55*, 150–158. [CrossRef]
17. Pikas, E.; Thalfeldt, M.; Kurnitski, J. Cost optimal and nearly zero energy building solutions for office buildings. *Energy Build.* **2014**, *74*, 30–42. [CrossRef]
18. Pulido Arcas, J.A.; Rubio-Bellido, C.; Perez-Fragallo, A.; Orpeza-Perez, I. Net zero energy buildings and low carbon emission, a case of study of Madagascar Island. In *Zero-Energy Buildings—New Approaches and Technologies*; IntechOpen: London, UK, 2020.
19. SPW Energie. Dépliant—Exigences PEB. 2017. Available online: https://energie.wallonie.be/fr/exigences-peb.html?IDC=9136 (accessed on 17 November 2020).
20. Les Criteres du Passif. Available online: https://www.maisonpassive.be/?Les-criteres-du-passif (accessed on 21 November 2020).
21. Comment le Tarif de L'électricité Est-Il Fixé en 2021? Available online: https://www.comparateur-energie.be/blog/prix-electricite-belgique/#evolution (accessed on 13 July 2021).
22. Wikipedia. The Free Encyclopedia. Available online: https://en.wikipedia.org/wiki/Washington (accessed on 13 July 2021).
23. Nematchoua, M.K.; Orosa, J.A.; Buratti, C.; Obonyo, E.; Rim, D.; Ricciardi, P.; Reiter, S. Comparative analysis of bioclimatic zones, energy consumption, CO_2 emission and life cycle cost of residential and commercial buildings located in a tropical region: A case study of the big island of Madagascar. *Energy* **2020**, *202*, 117754. [CrossRef]
24. DUALSUN. Available online: https://news.dualsun.com/co-en/12/2014/what-is-the-optimal-orientation-and-tilt-angle-for-solar-panels/ (accessed on 13 July 2021).
25. Nematchoua, M.K.; Tchinda, R.; Orosa, J.A. Adaptation and comparative study of thermal comfort in naturally ventilated classrooms and buildings in the wet tropical zones. *Energy Build* **2014**, *85*, 321–328. [CrossRef]
26. *ASHRAE Guideline 14-2002: Measurement of Energy and Demand Savings*; ASHRAE: Atlanta, GA, USA, 2002.
27. Nematchoua, M.K.; Ricciardi, P.; Buratti, C. Statistical analysis of indoor parameters an subjective responses of building occupants in a hot region of Indian ocean; a case of Madagascar island. *Appl. Energy* **2017**, *208*, 1562–1575. [CrossRef]
28. Nematchoua, M.K.; Noelson, J.C.V.; Saadi, I.; Kenfack, H.; Andrianaharinjaka, A.-Z.F.; Ngoumdoum, D.F.; Sela, J.B.; Reiter, S. Application of phase change materials, thermal insulation, and external shading for thermal comfort improvement and cooling energy demand reduction in an office building under different coastal tropical climates. *Sol. Energy* **2020**, *207*, 458–470. [CrossRef]
29. Nematchoua, M.K.; Mahsan, S.; Sigrid, R. Strategies and scenarios to reduce energy consumption and CO_2 emission in the urban, rural and sustainable neighbourhoods. *Sustain. Cities Soc.* **2021**, *72*, 103053. [CrossRef]
30. Nematchoua, M.K.; Roshan, G.R.; Tchinda, R.; Nasrabadi, T.; Ricciardi, P. Climate change and its role on forecasting the Energy demand in buildings, Case study of Douala City, Cameroon. *J. Earth Syst. Sci.* **2015**, *124*, 269–281. [CrossRef]
31. Nematchoua, M.K.; Nishimwe, A.M.-R.; Reiter, S. Towards Nearly Zero-Energy Residential Neighbourhoods in the European Union: A case study. *Renew. Sustain. Energy Rev.* **2021**, *135*, 110198. [CrossRef]
32. Fortman, D.J.; Brutman, J.P.; De Hoe, G.X.; Snyder, R.L.; Dichtel, W.R.; Hillmyer, M.A. Approaches to sustainable and continually recyclable cross-linked polymers. *ACS Sustain. Chem. Eng.* **2018**, *6*, 11145–11159. [CrossRef]
33. Zhu, Y.; Yan, Y.; Zheng, F.; Ge, J.; Gu, Y. Research on the Renovation of Historical Buildings and Improvement of the Residential Environment of Hangzhou Zhuyangxin Plaster Store. *J. Asian Arch. Build. Eng.* **2010**, *9*, 395–402. [CrossRef]
34. Aktas, Y.D.; Wang, K.; Zhou, Y.; Othman, M.; Stocker, J.; Jackson, M.; Hood, C.; Carruthers, D.; Latif, M.T.; D'Ayala, D.; et al. Outdoor Thermal Comfort and Building Energy Use Potential in Different Land-Use Areas in Tropical Cities: Case of Kuala Lumpur. *Atmosphere* **2020**, *11*, 652. [CrossRef]
35. Nematchoua, M.K.; Reiter, S. Evaluation of bioclimatic potential, energy consumption, CO_2-emission, and life cycle cost of a residential building located in Sub-Saharan Africa; a case study of eight countries. *Sol. Energy* **2021**, *218*, 512–524. [CrossRef]
36. Olonscheck, M.; Holsten, A.; Kropp, J. Heating and cooling energy demand and related emissions of the German residential building stock under climate change. *Energy Policy* **2011**, *39*, 4795–4806. [CrossRef]

energies

MDPI

Article

"Oil is the New Data": Energy Technology Innovation in Digital Oil Fields

HaeOk Choi and Hwanll Park *

Science and Technology Policy Institute, Sejong National Research Complex 370, Sicheng-daero, Sejong 30147, Korea; hochoi@stepi.re.kr

* Correspondence: hipark@stepi.re.kr; Tel.: +82-44-287-2461

Received: 11 August 2020; Accepted: 12 October 2020; Published: 23 October 2020

Abstract: Digital oil fields (DOFs) are built on data produced from energy technology innovation during the application of new technologies to oil resource development. In this study, this conversion is examined through the paradigm switch to "oil is the new data". An analysis of related patents shows that DOF technology is developing through convergence and close links with other industries, specifically the equipment, parts, and material industries. Additionally, it is conjectured that the strategic preemption of standards will emerge as an important policy issue because a standard must be established for the interoperability of the elemental technologies of DOFs. Furthermore, with the expansion of DOF-related technologies, device-related technologies have also been developed. Of these device-related technologies, sensor technology specifically provides new possibilities for the development of DOFs. The significance of this study is that it explains the evolution of DOFs over the course of 10 years, which is illustrative of technological innovation in the field of energy-related development and data collection.

Keywords: digital oil field; oil upstream; technology innovation; oil exploration and production; energy policy

1. Introduction

The expression, "data is the new oil" implies the considerable current economic value of big data in the digital era [1]. This indicates that data, like coal and oil used as raw material and fuels for industrial applications, may be used to develop other industries, creating new economic value. Recently, the oil exploration and production (oil E&P) industry has been generating, storing, and using vast amounts of data. The oil E&P industry has applied sensors throughout the entire oil production process to create data in real time, and uses artificial intelligence to analyze the collected data to make decisions efficiently. We describe the digital transformation of the oil E&P industry as the digital oil field (DOF). DOFs, which are preoccupied with value added data, have attracted significant attention recently with the introduction of new technologies. Since DOFs are oil fields that produce big data, we paraphrase the term "data is oil", to state that "oil is the new data". Furthermore, data have garnered attention as the crude oil of the 21st century [2]. To date, data have had a great impact worldwide, such that new economic drivers have been established owing to its applications throughout industry. Although the resource development industry has been dealing with vast amounts of material for a long time, the rapid development of computers, sensors, information and communication technologies (ICTs) in recent years has exponentially increased the amount of data being generated. Big data technologies can be used to analyze large amounts of unstructured data that hitherto could not be processed. As processing technologies, big data analysis methods are the basis for realizing intelligent services [3].

Recently, big data analysis has been utilized for innovation based on new DOF technology. DOF innovation is progressing rapidly because the combination of resource development and ICT can reduce the overall cost and increase profitability of the whole process. To develop a DOF, it is necessary to study the optimal search path by combining various data such as geographic information, field information, core data, and reservoir data. Therefore, related patents, which are being continuously monitored, have been filed in connection with these fields [4,5]. DOF technology primarily develops by adopting technologies from fields such as short-range and telecommunication infrastructure, cloud computing and integrated data management, digital sensor modules, and data analysis and visualization [6]. This is different from existing energy-related data conversion, which necessitates the periodic consideration of technology for comprehensive data collection. This study contributes toward finding ways to develop energy-related innovation capabilities through the analysis of DOF-related innovation capabilities.

Although the influence of oil in some areas in the world is diminishing due to the increasing deployment of renewable energy, oil remains important in the world economy. Recently, with production rates in traditional oil and gas fields reaching their peaks, or even depletion stages, exploration environments have become increasingly hostile owing to greater target depths, more extreme climates, or inaccessible locations. In the case of new petroleum resources such as shale gas, the required drilling per unit of leased area is typically higher than that of conventional petroleum resources, and so are the required work and management operations, such as in hydraulic crushing. DOF technology makes it possible to efficiently manage sites that require oil wells comprising several hundred to several thousand holes, such as those for shale gas, and to facilitate the unmanned or remote management of oil and gas development sites in extreme or remote locations. In addition, data generated in the oil production process are being leveraged to increase production and economic efficiency through risk prevention and demand forecasting. Therefore, this study aims to identify the implications of DOFs in the field of oil resource development using specific data. Various efforts have been made in the DOF-related hardware field to remotely monitor and control oil wells [7] by integrating ICT technologies such as data communication with servers and real time data processing. Research in this field is ongoing [8,9].

Several technological innovations have been designed recently in the field of oil exploration and production, through the adoption of new technologies [4]. DOFs are convergence technologies that manage and operate online [10,11]. New technologies that are adopted in the field include the extraction of limited petroleum resources. Examples of the extraction technology include extreme pole and gas hydrates, recovery/production by robots, AI-based exploration and evaluation technology, digital production optimization technology (DOF), and consumption prediction using big data [3,12]. This research reflects these new technologies based on data accumulated in the oil resource development field [13,14]. DOFs, in relation to petroleum resource development, have the potential to facilitate the creation of various business models through their roles as platforms for scaling and managing data. Dominant technology development companies in the petroleum field have a strong network in the value chain process, which makes it difficult for new startups to participate in this area. This is a real problem, as it is difficult for new companies to penetrate a niche market with well-established methods and strategies. As an opportunity for new companies to participate in the digital oil field, we make this field our focus. We believe that the results of this study can be used as reference material for future oil-related policies.

In this study, the structural characteristics of the innovation process in DOFs were investigated through the conversion of energy-related innovation capabilities focused on oil resource development stages. To this end, we analyzed changes in the innovation capabilities and structural characteristics of DOFs, which focus on using data to produce value addition. Previous studies on the digital oil field have focused on detailed element technology or business feasibility for specific cases [10,15–18], without discussions about the overall system perspective. In addition, there is little research that analyzes the contents of unstructured data using actual patent data in detailed technology. Furthermore, compared with the interest in the rapidly developing digital oil field owing to the grafting of new technologies, limited research has been conducted on case studies, and research on the structural characteristics related to technological innovations in this field is insufficient [10,17,19].

This study aims to solve major issues in technological innovation processes taking place in the digital oil field by using data on actual patent content. Specifically, by analyzing the innovation capabilities of DOFs, future research directions will be derived and proposed. The characteristics of the rapidly growing innovation capabilities of DOFs will be examined, and conclusions will be drawn. To overcome the limitations of existing research methods, we focused on innovative competency-related data (unstructured data in patents), and analyzed and derived structural implications through the analysis of structural characteristics over time.

2. Materials and Methods

The main activities of oil E&P are composed of exploration, reservoir characterization, development, and production, which collectively have the highest number of patents. We analyzed how DOF technologies evolved in the above mentioned areas. First, we focused on the development and production fields, which have the highest number of patents for E&P activities. We analyzed these patents using DOF-related methods. We collected and analyzed related data and derived future implications. During the study, relevant experts on patent-related search formulas were consulted (see Appendix A).

In oil E&P, securing core technologies is directly linked to profitability at each stage of the oil and gas field development; therefore, it is necessary to identify and continuously monitor core competencies. Despite the importance of these core competencies, they have so far hardly been organized and analyzed based on the number of patent applications and trends by country. In this study, we deployed a differentiated analysis method to overcome the limitations of existing research methods and to continuously monitor innovation capabilities in the field of oil resource development. For innovation capability analysis, we attempted to extract key content using patent document information such as unstructured and text data, and identify the key content and issues using cluster analysis and topic modeling.

Topic modeling is a machine learning technique that can extract the topics inherent in document data to classify documents or derive word clusters that constitute topics. The topic analysis modeling method extracts topics through the latent Dirichlet allocation (LDA) algorithm and visualizes the clustering of keywords and documents for each topic. The rationale for using the LDA algorithm in topic modeling is to discover the hidden semantic structure of the text body. In particular, LDA extracts topics by estimating the probabilities that a word exists in a specific subject and that a specific subject exists in a document as a combination probability [20]. This study uses scientific analysis tools to understand the knowledge-based network structure of the digital oil field, monitor the technological innovation process, and utilize the results of the analysis in policy development, supporting the entry of startups into the field.

The analysis steps (shown in Figure 1) are as described here.

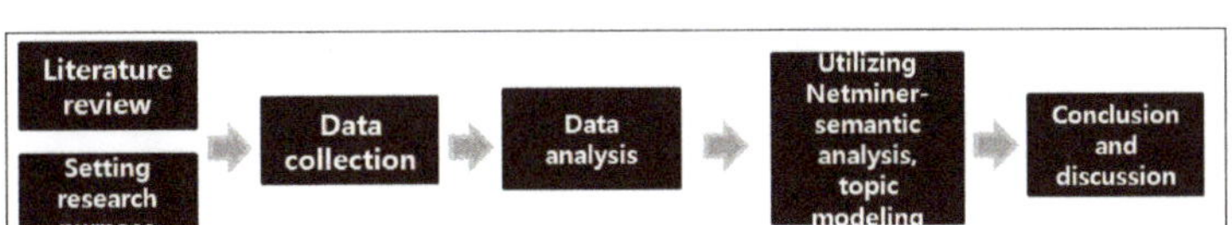

Figure 1. Research process.

First, the related literature was investigated to determine DOF research trends in the field of oil resource development.

Second, methods for collecting DOF-related data were designed in consultation with experts in the relevant fields, and a search formula for finding related patents was derived (Appendix A). Data from the last 9 years (2 June 2011~2 June 2020) were collected from the Korean Intellectual Property Office. A total of 15159 data points were collected. The Korean Intellectual Property Office database is a reliable site showing the current status of each country's patents by linking up with the patent databases of other countries.

Third, the cluster network analysis method and topic modeling technique were used to examine the structural characteristics. Through these methods, major issues and contents were analyzed to draw their implications. Unstructured patent data were used for this analysis. The patent number and applicant data were omitted, and the patent description and unstructured data related to the title were collected, analyzed, and processed.

Fourth, to analyze the yearly change in innovation capacity, each keyword was extracted by dividing the data for the 9 years into three-year ranges, such as 2011~2014, 2015~2017, and 2018~2020.

Finally, conclusions were drawn based on the analysis of the contents, and future research tasks were proposed.

This study attempts to understand the structural characteristics of network formation by first determining the knowledge-based network structure occurring in the digital oil field, and then monitoring, in detail, the changes that occurred in the last 9 years. This is meaningful in monitoring the innovation capacity of the digital oil field, which has been rapidly changing in recent years. Additionally, this study is different from other studies in that it contains an abundant, unseen knowledge-based network structure that focuses on the contents of patent data and attempts to analyze it so as to overcome the limitations of analysis methods based on patent frequency alone.

3. Results

In this study, semantic network analysis was utilized to examine the structural characteristics of the innovative capabilities of DOFs. First, by analyzing the entirety of the 9-year data, we attempted to analyze critical keywords and issues in the related topics. To examine the temporal changes in technological innovation, the data were split into three time divisions.

The following results were obtained by analyzing the structural characteristics of DOFs (Table 1). To interpret the results by group (Figure 2), an analysis was attempted by referring to a study that distinguished existing DOFs [21,22], and explained the characteristics of their related technologies at each stage. This study was used to interpret the high-ranking keywords and main contents of each group among the classification groups, with reference to existing studies (see Table A1).

The most common keywords from the first group, denoted by G1, include "method", "system", "composition", "material", and "use". This group represents the transmission role within the system in the DOF. It can be identified as the component serving as the communication infrastructure, which assembles, supports, and builds hardware-related technologies.

The second group, G2, mainly focuses on keywords such as "process", "apparatus", "control", and other terms related to process modeling in the DOF that involve interpretation and control of the collected data. The group is closely related to automation, which is important for optimizing the overall work processes. Traditional technologies in the field of oil resource development are being utilized to support the overall system efficiency improvement of DOFs by combining AI and machine learning, which are nontraditional technologies, to support decision making [23,24]. It was also confirmed that these technologies are closely related to those used in the equipment industry in terms of remote monitoring and control.

Among the main technologies in a DOF, process control can be made redundant by improving the efficiency of the oil field using methods such as prediction and production optimization through automated data collection and alarm systems. The management life cycle is divided into data processing, analysis, and modeling. Specifically, this process is used to make decisions with data obtained from petroleum resource development [25,26].

Table 1. Results on group network analysis of digital oil fields.

G1 (Communications Infrastructure)			G2 (Processing Modeling)			G3 (Sensor and Interface Support)			G4 (Control Hardware)		
Keywords	Frequency	Degree Centrality	Keywords	Frequency	Degree Centrality	Keywords	Frequency	Degree Centrality	Keywords	Frequency	Degree Centrality
method	4669	93	process	1046	59	device	718	39	plant	164	26
system	1972	73	apparatus	760	41	treatment	396	52	power	139	19
composition	1196	62	oil	750	49	production	392	56	acid	121	20
material	688	51	sand	655	57	fuel	204	25	unit	76	17
use	600	65	gas	459	46	compound	174	21			
product	333	39	fluid	375	36	particle	135	17			
surface	297	40	water	371	48	agent	129	18			
coating	225	26	hydrocarbon	312	40	mixture	119	27			
preparation	201	30	control	281	37	medium	109	22			
polymer	187	33	application	201	31	storage	106	23			
metal	186	31	recovery	195	35	bed	97	18			
structure	184	26	formation	183	17	soil	91	11			
assembly	178	24	catalyst	168	18	reactor	90	15			
Tool	173	21	processing	155	27	filter	83	22			
carbon	157	23	heat	153	26	chemical	81	25			
cement	149	18	energy	144	18						
mold	140	22	waste	143	29						
component	129	25	flow	136	23						
construction	120	22	operation	128	20						
vehicle	119	19	fracturing	123	22						
sanding	113	13	conversion	115	18						
core	104	21	bitumen	107	20						
casting	101	18	stream	103	16						
manufacture	100	18	removal	101	27						
article	98	14	extraction	100	21						
manufacturing	97	20	temperature	98	19						
proppant	97	11	slurry	96	20						
formulation	96	18	separation	91	19						
element	93	8	feedstock	89	19						
layer	90	18	tailing	84	11						
machine	90	12	pressure	83	13						
resin	90	15	biomass	82	18						
panel	89	13	liquid	81	15						
glass	86	16	screen	77	11						
concrete	85	12	well	76	13						
skin	83	6	steam	75	12						
fiber	81	11	proppants	74	8						
foam	79	22	field	72	14						
binder	76	12									
support	76	12									
body	75	11									
building	72	12									
fracture	72	12									

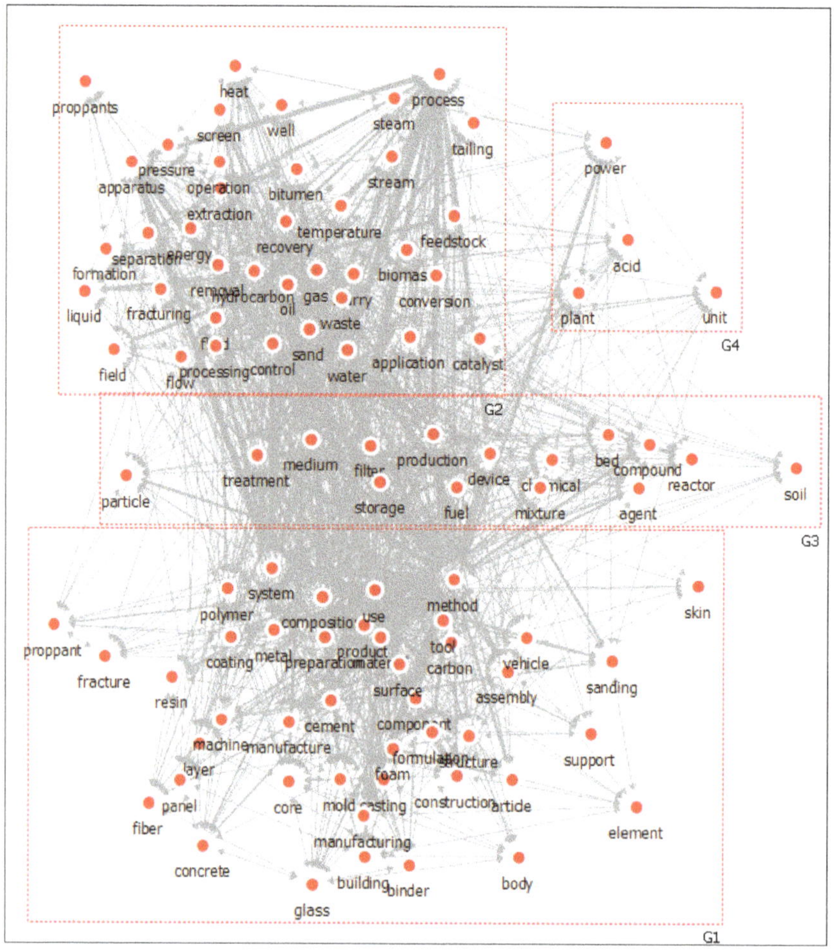

Figure 2. Group network analysis of digital oil fields.

The recently developed virtual field is a technology that can simulate the physical process by mathematically modeling a production network from oil and gas fields through a production line [27]. This is an important step, in which real time on-site data analysis, production optimization, and economic forecast analysis can be performed by incorporating risk factors into the simulation as well. Production optimization can be performed by applying each production classification; the optimization process utilizes additional drilling locations and plans at the field scale. Ultimately, this process can be integrated and used to predict oil price fluctuations or production volumes, thereby establishing an optimal management system for major oil fields [28].

> A device with one or more sensors for monitoring the effectiveness of sand compaction on a production line. The sensor measures the changes in sand compaction, which is affected by the mechanics of the vibration system, changes in the sand properties, and environmental changes—from patent #13204677, 2011.8.6. Moha**-.

The third group, G3, includes keywords such as "device", "treatment", and "production", and indicates major issues related to sensor and interface support. Device technologies include sensors and interfaces, and represent the processes of seismic exploration, 4D exploration, and monitoring, including remote sensing [29–31]. Processing and visualizing the obtained data can increase the efficiency of oil production by unifying complex data, such as geological and borehole data obtained from earthquake disasters, to increase the drilling and development efficiency.

G4, the last group, comprises major issues related to control hardware, such as the "plant", "power", "acid", and "unit". The major DOF companies characteristically develop technology mainly on the sea, where energy-related technology developments for plant management are being made. It has been confirmed that these companies are attempting to develop more efficient technology for situations where it is necessary to supply electricity to an offshore oil field.

A digital offshore plant field has also been analyzed. This field is used for discovering, drilling, and producing marine resources such as oil and gas, and is closely related to the DOF. Developments in this area can significantly reduce the cost of generating and producing marine resources in offshore plants by combining new technologies, such as ICT, with established ones [32,33].

In marine installations, a DOF is unmanned and remotely controlled. Owing to initial cost limitations, the technology for offshore oil fields tends to be developed mainly through collaboration with major companies that develop large-scale oil fields and major ICT companies. To disrupt the monopoly that major companies with experience in large-scale oil fields have, it is necessary to develop DOFs with differentiated strategies.

> A method for recovering gas in natural gas hydrate exploitation is disclosed, in which a gas–water mixture at the bottom of an exploitation well is delivered to an ocean surface platform through a marine riser by adopting the gas-lift effect of methane gas derived from the dissociation of natural gas hydrate, thus achieving a controllable flow production of marine natural gas hydrate. #15765652 2018.2.12. G* institute-.

Technological innovations in DOFs have evolved. Initially, in early 2010, keywords such as "oilfield", "gas system", "process composition", and "apparatus" ranked highly. A change was observed in the mid-2000s, and as the importance of devices that improved the process efficiency of DOFs increased, the ranking of keywords related to those devices also increased. In recent years, devices that are considered important in DOFs have been sensor, block chain, and predictive analytics technologies (Research and markets, 2018). Among these, sensor technology is central in data collection and processing, and plays an important role in the development of DOFs. To manage a large area, it is necessary to install a sensor, and collect and analyze its data; the economic feasibility of this can be determined based on the price of the sensor. As a matter of fact, sensor prices have been falling since 2010 ($0.66 per unit) and, at the time of writing, they have fallen to half their initial values [34]. It has been confirmed that DOFs are becoming more economically efficient owing to the drop in sensor prices and that the majority of the keywords are device-related issues. In the development of DOFs, sensors can collect different data and, by processing these data, increase the efficiency and predictability of the entire process, thus facilitating the development of a smarter oil field.

> A completion system for use in a well includes a first completion section and a second section. The first completion section has a sand control assembly to prevent passage of particulates, a first inductive coupler portion, and a sensor positioned proximate to the sand control assembly, which is electrically coupled to the first inductive coupler portion—from patent #14586375, S* cooperation, 2014.12.30.-.

Data integration, which involves collecting data from each step, is a comprehensive process that covers data collection and processing. Optimization, which improves business efficiency, is performed by varying the judgment and manipulation method according to the collection of information in the upstream stage. Intelligent drilling and completion is a process that extracts underground information in real time during drilling. It helps drilling technicians remove obstacles and optimize operations such as bending. Specifically, it helps in optimizing the productivity of boreholes during drilling through means such as monitoring the temperature and pressure through a fluid sensor. It also prevents accidents by facilitating the early detection of hazards.

Next, changes in the knowledge-based network structure were examined by extracting the frequency of keywords over time (Figure 3), based on the keywords extracted from the network by group. The variation of the DOF over time is as presented here. First, when considering the contents of related patents, the keywords related to "system" and "process" are located at the root. This has remained the same for almost a decade and it can be confirmed that, in the development of the DOF, technological developments related to improving the efficiency of these systems and processes are being made. Additionally, it has been confirmed that DOFs, which manage the entire cycle, tend to be geared toward the development of technologies for overall optimization rather than the development of specific technologies.

I(2011–2014)	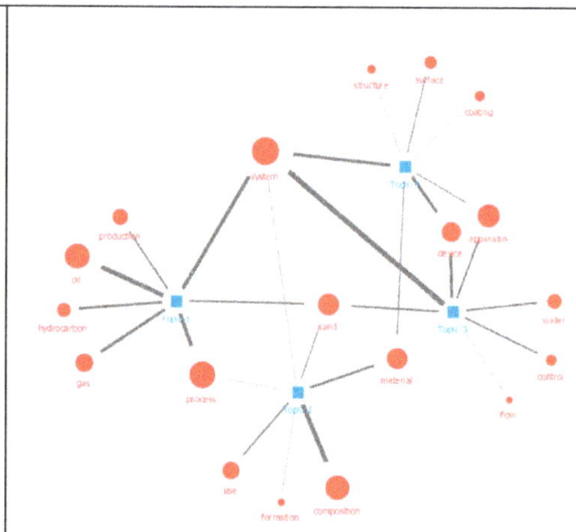
II(2015–2017)	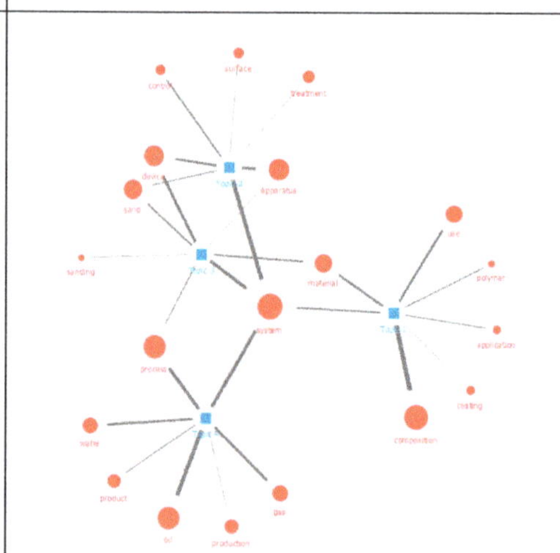

Figure 3. *Cont.*

III(2018–2020)	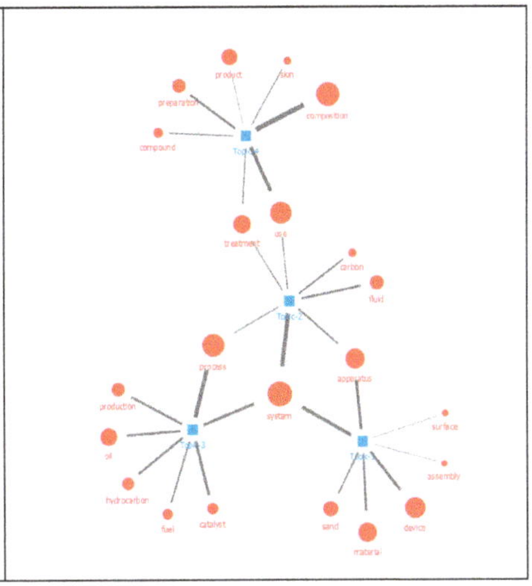

Figure 3. Result of topic modeling by time series.

In the development of DOF technology, "device" development is closely related to the sensor, communication base, decision support, remote operation, and control system fields. In the field of traditional petroleum resource development, which is mainly based on existing hardware device technology, it is possible to improve the economic efficiency of oil fields by adopting new ICT technology [35]. The core technology in this field is based on equipment components. Furthermore, hardware technology development is a field that has been developed in combination with AI and machine learning technologies that can support decision making [18].

The analysis results obtained from extracting the frequency of related keywords in the groups over time are detailed as follows.

The groups G1 (communications infrastructure), G3 (sensor and interface support), and G4 (control hardware) have become more important over time (Figure 4). The frequency of G2 (processing modeling), conversely, has reduced over time, indicating that the DOF has progressed beyond the initial stage of development, which requires overall process modeling. Furthermore, it can be observed that various technologies related to traditional oil resource development are evolving in combination with other innovative technologies as they undergo a scale-up process while being assimilated into the DOF. The technology related to G1 is advancing steadily, which suggests that the standards for maintaining interoperability by integrating different technologies are growing in importance over time.

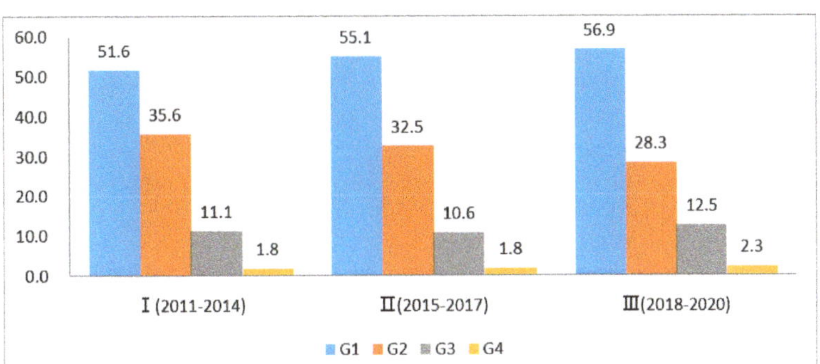

Figure 4. Characteristics of digital oil field (DOF) development by time series.

Groups G1, G3, and G4 have become increasingly important over time by percentage distribution. However, as seen in Table 2, the occurrence frequency of G1 key terms decreased from 5211 (2011–2014) to 4344 (2018–2020); the occurrence frequency of G2 key terms decreased from 3591 to 2158. The frequency ratio gradually decreases for all groups, but the distribution shows a tendency to increase gradually for groups G1, G3, and G4. This shows a tendency to decrease in quantity but increase in actual overall distribution, which is meaningful in that it shows qualitative growth of core technologies rather than a quantitative increase.

Table 2. Characteristics of DOF development by time series.

Group Division	I (2011–2014)		II (2015–2017)		III (2018–2020)	
	Frequency	%	Frequency	%	Frequency	%
G1 (Communications infrastructure)	5211	51.6%	4349	55.1%	4344	56.9%
G2 (Processing modeling)	3591	35.6%	2568	32.5%	2158	28.3%
G3 (Sensor and interface support)	1116	11.1%	839	10.6%	951	12.5%
G4 (Control Hardware)	178	1.8%	141	1.8%	177	2.3%
Total	10096	100%	7897	100%	7630	100%

4. Discussion

To develop a DOF, a strategic innovation transitioning from "data is the new oil" to "oil is the new data" is necessary. In the past, petroleum resource development involved oil production through the process of drilling, reservoir characterization, development, and the production of traditional oil. The oil industry, in particular, is characterized by the pervasive presence of cartels in the technology sector, as major corporations around the world dominate and lead technology development. Entering the petroleum sector, unlike other industries, requires high initial capital and experience. The availability of empirical data can reduce the opportunity cost and improve the economic efficiency of the DOF.

In DOFs, a technology that can increase the overall optimization efficiency is being primarily developed. In the petroleum field, petroleum technology developments so far can be mainly attributed to innovations in elemental technologies. However, the digital oil field is developing into a platform that enhances overall efficiency, focusing on the interoperability of element technologies [36]. Additionally, by using the data generated by the DOF platform, we are trying to realize not only the optimization of the entire system but also a reduction in management costs through the optimization management of the oil field. Therefore, in establishing policies related to DOFs, rather than focusing on specific technologies, it is necessary to focus on the overall system or control field and seek data-based support measures. In this field, the integration of different elementary technologies will emerge as an important issue in the future; technological innovation policy support is needed.

In the existing oil industry, the participation of venture companies and small and medium enterprise (SMEs) is restricted, owing to the monopoly of certain dominant companies. However, a situation where new technology is inevitable in the petroleum industry can provide a new opportunity for venture companies and SMEs with differentiated technologies to incorporate such new high-tech technologies into the market. Therefore, this study is meaningful in the sense that it offers a technological innovation direction for startups to enter into the oil industry, by providing the structural characteristics of the knowledge base of the digital oil field.

Digitalization of the oil resource development field is closely related to industrial development. Regardless of how good a technology is, if it does not yield industrial profits, it may be obliterated. Furthermore, technology in the field of petroleum resource development has developed under the monopoly of a few leading companies over the past few years; but there has hardly been any external technical impetus. A change in this aspect can provide an opportunity for the industrial revolution to shift to an oil-oriented development strategy centered on economics, especially against the backdrop of the surge in oil prices that occurred in the mid-2000s. A strategic approach is needed to shift the paradigm of energy-related technological development.

DOFs are being developed in tandem with various industries such as the parts, materials, equipment, and device industries; it goes beyond merely adopting ICT technology. Future studies could focus on confirming the development trend of DOFs in relation to the various industries in a development strategy focused on early productivity improvement methods.

5. Conclusions

In this study, using DOF-related patents, the technological innovations in DOFs, which have recently attracted significant attention in energy-related technology development, were analyzed. This study overcame the limitations of the existing patent analysis by focusing on the contents of the technology. Furthermore, research methodologies, such as the evolution of the technology field and cluster analysis through semantic network analysis, were utilized.

First, as can be seen in the DOF-related patent analysis, technological development is achieved through fusion with other industries, specifically the equipment, parts, and material industries. DOFs develop closely with the equipment, parts, and materials industries; they are not created by merely grafting new ICT into the traditional oil industry. It can be observed that developments are being made through innovations that go beyond simply monitoring the entire system; these developments do not involve merely deploying an ICT-based monitoring system in the existing oil resource development field. Therefore, to monitor the situation of oil and gas fields in real time and build an online management system by automating the entire process from exploration to production, it is necessary to go through the scale-up process of all related technologies and confirm that the overall technology optimization is taking place.

It is conjectured that the strategic preemption of the standard will emerge as an important policy task because standards must be established to ensure the interoperability of the elemental technologies in DOF technology optimization. In developing DOFs, it was confirmed that the main systems and processes are central to the interoperability of convergence technologies and closely related to standard preemption; technology development is being geared toward improving the efficiency of the component systems and processes of DOFs.

Several device-related technologies have also been developed. Among these device-related technologies, sensor technology provides new possibilities for the development of DOFs. This is because the development of sensor technology can lead to a qualitative growth in data collection and processing, as various data are produced and collected through sensors in DOFs. In terms of economics, the decline in the price of core devices plays an important role in the diffusion of DOFs.

Hardware technologies in the traditional oil resource development field are rapidly developing DOF devices, which is a deviation from traditional preoccupation, by combining AI and machine learning, data collection, and processing technologies. The core significance of this study is that it explains the evolution of DOF technology, which is illustrative of technological innovation in the field of energy-related data development and collection.

Author Contributions: Conceptualization, H.C.; methodology, H.C.; validation, H.C.; formal analysis, H.C.; investigation, H.C.; data curation, H.C.; writing—original draft preparation, H.C. and H.P.; writing—review and editing, H.C. and H.P. supervision, H.C. and H.P.; All authors have read and agreed to the published version of the manuscript.

Funding: The publication fee for this manuscript was funded by the Science and Technology Policy Institute (project number: P0200300 and B0200502).

Acknowledgments: This research is supported by STEPI (project number: P020030, title: Global tech-knowledge of strategic resources competitiveness for securing Korea's competitiveness). Furthermore, this research is related to "Research on Data-based policy priorities for technology innovation-rediscovery of dark data" (project number: B0200502).

Conflicts of Interest: The authors declare no conflict of interest.

Appendix A

This research focused on Oil resource development field development and production stage digital oil fields.

Table A1. Network analysis results of topic modeling analysis by time series.

No.	2011–2014			2015–2017			2018–2020		
	Keywords	Frequency	%	Keywords	Frequency	%	Keywords	Frequency	%
1	system	704	16.9	system	645	20.0	system	616	18.6
2	process	403	9.7	composition	300	9.3	composition	550	16.6
3	oil	366	8.8	process	266	8.2	process	373	11.3
4	composition	340	8.2	oil	250	7.8	use	240	7.3
5	apparatus	315	7.6	apparatus	229	7.1	device	215	6.5
6	sand	314	7.5	device	223	6.9	apparatus	209	6.3
7	material	288	6.9	sand	222	6.9	material	185	5.6
8	device	278	6.7	material	215	6.7	treatment	145	4.4
9	gas	211	5.1	use	168	5.2	oil	134	4.1
10	use	189	4.5	gas	166	5.1	product	131	4.0
11	production	176	4.2	water	124	3.8	sand	119	3.6
12	water	161	3.9	production	118	3.7	fluid	102	3.1
13	hydrocarbon	154	3.7	product	105	3.3	preparation	96	2.9
14	surface	145	3.5	treatment	101	3.1	production	96	2.9
15	control	124	3.0	surface	93	2.9	hydrocarbon	95	2.9
total		4168	100%		3225	100%		3306	100%

References

1. Haupt, M. "Data is the New Oil"—A Ludicrous Proposition. Available online: https://medium.com/project-2030/data-is-the-new-oil-a-ludicrous-proposition-1d91bba4f294 (accessed on 7 July 2020).
2. Hirsch, D.D. The glass house effect: Big Data, the new oil, and the power of analogy. *Maine Law Rev.* **2013**, *66*, 373.
3. Korea Institute of Geoscience and Mineral Resources. *Smart Evaluation Solution for Oil and Gas Field Based on Integration of Static and Dynamic Data*; KIGAM: Daejeon, Korea, 2019.
4. Khan, M.Y.; Chetri, H.; Saputelli, L.; Singh, S. Waterflood Optimization and its Impact Using Intelligent Digital Oil Field (Idof) Smart Workflow Processes: A Pilot Study in Sabriyah Mauddud, North Kuwait. In Proceedings of the IPTC 2014: International; North Kuwait Petroleum Technology Conference, Doha, Qatar, 19–22 January 2014; pp. 1–15.
5. Hussain, A.; Vega, J.C.; Hassane, M.A.S.; Yusaf, S.A.; Abdul-Halim, A.A. Enhancing Smart Completion Capabilities by Integration with Digital Oil Field Real Time Monitoring System in a Green Field of ADMA-OPCO. In Proceedings of the Abu Dhabi International Petroleum Exhibition & Conference, Abu Dhabi, UAE, 7–10 November 2016.
6. HyunTae, K. The Technology Trend of Upstream Field in Oil. Presented at STEPI Presentation Material, Sejong, Korea, 22 July 2020.
7. Qin, H.; Han, Z. Stochastic Resource Allocation for Well Control With Digital Oil Field Infrastructure. *IEEE Syst. J.* **2016**, *12*, 1295–1306. [CrossRef]
8. Burda, B.; Crompton, J.; Sardoff, H.M.; Falconer, J. Information Architecture Strategy for the Digital Oil Field. In Proceedings of the Digital Energy Conference and Exhibition, Huston, TX, USA, 11–12 April 2007.
9. Tao, Z. IOT's application in the "Digital Oil Field". *Telecommun. Sci.* **2010**, *4*, 25–32.
10. Holland, D. Managing the journey to the digital oil field. *Lead. Edge* **2004**, *23*, 1137–1138. [CrossRef]
11. Chanana, P.; Soni, T.M.; Bhakne, U. Emerging Technologies and Workflows in Digital Oil Field. In Proceedings of the Offshore Technology Conference Asia, Kuala Lumpur, Malaysia, 22–25 March 2016.
12. Taneja, P.; Wate, P. Big data enabled digital oil field. *CSI Commun.* **2013**, *37*, 18–20.
13. Holland, D. Digital Oil Field 2.0: Maturity and Paradigms. *J. Pet. Technol.* **2009**, *61*, 56–57. [CrossRef]
14. Rajan, S. Future Technologies of the Oil Field. *Way Ahead* **2011**, *7*, 11–14. [CrossRef]
15. Records, L.R.; Shimbo, D.T. Petroleum enterprise intelligence in the digital oil field. In Proceedings of the SPE Intelligent Energy Conference and Exhibition, Utrecht, The Netherlands, 23–25 March 2010.

16. Beckwith, R. Apps and the Digital Oil Field. *J. Pet. Technol.* **2012**, *64*, 40–46. [CrossRef]
17. Ali, Z.; Al-Jasmi, A.K.; Qiu, F. Digital Oil Field Experience: An Overview and a Case Study. In Proceedings of the SPE Digital Energy Conference, The Woodlands, TX, USA, 5–7 March 2013.
18. Anderson, R.N. 'Petroleum Analytics Learning Machine' for optimizing the Internet of Things of today's digital oil field-to-refinery petroleum system. In Proceedings of the 2017 IEEE International Conference on Big Data (Big Data), Boston, MA, USA, 11–14 December 2017; pp. 4542–4545.
19. Cramer, R.; Gobel, D.; Mueller, K.; Tulalian, R. A measure of the digital oil field status-is it the end of the beginning? In Proceedings of the SPE Intelligent Energy International, Utrecht, The Netherlands, 27–29 March 2012.
20. Blei, D.M.; Ng, A.Y.; Jordan, M.I. Latent Dirichlet allocation. *J. Mach. Learn. Res.* **2003**, *3*, 993–1022.
21. BP, "The Field of the Future technology flagship," 2014.03. Available online: www.bp.com (accessed on 2 July 2020).
22. Crompton, J. The Digital Oil Field Hype Curve: A Current Assessment the Oil and Gas Industry's Digital Oil Field Program. In Proceedings of the SPE Digital Energy Conference and Exhibition, The Woodlands, TX, USA, 3–5 March 2015.
23. Devries, S.G. Production Management Information Challenges of the Digital Oil Field. In Proceedings of the SPE Annual Technical Conference and Exhibition, Dallas, TX, USA, 9–12 October 2005.
24. Jafarizadeh, B.; Bratvold, R.B. Strategic Decision Making in the Digital Oil Field. In Proceedings of the SPE Digital Energy Conference and Exhibition, Huston, TX, USA, 7–8 April 2009.
25. Bonham, G.M.; Heradstveit, D.; Narvesen, O.; Shapiro, M.J. A Cognitive Model of Decision-making: Application to Norwegian Oil Policy. *Coop. Confl.* **1978**, *13*, 93–108. [CrossRef]
26. De Oliveira, V.L.C.; Tanajura, A.P.M.; Lepikson, H. A Multi-agent System for Oil Field Management. *IFAC Proc. Vol.* **2013**, *46*, 35–40. [CrossRef]
27. Lin, L.; Xianmei, L. Design of training system of oil field simulation based on virtual reality technology. *Comput. Technol. Dev.* **2012**, *22*, 205–208.
28. Lim, J.-T.; Park, H.-W.; Lim, J.-S. Classification and Application of Digital Oil Field System. *J. Korean Soc. Miner. Energy Resour. Eng.* **2014**, *51*, 750–756. [CrossRef]
29. Kwarteng, A.Y. Multitemporal remote sensing data analysis of Kuwait's oil lakes. *Environ. Int.* **1998**, *24*, 121–137. [CrossRef]
30. Landro, M.; Solheim, O.A.; Hilde, E.; Ekren, B.O.; Stronen, L.K. The Gullfaks 4D seismic study. *Pet. Geosci.* **1999**, *5*, 213–226. [CrossRef]
31. Hatchell, P.; Kawar, R.; Savitski, A. Integrating 4D Seismic, Geomechanics and Reservoir Simulation in the Valhall Oil Field. In Proceedings of the 67th EAGE Conference & Exhibition, Madrid, Spain, 14–17 June 2005; p. cp-1-00585.
32. Zhang, X.; Jiang, X.; Shi, S.; Chi, T. Digital ocean technological advances. In *Remote Sensing and Modeling*; Springer: Cham, Switzerland, 2014; pp. 195–214.
33. Zhang, X.; Wang, L.; Jiang, X.; Zhu, C. *Modeling with Digital Ocean and Digital Coast*; Springer International Publishing: Basel, Switzerland, 2017.
34. Statista, 2020. Smart Sensors: Global Average Sales Price 2010–2020. Available online: https://www.statista.com/statistics/736563/global-average-sales-price-of-smart-sensors/ (accessed on 3 July 2020).
35. Holdaway, K.L. Enhance Digital Oil Fields by Plugging the Technological Capability Gap. In Proceedings of the North Africa Technical Conference and Exhibition, Cairo, Egypt, 14–17 February 2010.
36. Janakiraman, S. Digital oil fields—Intelligent wells and platforms. *Pet. Eng.* **2018**, *16*, 24. [CrossRef]

energies

MDPI

Article

Low-Carbon Energy Governance: Scenarios to Accelerate the Change in the Energy Matrix in Ecuador

Flavio R. Arroyo M. [1,2,*] and Luis J. Miguel [1,3]

1 Systems Engineering and Automatic Control, School of Industrial Engineering, Paseo del Cauce s/n, University of Valladolid, 47011 Valladolid, Spain; ljmiguel@eii.uva.es

2 Faculty of Engineering, Physical Sciences and Mathematics, Av. Universitaria, Central University of Ecuador, Quito 170129, Ecuador

3 Research Group on Energy, Economy and System Dynamics, School of Industrial Engineering, Paseo del Cauce s/n, University of Valladolid, 47011 Valladolid, Spain

* Correspondence: flavio.arroyo@gmail.com

Received: 5 August 2020; Accepted: 9 September 2020; Published: 11 September 2020

Abstract: This article describes the results of a study of Ecuador's energy status, using the system dynamics methodology to model supply, demand and CO_2 emissions scenarios for the year 2030. Primary energy production increased in the different projected scenarios, with oil as the most important source of energy. The increase observed in final energy consumption was mainly associated with the transport and industry sectors. A reduction in energy intensity was projected for the different scenarios, which could be associated with the projected economic growth. The results obtained were used to build a proposal for energy policies aimed at mitigating emissions. The proposed changes to the national energy matrix could be the factors that will contribute most to the achievement of carbon emission reductions projected by the different scenarios; changes in the energy matrix are mainly associated with the development of projects to replace fossil fuels with renewable energies, mainly hydropower.

Keywords: climate change; environmental sustainability; low-carbon energy; energy governance; energy policy; renewable energy

---

1. Introduction

As more and more extreme weather events occur across the world, global climate risks are intensifying. In recent years, most meteorological phenomena have been blamed for global warming [1]. The impacts of global warming generate large investments in the global economy, mainly for developing countries with little capacity to generate foreign exchange [2,3].

To prevent substantial changes in ecosystems from occurring, global CO_2 emissions must decrease significantly by 2030. Although the intensity of the production of greenhouse gases has been reduced to some extent, this decrease is not taking place quickly enough to achieve a transition towards environmentally sustainable production and consumption, which favors an increase in the level of carbon emissions and the acceleration of climate change [1].

According to information from the United Nations, the latest estimates attribute seven million deaths to air pollution, which is currently one of the main general risk factors for human health worldwide, overtaken by problems of high blood pressure, diabetes and smoking [4].

Global energy-related CO_2 emissions increased by 1.7% in 2018 and reached a historical maximum of 33.1 Gt [5]; this growth contrasts with reduction commitments to meet the objectives of the Paris Agreement on climate change. The use of renewable energy in the production of electrical energy in the United States has contributed to a reduction in emissions [6].

Between 1989 and 2014, the use of fossil fuels remained constant, the demand for oil decreased by 6%, while the consumption of natural gas and coal increased by 2% and 3%, respectively. Bioenergy covers 10% of the world's energy demand [7].

The global energy demand worldwide grew by 2% in 2017, according to preliminary estimates by the International Energy Agency (EIA), more than double the growth rate in 2016. According to the 2019 edition of BP's Energy Outlook which analyzed a scenario of energy consumption by 2040, the world will demand 17,865.82 Mtoe [5]. The growth in energy consumption from 1996 to 2007 reached almost 58% and the projection made by BP foresees an increase of 32% by 2040.

India and China are the countries with the highest growth in energy demand in the world [8]. Transparent energy policies regulate the power generation capacity of nations [9]. The increase in energy demand is focused on improving living conditions and global economic development.

In 2018, world economic growth remained stable at around 3% [1]; between 2017 and 2018, high-income economies expanded at a constant rate of 2%, while upper–middle income economies in the East and South Asian regions followed a relatively strong growth path with an average growth of 5.7% [10].

Prospects for global macroeconomic development must urgently take concrete political action to achieve sustained economic growth sustainably. The investigations of [11–20] establish that energy consumption is related to economic growth.

The efforts of an international collective action aimed at the administration and distribution of energy resources, as well as the provision of energy services in the form of global energy governance, have been proposed to provide a clear picture of the challenges related to energy use [21].

In recent years, energy governance has become an important and new field of research in international studies. It has sought to understand how the energy sector is regulated, for whom and with what consequences. Energy security uses indicators for governance. The related research has used quantitative indicators for topics such as quality of government, competition, transparency among others [9,22–24]. Indicators have been established in the energy sector to be used in energy governance. [25] This establishes energy policies related to planning and development, energy tariffs and subsidies as effective indicators of governance.

There are other important quantitative indicators to understand the context of the energy system and the options available to reduce future carbon emissions. These include basic factors such as the population wealth and the climate of the city along with energy indicators: the price of energy in each city and the relative proportion of any tax or levy; the carbon intensity of fuels the objectives established by each city for the reduction in greenhouse gas emissions; the annual energy demand [26].

An analysis of energy consumption and economic growth in Ecuador conducted by [27] indicated that, until the 1970s, the country mainly had an agro-export model; by 1937, a transition had begun that led to an oil-producing model in the country. At that time, oil production was more than three times higher than oil consumption which led to significant changes in the country's technification.

Ecuador is an oil-producing country; as its financial resources are limited, investment in renewable energy has focused on hydroelectric energy. Ecuador has important challenges regarding the exploitation of its natural resources and the conservation of its ecosystems [8].

The main oil fields in Ecuador are in the Amazon region, where biodiversity and the fragility of ecosystems impose technical and economic challenges in relation to their exploitation. For a long time, the natural gas associated with oil production was ventilated and burned instead of being utilized, releasing significant amounts of CO_2 and other pollutants into the atmosphere [8].

By world standards, Ecuador's oil industry is relatively modest; its proven oil reserves are approximately 8000 million barrels [28]. Despite this modest classification, the oil sector plays a prominent role in the country's economic policy and welfare, representing about 50% of export earnings and about a third of all fiscal revenues. Clean energies have had a minimal contribution in the primary energy matrix, except for hydroelectric energy production, which increased by 115% between 2000 and 2016 [29]. The total final energy production between 2000 and 2016 maintained a regularity of around 71,000 KBOE. It is important to highlight how, from 2014, electricity became the main final energy source in Ecuador [30].

Global warming is palpable in Ecuador; in recent years, there has been a reduction of about 40% in the glaciers of the Chimborazo, Cotopaxi and Antisana volcanoes [8]. To address this problem, since 2012, the 'Climate Change Strategy', developed by the Ministry of Environment and socialized within municipalities, has been implemented. The plan includes three aspects: adaptation, mitigation and a reduction in emissions from deforestation [31].

Renewable energies have become the fastest growing energy source, representing about a quarter of the global electricity generation [32]. Their participation is expected to grow as there is an increase in the electrification of the end-use sectors [33]. Since 2016, electricity has generated greater investments than the oil and gas sectors [34].

The costs of producing renewable energy have decreased, becoming a major driver of change. In the last 10 years, the average cost of electricity generated from photovoltaic and wind solar energy decreased by more than 70% and 20%, respectively. By 2020, the average cost of electricity generated by solar and wind sources will be the lowest on the market [35]. This competitive advantage is the result of technological progress and increased investment in this sector. Hydropower and geothermal energy have also shown competitive costs since they have started to operate [36].

Several countries are switching to renewable energy because they want to be less dependent on imports of energy from oil and gas. Several of the major oil-producing countries have set targets to increase renewable energy generation. The United Arab Emirates, by 2050, expect 44% of their energy mix to be made up of renewable energy and a 70% reduction in their carbon emissions [36].

Renewable energies bring macroeconomic advantages; the expectations for 2050 are that the cost of production will go from 5% to 2% of the world GDP [37]. To achieve energy and decarbonization goals, countries promote the use of renewable energy and energy efficiency [38,39].

Energy and climate change policies should not be complicated and clean energy must be provided at a competitive cost aimed at reducing emissions [40]; it is necessary to review and reorient fuel subsidies and direct funds towards developing new technologies to mitigate environmental problems, and countries must plan energy transition policies that promote the use of renewable energy [41]. The European Union has proposed policies for an energy transition focused on reducing emissions, developing clean energy and energy efficiency [42].

It is necessary to develop a coherent and comprehensive governance system that can address general sustainability constraints and implement long-term objectives [43]. Designing a coherent energy and climate policy will require a clear political objective that aims to reduce carbon emissions and adapt the energy system to the impacts of climatic variability and change [27,44]; this energy policy must be based on research to enable the proper planning, decision making, implementation and evaluation of its objectives.

This article describes and tests an approach for estimating energy production, final energy demand and proposed energy policing aimed at reducing CO_2 emissions in a newly integrated assessment framework that focuses on the biophysical and economic dimensions and interactions that arise during energy transitions. System dynamics is a perspective and set of conceptual tools that enable us to understand the structure and dynamics of complex systems, as well as a modelling method that enables us to build formal computer simulations of complex systems [45].

2. Materials and Methods

This section describes the methodology we developed to estimate the final energy supply and demand for the sector in Ecuador. A method has been developed based on the top-down projection of the evolution of final energy demand by the economic sector. The input–output analysis is based on a matrix that describes the supply and demand of energy between different sectors. Its combination with environmental aspects allow us to assign a specific final energy consumption per unit of monetary production to each sector in the search for low-carbon energy governance.

2.1. Model Description

This research has a two-part methodology that includes a literature review and evaluation and a scenario analysis using system dynamics (SD). System dynamics originated in control engineering and automation to determine a structure with different, interrelated input and output flows to fill or drain (respectively) different stocks. Two variables, connected by a causal link, can change in the same direction (positive relationship) or in opposite directions (negative relationship). In a positive—or reinforcing—feedback loop, growth in the first variable causes growth in the second variable, but growth in the second variable also causes growth in the first variable [46].

SD analyzes complex socio-ecological systems and is considered a feasible resource for thinking about sustainable futures [45]. SD models are used to simulate dynamic behavior over time [46]. SD is recreated by the explicit mapping of information transfers between stocks and flows to model feedback interactions [47]. The energy perspective is an indispensable tool to determine and analyze the most probable future scenarios for an energy system. Future energy demand and composition scenarios have implications for political decisions [48].

The system dynamics methodology has been widely used to model complex systems in which feedback, delays, and nonlinearities are frequent [47,49]. Some of these applications have been aimed at modelling energy and environmental systems [45,50,51], as well as integrated evaluation models [5,6]. From all the analyses of the literature review, it is understood that low-carbon energy governance will help accelerate the shift to a sustainable energy matrix in Ecuador. The historical data series used in the research was taken from the information collected from the national energy balances [29,52–55].

2.2. Scenario Analysis

Our design shows the energy perspective and the production of CO_2 emissions in Ecuador. Three scenarios were developed to show the relationship between the energy mix, economic growth and emissions in Ecuador. The design of these scenarios allows us to predict the evolution of the variables and, in turn, helps to generate energy policies and project the reduction in CO_2 emissions.

The Business as Usual (BAU) scenario projects past trends and those that will continue [49]. The National Policies (NP) scenario proposes the prioritization of the use of renewable energy sources, promotes the use of hydroelectric energy and replaces inefficient thermal generation. The massive implementation of efficient lighting in homes and public roads is proposed. This scenario also suggests the replacement of equipment with newer models that have a higher energy consumption. The replacement of LPG with electricity is proposed, with the implementation of induction cookers and execution of energy management systems in the main industries. The scenario also promotes the realization of a sustainable transport system that uses electricity and not hydrocarbons as the main source of energy; it also raises energy sovereignty as one of the pillars of a new energy matrix. The Global Policies and Trends (GPAT) scenario considers multilateral environmental agreements and global macroeconomic prospects for sustainable development, and replacement strategies for clean energy and energy efficiency. It considers the projections or trends of the reports of organizations such as the Intergovernmental Panel on Climate Change (IPCC), International Energy Agency (IEA), and BP, among others.

2.3. Modelling and Simulation

The model is divided into four main modules: economy, energy demand, energy availability, and climate/emissions. The conceptual schematic description includes the main relationships between the different modules. The main characteristics of each module are:

- Economy: the economy is modeled, assuming demand-driven growth and the complementarity of the sector. Therefore, production is determined by final demand and economic structure, combined with supply-side constraints, such as energy availability.

- Energy demand: final energy demand by sector is estimated through the projection of sector economic production and sector final energy demand, considering efficiency improvements and inter-final energy replacements driven by policies and physical scarcity.
- Energy availability: this module includes the potential and availability of RES and non-renewable energy resources, considering the biophysical and temporal limitations. In particular, the availability of non-renewable energy resources depends on both stock and flow limitations. In total, 5 energy sources and technologies are considered, and 9 final energies (electricity, solids, gases and liquids), with a technological breakdown. The modeling of energy availability is mainly based on the previous WoLiM model [56].
- Climate/emissions: the global model calculates the levels of CO_2 emissions generated by the final amount of energy used in Ecuador.

The evolution and integration of the model across modules requires, for each time step, the energy demand and supply to be balanced dynamically. The system was designed to establish energy policies and CO_2 emissions in Ecuador in 2030. The traditional energy resources of the country are also considered. The impact of economic growth on energy consumption and CO_2 emissions is analyzed. The flowchart of the economic–energy–carbon emissions system is shown in Figure 1. The energy consumption of each of the economic sectors generates CO_2 emissions.

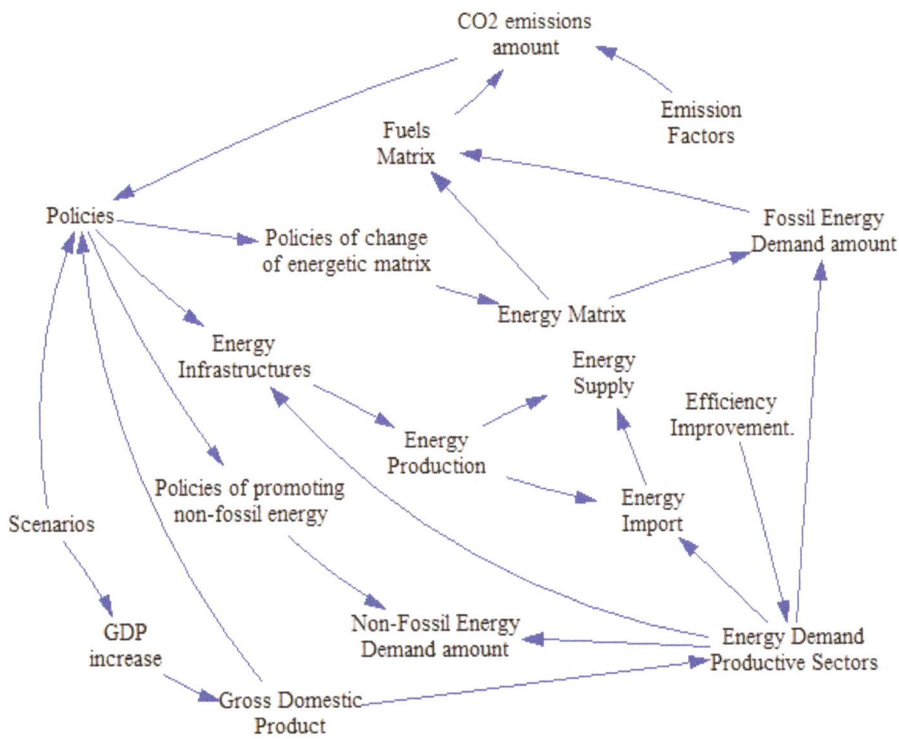

Figure 1. Flow diagram of the energy system and production of carbon emissions in Ecuador (adapted from [18]).

3. Results

The results obtained in order to simulate the scenarios of supply and demand for energy and the production of CO_2 emissions in 2030 are shown in Table 1. In all the projected scenarios, an increase in primary energy production is observed; this increase is greater in the NP and GPAT scenarios because these scenarios consider energy exploitation and production policies. In the NP scenario, energy policies continue with oil exploitation and the construction of new hydroelectric projects [57].

Table 1. Results obtained through the dynamics model for different energy variables and CO_2 emissions for 2016 and the Business as Usual (BAU), National Policies (NP) and Global Policies and Trends (GPAT) scenarios for 2030.

Variables		Simulated Scenarios to 2030		
	2016	**BAU**	**NP**	**GPAT**
Primary energy production (KBOE)	117,987.00	151,219.00	225,901.00	195,145.00
Final energy demand (KOB)	86,400.80	184,345.00	134,193.00	103,521.00
PIB per capita (USD 2007 per habitant)	4355.61	6223.09	6648.99	6283.12
CO_2 emissions (KT CO_2)	36,073.50	75,182.60	43,938.30	42,191.40
Carbon emissions per capita (TCO_2)	2.20	3.89	2.33	2.46
Carbon emissions/GDP ($kgCO_2$/USD 2007)	0.51	0.63	0.35	0.39
Energy intensity (BEP thousands USD 2007)	1.21	1.51	1.07	0.96

A breakdown of primary energy production by sector can be seen in Figure 2, revealing that obtaining energy from oil is the most important source in all the scenarios considered. However, a significant increase in the contribution of energy from hydroelectric sources and natural gas can be observed. Ecuador's energy mix was mainly based on the use of non-renewable sources; the use of renewable energy generation technologies is the least developed and is related to the launch of large-capacity hydroelectric plants, as their greater participation is expected.

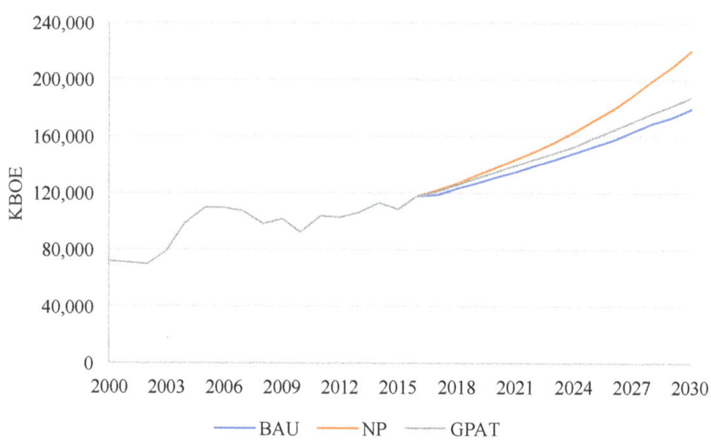

Figure 2. Primary energy production, Business as Usual (BAU), National Policies (NP) and Global Policies and Trends (GPAT) scenarios for 2030.

An increase in final energy demand by 2030 can be observed (Table 1). This will increase by 3.3 times in the BAU scenario if the current characteristics of energy consumption are maintained, which projects high and inefficient consumption. Maintaining current consumption habits would require Ecuador to significantly increase energy production; one should even consider importing hydrocarbons due to the incredibly high national energy demand. The lower projections raised by the PN (National Policies (NP) scenario) and PYTM (Global Policies and Trends scenario (GPAT)) scenarios, reducing the energy demand by 2.6 and 1.8 times, respectively, could be associated with a government policy to replace energy sources and improve energy efficiency. The BAU scenario projection shows a final energy demand of 184,345 KBOE. The NP and GPAT scenarios project an energy demand of 134,193 and 103,521 KBOE, respectively; this is due to the higher efficiency of electrical energy (Figure 3).

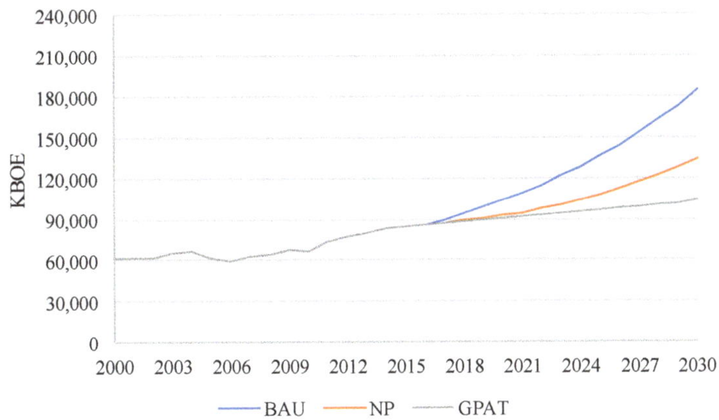

Figure 3. Energy demand, Business as Usual (BAU), National Policies (NP) and Global Policies and Trends (GPAT) scenarios for 2030.

A more exhaustive analysis of final energy consumption indicates, for the National Policies scenario (NP) and Global Policies And Trends scenario (GPAT), that 20% and 54%, respectively (Figure 4), of the total use would be in the transport sector, a sector that has been characterized by its inefficiency in the use of energy, mainly due to its use of fossil fuels.

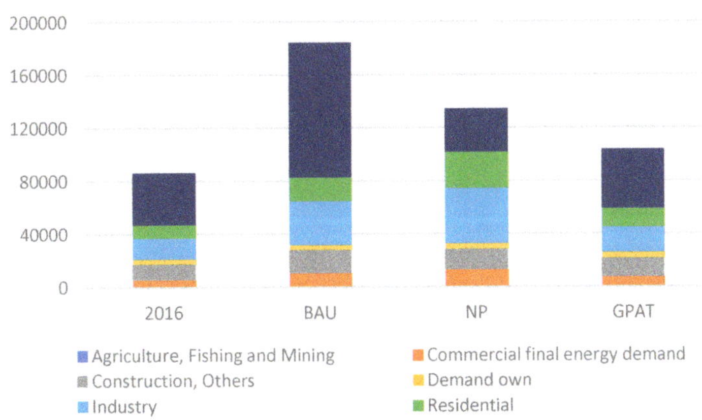

Figure 4. Distribution of energy demand by sector for 2016 and the Business as Usual (BAU), National Policies (NP) and Global Policies and Trends (GPAT) scenarios for 2030.

The projected reduction in relation to the BAU scenario is probably the result of the promotion of the use of biofuels and the use of electricity in mass passenger transport nationwide. Another sector that shows an increase in final energy consumption is the industrial sector, for which an energy transition has been proposed—a change in production equipment.

The projection of the model for GDP per capita foresees increases of 128% and 123% for the National Policies scenario (NP) and Global Policies and Trends scenario (GPAT) (Table 1); however, economic growth would be relatively low, even for a region, as currently the average region value is

USD 9270, and Ecuador barely reaches half that value. Currently, the GDPs of North America exceed USD 58,000 and the GDPs of the European Union exceed USD 33,000. The projections for 2030 foresee a moderate growth worldwide and this would only amplify the gap that Ecuador must solve; therefore, a structural change in the economy focused on sustained and equitable long-term economic growth is essential [58].

The proposed changes in the national energy matrix and the substitution of the type of energy could be the factors that contribute most to the achievement of a reduction in carbon emissions projected by the National Policies (NP) and Global Policies And Trends (GPAT) scenarios (Table 1). In relation to the maintenance of exaggerated current trends, emissions in the BAU scenario will increase by 2.7 times by 2030. If Ecuador does not implement measures in relation to consumption and energy sources, CO_2 emissions will continue to increase, especially if it does not adopt the use of more efficient and less pollutant energies, not like those currently used, which are mostly of fossil origin. The implementation of measures for the proposed scenarios would favor reductions of 44% and 38% in the ratio of CO_2 emissions to GDP for the National Policies (NP) and Global Policies And Trends (GPAT) scenarios compared to the increase observed for the BAU scenario by 2030 (Table 1).

The energy intensity projections for the National Policies (NP) scenario and Global Policies and Trends scenario (GPAT) project a reduction in relation to the current values as well as an expected increase for the BAU scenario (Table 1). [59] It has been proposed that economic growth goes hand in hand with the efficient use of energy and, in the Ecuadorian case, changes in the energy matrix have been contemplated mainly due to the development of projects for the replacement of fossil fuels, with renewable energy being represented by hydroelectric energy. Ecuador mainly presents inefficiency indicators in the transportation and industry sectors [29]; if current conditions are maintained, Ecuador will continue to be an inefficient nation in relation to energy use.

4. Discussion

Considering the simulation results, Ecuador shows an energy inefficiency in the transportation and industry sectors. Using the BAU, National Policies (NP) and Global Policies and Trends (GPAT) scenarios for 2030, we can analyze the energy policies that will serve as the basis for the design of energy governance aimed at mitigating emissions and improving national energy efficiency. These three scenarios allow us to establish the effects of energy governance and environmental sustainability on Ecuador through appropriate energy policies.

After examining the role of the world's leading institutions in the area of energy governance, intergovernmental organizations create multilateral banks and global action networks. Existing forms of global governance coincide with the scope of the global energy challenges. The urgency of effectively implementing global energy policies to address nations' geopolitical tensions and economic changes compels researchers to investigate global energy governance [21,60]. The efforts of international collective action aimed at the administration and distribution of energy resources, as well as at the provision of energy services in the form of global energy governance, could provide a clear picture of the energy-related challenges.

Energy efficiency seeks to reduce energy waste and expenses along the energy chain, reduce dependence on energy imports, mitigate damage to the environment and improve a country's productive efficiency. Effective energy efficiency intervention requires several strategic actors in the energy industry, including consumers and authorities [61].

When considering scenarios with better economic prospects and lower energy consumption, energy intensity tends to be reduced, which reduces the amount of CO_2 emissions. By increasing the proportion of renewable energy, we can see how this reduces CO_2 emissions.

Fossil fuels are the main source of energy consumption in Ecuador. Efforts to improve Ecuador's energy efficiency have not achieved the expected results, making it urgent to improve the energy efficiency of different sectors, which would contribute to reducing expenses throughout the energy chain, reducing emissions and improving the country's productivity. A reduction in CO_2 emissions is feasible if the use of an energy mix is considered, utilizing Ecuador's hydroelectric potential and other renewable sources.

Ecuador has begun a transformation in the energy sector, departing from a dependence on fossil fuels for electricity generation to complete self-sufficiency through hydroelectric power. The country is also improving its energy transfer infrastructure to allow the more efficient transportation of massive high-voltage loads.

Considering Ecuador's hydroelectric potential, the combination of a higher proportion of hydroelectric power and other renewable sources would allow a smaller amount of CO_2 emissions to be generated. Finally, there are projections that consider that increases in the share of hydroelectric power by 50% and 70% would result in a decrease in carbon emissions to 27.36 and 15.03 $MtCO_2$, respectively, instead of the 37.10 MTons currently projected, which shows that the road to follow is the one with the greatest use of renewable energy [57].

Transportation growth is inevitable; therefore, transportation management must be optimized. In this sector, it is necessary to improve the quality of fuels. The replacement of inefficient or obsolete technologies under new projects by using energy from renewable sources is essential. The use of passenger transport should be promoted, as well as the substitution of fuels to modify energy intensity due to its higher level of efficiency.

In the industrial sector, it is necessary to change the energy sources used, and this sector must migrate towards renewable energies. Inefficient equipment such as motors, pumps or boilers must be replaced; cogeneration systems must be included, and energy management systems must be applied.

National policies for the efficient use of energy in buildings must be regulated and created, based on habitability criteria, to promote the replacement of high-energy consumption equipment, optimize public lighting, and strengthen the energy efficiency program for induction cooking and heating water, which does not use fossil fuels.

A variety of responses have been noted in the literature when analyzing the relationship between economic growth and CO_2 emissions for different countries; some studies have indicated that economic growth has a causal relationship with the variation in CO_2 emissions [17,18,20,62–67]. Other studies indicate that there is a two-way relationship between economic growth and CO_2 emissions. In the last two years, the global economy has grown, according to a report by the International Energy Agency (IEA); however, CO_2 emissions have not increased at all, indicating that there is no clear link between economic growth and the growth in CO_2 emissions [1,68].

The relationship between CO_2 emissions and economic growth makes it possible to formulate energy policies and promote the use of renewable energy resources [69–71]. Research has allowed us to realize that, currently, in developed economies, the influence of economic growth on CO_2 emissions is lower. On the other hand, in developing countries, this relationship remains strong because economic growth is prioritized over environmentally friendly production processes, but this approach does not consider the type of resources used or the effects that will appear in the future.

5. Conclusions

Greater investment in the development of renewable energy projects would contribute to the mitigation of CO_2 emissions and lead to better care of our environment for future generations [27]. After performing an energy analysis, we found that, between 1979 and 2015, in Ecuador, economic growth was promoted by energy consumption, particularly by economic growth in the primary, secondary and tertiary economic sectors, and that there is a causal relationship that associates the consumption of oil and hydroelectricity with economic growth.

The increase in energy consumption in Ecuador affects the economic situation of the country; large subsidies are granted in relation to the consumption of electrical energy. Ecuador is an exporter of crude oil and an importer of petroleum derivatives, both of which generate expenses for the national government. To achieve a more sustainable development dynamic, the price of petroleum-based energy should be increased, especially when used for unproductive activities, while the consumption of renewable energy, especially oriented to the industrial sector, should be subsidized.

A scenario of significant economic growth would allow for investment in projects related to the mitigation of CO_2 emissions. In order to achieve sustainable economic growth and maintain a good relationship with our ecosystem, the growth of production and the commercialization of products must be accompanied by investments in constant improvements in all economic activities to reduce pollution to the minimum amount possible in Ecuador.

Economic projections do not predict encouraging scenarios for Ecuador, so it is necessary to establish energy policies that improve the wellbeing of Ecuadorians. On the other hand, if we consider the scenarios that preview greater economic growth and the use of renewable energy, we will realize that, over time, this relationship is not as close as in the past. The economic recovery of Ecuador and the promotion of greater investments in innovation and technological development will be necessary in the fight to mitigate CO_2 emissions and take care of our environment for the future generations.

To achieve sustainable economic growth and maintain a good relationship with our ecosystem, the growth of production and the marketing of products must be accompanied by investments in constant improvements in all economic activities to reduce pollution to the minimum amount possible. Maintaining the policies that have been implemented and incorporating others will allow Ecuador to maintain progressive economic development, which currently aims to invest in new technologies and in projects focused on the use of clean energy and caring for the environment. Replacing non-renewable energy with clean energy and designing energy policies focused on climate change must go hand in hand, projecting a more encouraging image for the future of CO_2 emissions in Ecuador.

Author Contributions: Investigation F.R.A.M.; conceptualization, F.R.A.M. and L.J.M.; methodology, F.R.A.M. and L.J.M.; software, F.R.A.M. and L.J.M.; writing—original draft preparation F.R.A.M.; writing—review and editing F.R.A.M. and L.J.M.; supervision L.J.M.; validation L.J.M. All authors have read and agreed to the published version of the manuscript.

Funding: This research received no external funding.

Conflicts of Interest: The authors declare no conflict of interest.

References

1. United Nations Environment Programme. *Global Environment Outlook GEO-6: Healthy Planet*; UN Environment: Nairobi, Kenya, 2019.
2. Claussen, E.; Peace, J. Energy myth twelve-climate policy will bankrupt the US economy. In *Energy and American Society-Thirteen Myths*; Springer: Dordrecht, The Netherlands, 2007; pp. 311–340.
3. Stern, N. *The Economics of Climate Change*; Cambridge University Press: Cambridge, UK, 2007.
4. International Energy Agency. *Energy and Air Pollition*; International Energy Agency: Paris, France, 2016.
5. BP. *BP Statistical Review of World Energy*; Pureprint Group Limited: London, UK, 2019.
6. International Energy Agency. *Global Energy & CO2 Status Report*; IEA: Paris, France, 2018.
7. International Energy Agency. *World Energy Outlook*; IEA: Paris, France, 2016.
8. Arroyo, F.R.; Miguel, L.J. Analysis of energy demand scenarios in Ecuador: National government policy perspectives and global trend to reduce CO_2 emissions. *Int. J. Energy Econ. Policy* **2019**, *9*, 364–374.
9. Ren, J.; Sovacool, B.K. Quantifying, measuring, and strategizing energy security: Determining the most meaningful dimensions and metrics. *Energy* **2014**, *76*, 838–849. [CrossRef]
10. World Bank Group. *Global Economic Prospects*; The World Bank: Washington, DC, USA, 2019.
11. Chang, C.-C. A multivariate causality test of carbon dioxide emissions, energy emissions, energy consumption and economic growth in China. *Appl. Energy* **2010**, *87*, 3533–3537. [CrossRef]
12. Zhang, X.-P.; Cheng, X.-M. Energy consumption, carbon emissions, and economic growth in China. *Ecol. Econ.* **2009**, *68*, 2706–2712. [CrossRef]
13. Saboori, B.; Sulaiman, J. CO_2 emissions, energy consumption and economic growth in Association of Southeast Asian Nations (ASEAN) countries: A cointegration approach. *Energy* **2013**, *55*, 813–822. [CrossRef]
14. Pao, H.-T.; Tsai, C.-M. CO_2 emissions, energy consumption and economic growth in BRIC countries. *Energy Policy* **2010**, *38*, 7850–7860. [CrossRef]

15. Mirzaei, M.; Bekri, M. Energy consumption and CO_2 emissions in Iran, 2025. *Environ. Res.* **2017**, *154*, 345–351. [CrossRef] [PubMed]
16. Hasseb, M.; Azam, M. Energy consumption, economic growth and CO_2 emission nexus in Pakistan. *Asian J. Appl. Sci.* **2015**, *8*, 27–36. [CrossRef]
17. Saidi, K.; Hammami, S. The impact of CO_2 emissions and economic growth on energy consumption in 58 countries. *Energy Rep.* **2015**, *1*, 62–70. [CrossRef]
18. Arroyo, F.; Miguel, L.J. The Trends of the Energy Intensity and CO_2 Emissions Related to Final Energy Consumption in Ecuador: Scenarios of National and Worldwide Strategies. *Sustainability* **2019**, *12*, 20. [CrossRef]
19. Gazheli, A.; Van den Bergh, J.; Antal, M. How realistic is green growth? Sectoral-level carbon intensity versus productivity. *J. Clean. Prod.* **2016**, *129*, 449–467. [CrossRef]
20. Arroyo, M.F.R.; Miguel, L.J. The Role of Renewable Energies for the Sustainable Energy Governance and Environmental Policies for the Mitigation of Climate Change in Ecuador. *Energies* **2020**, *13*, 3883. [CrossRef]
21. Florini, A.; Sovacool, B.K. Who governs energy? The challenges facing global energy governance. *Energy Policy* **2009**, *37*, 5239–5248. [CrossRef]
22. Sovacool, B.K. An international assessment of energy security performance. *Ecol. Econ.* **2013**, *88*, 148–158. [CrossRef]
23. Sovacool, B.K.; Mukherjee, I.; Drupady, I.M.; D'Agostino, A.L. Evaluating energy security performance from 1990 to 2010 for eighteen countries. *Energy* **2011**, *36*, 5846–5863. [CrossRef]
24. Saunders, H.; Sovacool, B.K. Competing policy packages and the complexity of energy security. *Energy* **2014**, *67*, 641–651.
25. Ang, B.; Choong, W.; Ng, T. Energy security: Definitions, dimensions and indexes. *Renew. Sustain. Energy Rev.* **2015**, *42*, 1077–1093. [CrossRef]
26. Morlet, C.; Keirstead, J. A comparative analysis of urban energy governance in four European cities. *Energy Policy* **2013**, *61*, 852–863. [CrossRef]
27. Pinzón, K. Dynamics between energy consumption and economic growth in Ecuador: A granger causality analysis. *Econ. Anal. Policy* **2018**, *57*, 88–101. [CrossRef]
28. BP. *Statistical Review of World Energy*; BP: London, UK, 2017.
29. Coordinating Ministry of Strategic Sectors. *National Energy Balance 2016*; Coordinating Ministry of Strategic Sectors: Quito, Ecuador, 2017.
30. Coordinating Ministry of Strategic Sectors. *National Energy Agenda 2016–2040*; Coordinating Ministry of Strategic Sectors: Quito, Ecuador, 2016.
31. Sosa, G.F. El Cuidadano Períodico Oficial. 15 Noviembre 2017. Available online: http://www.elciudadano.gob.ec/ecuador-implementa-programas-para-reducir-las-emisiones-de-gases-contaminantes/ (accessed on 30 June 2019).
32. IRENA. *Renewable Energy Statistics 2018*; International Renewable Energy Agency: Masdar City, UAE, 2018.
33. International Energy Agency. *World Energy Outlook 2018*; IEA Publications: Paris, France, 2018.
34. International Energy Agency. *World Energy Investment*; IEA Publications: Paris, France, 2018.
35. IRENA. *Renewable Power Generation Costs in 2017*; International Renewable Energy Agency: Masdar City, UAE, 2018.
36. Global Commission on the Geopolitics of Energy Transformation. A new world. In *The Geopolitics of the Energy Transformation*; International Renewable Energy Agency: Masdar City, UAE, 2019.
37. DVG.GL. *Energy Transition Outlook 2018*; DVG.GL: Oslo, Norway, 2018.
38. Downie, C. Business actors, political resistance, and strategies for policymakers. *Energy Policy* **2017**, *108*, 583–592. [CrossRef]
39. Simsek, Y.; Lorca, Á.; Urmee, T.; Bahri, P.A.; Escobar, R. Review and assessment of energy policy developments in Chile. *Energy Policy* **2019**, *127*, 87–101. [CrossRef]
40. Helm, D. The European framework for energy and climate policies. *Energy Policy* **2014**, *64*, 29–35. [CrossRef]
41. Chang, C.-C.; Carballo, C.F.S. Energy conservation and sustainable economic growth: The case of Latin America and the Caribbean. *Energy Policy* **2011**, *39*, 4215–4221. [CrossRef]
42. Gao, M.-Z.A.; Fan, C.-T.; Liao, C.-N. Application of German energy transition in Taiwan: A critical review of unique electricity liberalisation as a core strategy to achieve renewable energy growth. *Energy Policy* **2018**, *120*, 644–654. [CrossRef]

43. Hildingsson, R.; Johansson, B. Governing low-carbon energy transitions in sustainable ways: Potential synergies and conflicts between climate and environmental policy objectives. *Energy Policy* **2016**, *88*, 245–252. [CrossRef]
44. Emodi, N.V.; Chaiechi, T.; Rabiul, A.B.M.; Beg, A. A techno-economic and environmental assessment of long-term energy policies and climate variability impact on the energy system. *Energy Policy* **2019**, *128*, 329–346. [CrossRef]
45. De Blas, I.; Miguel, L.J.; Capellán-Pérez, I. Modelling of sectoral energy demand through energy intensities in MEDEAS integrated assessment model. *Energy Strategy Rev.* **2019**, *26*, 100419. [CrossRef]
46. Nieto, J.; Carpintero, Ó.; Miguel, L.J.; De Blas, I. Macroeconomic modelling under energy constraints: Global low carbon transition scenarios. *Energy Policy* **2020**, *137*, 111090. [CrossRef]
47. Sterman, J.D. Business dynamics. In *Systems Thinking and Modeling for a Complex World*; McGraw-Hill Higher Education: New York, NY, USA, 2000.
48. Mondal, M.A.H.; Bryan, E.; Ringler, C.; Mekonnen, D. Ethiopian energy status and demand scenarios: Prospects to improve energy efficiency and mitigate GHG emissions. *Energy* **2018**, *149*, 161–172. [CrossRef]
49. Meadows, D.H.; Meadows, D.L.; Randers, J.; Behrens, W.W. *The Limits to Growth*; Universe Books: New York, NY, USA, 1972.
50. Ford, A. *Modeling the Environment*, 2nd ed.; Island Press: Washington, DC, USA, 2010.
51. Sterman, J.; Fiddaman, T.; Franck, T.R.; Jones, A.; McCauley, S.; Rice, P.; Sawin, E.; Siegel, L. Climate interactive: The C-ROADS climate policy model. *Syst. Dyn. Rev.* **2012**, *28*, 295–305. [CrossRef]
52. Coordinating Ministry of Strategic Sectors. *National Energy Balance 2013*; Coordinating Ministry of Staging Sectors: Quito, Ecuador, 2013.
53. Coordinating Ministry of Staging Sectors. *National Energy Balance 2014*; Coordinating Ministry of Staging Sectors: Quito, Ecuador, 2014.
54. Coordinating Ministry of Strategic Sectors. *National Energy Balance 2015*; Coordinating Ministry of Staging Sectors: Quito, Ecuador, 2015.
55. Ministry of Energy and Non-Renewable Natural Resources. *National Energy Balance 2017*; Ministry of Energy and Non-Renewable Natural Resources: Quito, Ecuador, 2018.
56. Capellán-Pérez, I.; Mediavilla, M.; De Castro, C.; Carpintero, Ó.; Miguel, L.J. Fossil fuel depletion and socio-economic scenarios: An integrated approach. *Energy* **2014**, *77*, 641–666. [CrossRef]
57. CONELEC. *Electrification master plan 2013–2022*; CONELEC: Quito, Ecuador, 2013.
58. Calderón, Á.; Dini, M.; Stumpo, G. *Los Desafíos del Ecuador Para el Cambio Estructural con Inclusión Social*; Naciones Unidas: Santiago, Chile, 2016.
59. Everett, T.; Ishwaran, M.; Ansaloni, G.P.; Rubin, A. *Economic Growth and the Environment*; Department for Environment, Food and Rural Affairs: London, UK, 2010.
60. Kunchornrat, J.; Phdungsilp, A. Multi-Level Governance of Low-Carbon Energy Systems in Thailand. *Energies* **2012**, *5*, 531–544. [CrossRef]
61. Limaye, D.R.; Heffner, G.C.; Sarkar, A. *An Analytical Compendium of Institutional Frameworks for Energy Efficiency Implementation*; The World Bank Group: Washington, DC, USA, 2008.
62. Munir, Q.; Lean, H.H.; Smyth, R. CO_2 emissions, energy consumption and economic growth in the ASEAN-5 countries: A cross-sectional dependence approach. *Energy Econ.* **2020**, *85*, 104571. [CrossRef]
63. Acheampong, A.O. Economic growth, CO_2 emissions and energy consumption: What causes what and where? *Energy Econ.* **2018**, *74*, 677–692. [CrossRef]
64. Xue, B.; Geng, Y.; Muller, K.; Lu, C.; Ren, W. Understanding the Causality between Carbon Dioxide Emission, Fossil Energy Consumption and Economic Growth in Developed Countries: An Empirical Study. *Sustainability* **2012**, *6*, 1037–1045. [CrossRef]
65. Nain, M.Z.; Ahmad, W.; Kamaiah, B. Economic growth, energy consumption and CO_2 emissions in India: A disaggregated causal analysis. *Int. J. Sustain. Energy* **2015**, *36*, 807–824. [CrossRef]
66. Hossain, M.S. Panel estimation for CO2 emissions, energy consumption, economic growth, trade openness and urbanization of newly industrialized countries. *Energy Policy* **2011**, *39*, 6991–6999. [CrossRef]
67. Piłatowska, M.a.G.A.; Włodarczyk, A. The Effect of Renewable and Nuclear Energy Consumption on Decoupling Economic Growth from CO_2 Emissions in Spain. *Energies* **2020**, *13*, 2124. [CrossRef]
68. Kosow, H. *Methods of Future and Scenario Analysis: Overview, Assessment, and Selection Criteria*; German Development Institute (DIE): Bonn, Germany, 2008.

69. Ahmed, M.M.; Shimada, K. The Effect of Renewable Energy Consumption on Sustainable Economic Development: Evidence from Emerging and Developing Economies. *Energies* **2019**, *12*, 2954. [CrossRef]
70. Mardani, A.; Streimikiene, D.; Nilashi, M.; Aranda, D.A.; Loganathan, N.; Jusoh, A. Energy consumption, economic growth, and CO_2 emissions in G20 countries: Application of adaptive neuro-fuzzy inference system. *Energies* **2018**, *11*, 2721. [CrossRef]
71. Mardani, A.; Streimikiene, D.; Cavallaro, F.; Loganathan, N.; Khoshnoudi, M. Carbon dioxide (CO_2) emissions and economic growth: A systematic review of two decades of research from 1995 to 2017. *Sci. Total. Environ.* **2019**, *649*, 31–49. [CrossRef]

energies

MDPI

Article

The Role of Electrification in the Decarbonization of Central-Western Europe

Gauthier de Maere d'Aertrycke [1,*], Yves Smeers [2], Hugues de Peufeilhoux [3] and Pierre-Laurent Lucille [3]

1. ENGIE Impact, Boulevard Simon Bolivar 34-36, 1000 Brussels, Belgium
2. CORE and Louvain School of Engineering, Université catholique de Louvain, 1348 Louvain-la-Neuve, Belgium; yves.smeers@uclouvain.be
3. ENGIE, Strategy Division, 1 Place Samuel de Champlain, 92400 Courbevoie, France; hugues.depeufeilhoux@engie.com (H.d.P.); pierre-laurent.lucille@engie.com (P.-L.L.)

* Correspondence: gauthier.demaeredaertrycke@engie.com; Tel.: +32-484-91-11-64

Received: 27 July 2020; Accepted: 15 September 2020; Published: 19 September 2020

Abstract: Scenario studies of energy transition generally point to the central role of electricity. This notion is ambiguous as its interpretation can range from an electricity-only policy to portfolios of different energy vectors with a dominance of electricity. This ambiguity adds to the uncertainty that already pervades today's investment environment. This paper examines the centrality of electricity through a so-called "variational scenario" analysis with policies obtained by a mix of electricity-only and green gas penetration while maintaining constant decarbonization objectives. Electricity is a complex product that can only be further complicated by the high penetration of renewables and its interaction with the production and use of synthetic fuels. The variational scenario analysis is conducted with sufficiently fine (hourly) granularity to produce an adequate representation of these phenomena. It shows that tilting the central role of electricity to a mix of electricity and green gas offers several advantages in terms of efficiency, flexibility of investment strategies, and robustness with respect to major uncertainties. It shows that the variational scenario analysis can be extended to more complex mixes of policies.

Keywords: energy transition; scenario quantification; linear programming; net zero carbon emissions; electrification; green gases

1. Introduction

Electricity is central to energy transition. The reasons for this are obvious: electricity is the easiest and cheapest sector to decarbonize and it enables the production of synthetic zero-carbon fuels that can substitute the fossil materials used in combustion today. This argument has been developed in the industry [1,2] and in scientific literature [3,4]. A stronger version of the statement is the claim that full decarbonization can only be reached using a 100% renewable electricity system even without nuclear or carbon sequestration [5]. Alternatively, electricity may be considered as just one in a set of important clusters of sub-policies [6]. The transition would then combine sub-policies taken from different clusters; this is the result commonly found in "ambitious" scenarios. However, this combination should also offer stability against uncertainty. The 100% renewable policy should be robust, at least locally, with respect to changes in the accompanying sub-policies. These are challenging goals, especially when considering results published in other parts of the literature. As an example, Resources for the Future's Global Energy Outlook [7] reviews 14 recent scenarios produced by important organizations, among which 10 of the 14 are based on continuations or developments of current policies. These 10 do not achieve the energy transition in 2050 even when allowing for nuclear and carbon sequestration. Only four of these scenarios, referred to as "ambitious", show a promise for reaching global decarbonization, but with nuclear and carbon sequestration.

This paper discusses an analysis of a 100% renewable policy for energy decarbonization. In contrast to most studies, this scenario was not selected from a discrete set of assumptions; it was derived from a local analysis of a continuous set of policies that balance storable and non-storable renewable energy. The approach enabled a comparative assessment of the retained policy with respect to more extreme (less diversified) policies and an analysis of the robustness of this assessment to major assumptions on some other possible policies (clusters). The problem treated in the paper is limited but relevant, as it is directly inspired by current French low-carbon strategy [8]. The analysis method is a modified scenario approach that can be extended to more complex issues that would involve several clusters of policies [6].

Opinions diverge on the technical and economic feasibility of the 100% renewable objective. Due to cost and availability of crops surfaces, the recourse using biomass is limited, which implies that intermittent solar and wind constitutes the bulk of the system. Heard et al. [9] questioned the technical feasibility of this objective on the basis of extensive literature. Brown et al. [10] contested this argument, also referring to extensive literature. The authors claimed that previously raised objections [9] have already been overcome and that the 100% objective is known to be technically feasible and economically viable. This discussion goes beyond the scope of this paper, but some of its conclusions directly apply to the discussion herein. Specifically, a study of 100% renewable penetration requires simulations of the system to be conducted with a temporal granularity not coarser than one hour. Zappa et al. [11] returned to that discussion, treating with a very detailed model. They concluded that the 100% objective is technically feasible but at the cost of significant additional investment and large contributions from biomass and biogas. The problem is that the latter condition is generally considered difficult to satisfy. The question is then to find a substitute for the insufficient biomass.

The above is related to short-term operations. The 100% renewable penetration raises other questions, both institutional and financial, that have seldom been addressed in the literature but may be important. Blazquez et al. [12] explained that 100% renewable penetration questions the current organization of the competitive electricity system. The claim is easy to understand and hard to contest: the current power design and its bidding system are based on the merit order, which ranks plants according to fuel cost. The notion loses significance in a renewable system where the intermittent plants have no fuel cost; thus, there is a coordination problem that the current market design may find difficult to solve. Polzin et al. [13] considered an investment issue; the authors argued that the many expenses due to the transition can easily be accommodated by the financial system but that risk may be the problem. Specifically, financing a 100% renewable system with a mix of equity and investment-grade corporate bonds may be difficult in the current highly leveraged market (and more so in a market that will have to redeem the debt incurred because of the Covid-19 pandemic). The penetration of renewables has mainly developed with public help and this may still be necessary in the future to accommodate risk. These questions appear when simulating the 100% renewable market.

Scenario analysis is the most commonly used tool for exploring the energy transition. The method involves the construction of a set of possible futures from which plausible (descriptive) or optimal (normative) technological paths are derived to achieving a target such as the 2015 Paris Agreement (see, for instance, Newell et al. [7], who discussed both types of scenarios). As shown in the brief literature survey that follows this introduction, the process rarely leads to precise policy recommendations; scenario results are documented using various quantitative methods, but the application of this information to policy remains vaguely formulated. This paper turns the standard approach upside down: it begins with a single (mainly electricity) policy, from which it studies variations to enrich the analysis. A variational scenario is one that differs on a policy objective (here, the penetration of biomass and green gas) from the reference scenario. All reference and variational scenarios achieve the same decarbonization and renewable objectives—they only differ by the penetration of electricity and green gas. The objective was to explore the impact of this variation on the rest of the system. Apart from standard costing and pricing analysis (through dual variables), this also implied examining how the introduction of green gas can help to deal with the intermittency of other renewable energy forms (as treated in Zappa et al. [11]). In summary, it is hoped that this structured local analysis would more directly produce information usable for understanding, valuing, and, if possible, improving initial policy. Risk is present regardless of how the transition problem is posed; it is handled here by testing the robustness of the considerations obtained by the analysis to changes in some cost or efficiency data. The following summarizes the implementation of this philosophy in this paper.

The analysis starts with a massive electricity (ME) policy, defined by its penetration in the energy system and a decarbonization target. The policy relies on significant penetration of intermittent renewable resources and some biomass development deemed compatible with domestic resources; it is directly inspired by an official policy. Policies concentrating on a given technology track have advantages in terms of economies of scale, communication clarity, and, hence, acceptability. As discussed in the controversy on the feasibility of the 100% renewable transition, they may also produce unexpected difficulties because of the implied tightness of the energy system. ME is then analyzed by considering variations that increase the contribution of biomass and green gas within given restrictions on their availability. Renewable and decarbonization objectives remain identical but intermittency is reduced. This may produce economic advantages: power systems are known to encounter peak management problems for which capacities with low use must be built. A massive penetration of the less-controllable solar and wind resources exacerbates these problems even if they can be alleviated by storage and demand-side actions. The penetration of green gas is a natural method to add storability. As a by-product, it stabilizes the operation of the equipment and hence reduces the volatility of cash flows, which in turn facilitates investment. Last, risk is overwhelming and investment costs for the energy transition are reported to be high. Diversification of assets is well-known to mitigate risk and the substitution of some electricity by green gas is a diversification of the risk bearing on these costly assets.

In short, exploring the consequences of policy variations may add insight that is usually absent from sole comparisons of technical paths derived from the standard approach. The variational scenario analysis thus does not intend to substitute the usual scenario method, which informs policies on a broad range of technological paths; its objective is to enrich the analysis of retained technological paths by looking at variations thereof. While this paper focuses on the centrality of electricity in its cluster of measures, the underpinning methodology is quite general and applies to a mix of clusters. A section at the end of the paper elaborates on how more complex investigations can be conducted using the same approach.

The paper is organized as follows: a comparison of the proposed approach to a more usual analysis of energy transition is presented in the literature review in Section 2. Scenario analysis relies on an evaluation model; Section 3 summarizes the main features of the model used in this study. The analysis of the centrality of electricity is described in Section 4. Starting from an ME strategy, two alternative policies are introduced: low electricity (LE) and mixed energy carriers (MECs), which result from partially substituting electricity with green gas and biomass. General results are provided that suggest that a moderate penetration of green gas and biomass presents several advantages with respect to massive (ME) or low (LE) electrification. This changes the view of the central role of electricity toward a mixed electricity and green gas strategy. This policy is further analyzed by first exploring its implications in terms of the decarbonization of the power sector in Section 5 before considering the decarbonization of energy fuels through synthetic gas or hydrogen. The findings suggest the advantages of MEC compared to ME and LE. Section 5 shows that these advantages are robust with respect to the uncertainty of some crucial parameters of the model. Section 6 briefly explains how the analysis can be extended to more complex policy problems. The conclusions are presented at the end of the paper.

2. Literature Survey and Contribution

The literature on scenario analysis and its application to energy and energy transition is extensive. Recent global analysis of Resources for the Future [7] and its systematic comparison of several important scenarios were mentioned in the introduction. The following brief review is limited to a few recent papers that are directly relevant to provide context to this work. A first set of papers, taken from a 2020 issue of the *Economics of Energy & Environmental Policy* journal, reflects the current state of the field and refers to numerous existing surveys on scenario analysis. Recent advances in the now-long tradition of scenario construction and use in the European Union (EU) are also briefly summarized.

Ansari et al. [14] summarized the different concepts underpinning scenarios and their use. They also analyzed some major energy transition outlooks issued by different organizations, to which they added their own scenario produced with their model. Most of the discussion involved descriptive scenarios (only three normative scenarios were analyzed), where the authors illuminated the wide differences between their results. They also emphasized the importance of story lines for motivating the scenarios and helping to explain the diversity of results. In addition to this important, but still rather general advice, the authors adhered to the standard approach of comparing the results of different outlooks without inferring any particular policy conclusions from this comparison. Paltsev [15] explored the value of scenarios for decision making by elaborating on the type of results and their forecasting quality. The scenario results were compared as a method for gaining insight. Paltsev [16] analyzed the use of scenarios in-depth for exploring possible paths to decarbonization by first recalling the IPPC systematic use of scenarios produced by important organizations and Members of the Integrated Assessment Model Consortium (IAMC). The stated objective of scenario comparison is to analyze similarities and reasons for diversity between the scenarios. Paltsev's paper first treats several descriptive scenarios and finds relatively good consistency among them in the sense that they all suggest that decarbonization goals will not be reached by current policies. The author then turns to prescriptive scenarios that explicitly target decarbonization. He notes that several technological paths exist that contribute toward the objective and concludes that the outlooks do not reveal a general agreement on the path to follow. The paper does not provide any recommendation based on that diversity but gives some insights on the linear programming (LP) modeling, which will be referred to in Section 6. Mohn [17] conducted an in-depth analysis of the three scenarios produced by the IEA's World Energy Outlook, exploring a possible bias against the development of renewables in these scenarios but not commenting on the insight that can be derived from the scenarios for decision purposes.

These papers do not elaborate much on modeling technology. In contrast, Oei et al. [18] focused on the use of an LP model (GENeSYS-MOD) for constructing 100% renewable scenarios. They emphasized the role of the power sector and noted the need to work with sufficiently refined temporal and spatial granularity to handle high penetration of renewables properly. Notwithstanding a diversified experience with the model, the authors do not arrive at a recommendation on this granularity issue. They do not discuss the use of different scenarios. Crespo del Granado et al. [19] also focused on the need to rely on a good description of the power sector in the context of the energy transition. In contrast to a study that used a single LP model [18], they treated the question through a set of interacting sectoral models. The authors argued that this allows to capture the full complexity of the problems but did not elaborate on the mathematical structure of those interactions or on the economic properties of the global results. The authors treated four different scenarios that could best be classified as descriptive as they considered different institutional arrangements. It is hard to see how such diversified scenarios could provide a guide to investors and the authors did not attempt to do so.

The European Union (EU) has a long tradition of energy scenarios modeling, which was initiated in the aftermath of the energy crisis of the 1970s, later expanded to climate issues, and is now applied to energy transition. The PRIMES model [20], which was progressively developed over more than 30 years, plays a central role in these studies. PRIMES is a partial equilibrium model; it interacts with some satellite models but is a monolithic model that can include policy targets. Recent illustrations of the construction of a scenario by the use of these models have been published [21,22]. Capros et al. [23] expanded upon this experience and provided the basis for an analysis of energy transition, which is particularly interesting in terms of scenario construction. As in other organizations, European Commission's communication [24], which extend beyond Capros et al. [23], started from a basic scenario that was essentially the extension of the "EU Reference Scenario 2016" [22] and embedded the already-decided policies. This first scenario served as counterfactual to other scenarios, constructed as follows: a set of "category 1" scenarios was constructed, where each one added a specific technological path to the reference scenario and examined how this combination contributed to a given decarbonization target set at 80% greenhouse gas (GHG) reduction in 2050. A "category 2" scenario (known as COMBO) combined these technological pathways and added them to the basic scenario to examine the progress in decarbonization that could be achieved. Finally, two "category 3" scenarios considered disruptions compared to COMBO: additional actions to achieve negative emissions and to change consumer choice. The work was based on abundant technological documentation. PRIMES relies on a rather coarse description of the load curve with 48 and 216 load segments per year in the reduced or extended model versions, respectively [20]. The policy analysis consisted of a comparison of the results of the different scenarios without recommendations on what to extract from their diversity for decision making.

The contribution of this paper can be stated in reference to the literature survey and the discussion in the introduction. As explained in the introduction, the analysis proceeds with scenarios that represent policies and variations on policies (variational scenario analysis). This was performed inside a single linear programming (LP) model that embeds all the necessary technological data of the problem and adds policies in the form of constraints that reflect targets or combinations thereof. Varying the targets allows for an easy exploration of a given domain of policies. One can then search for improved mixed policies and test their stability (smooth reaction to changes of targets) through dual variables of these targets. The approach is applied to the robustification of a given massive electricity policy through the penetration of biomass and green gas, which, following Zappa et al. [11], plays a crucial role in determining the feasibility of the 100% renewable policy. As explained in the core of the text, this reveals the key role played by hydrogen produced from renewable energies when biomass and green gas availability are limited. This requires the analysis to be conducted with an hourly granularity to be credible in terms of technical feasibility [9,10]. Hourly granularity avoids simplifications commonly found in representations of the power system that distort price signals (implicit value of constraints or dual variables in LP parlance) sent to investment.

In practical terms, an hourly granularity increases the size of the model, making it more time consuming to solve. This problem can be overcome by limiting its scope to the electricity sector and the direct connections to decarbonization through the replacement of fossil fuels by methane or hydrogen supply. Avoiding distortions of price signals is of the essence for accurately capturing the interactions between electricity, hydrogen, and methane through storage and their pricing consequences during long periods of intermittency of renewable plants. Globally, this type of analysis contributes to filling the gap between the production of a discrete set of technical paths associated with possible futures in the usual scenario studies and detailed policy analysis.

3. Model Environment

3.1. The Model

The analysis was conducted for Central–Western Europe (CW-E) using a proprietary model operated on a succession of five-year periods. The results were obtained by an intertemporal optimization conducted over five-year intervals. They are only reported for the years 2020, 2030, 2040, and 2050 for space reasons. Each of those years was modeled with hourly time granularity to accurately capture the operations of electricity, methane, and hydrogen supply and storage that are central to the analysis. The model was formulated as a classic linear program similar to TIMES [25], covering investment and operations. LP energy models can be interpreted in terms of competitive markets, possibly completed with certain market instruments (such as emission trading or excise taxes) that do not destroy the usual reasoning in terms of marginal costs. An LP model can also be interpreted as representing a market design implementing co-optimization of operations, as in many U.S. electricity markets. Perfect competition or co-optimization assumptions are rarely satisfied in practice. A useful outcome of an LP model is revealing phenomena where these ideal assumptions may cause difficulties that should be further explored by the analyst, possibly with other models. Tight situations, which lead to high marginal costs in an ideal market model, are likely to become system stresses and possibly lead to market interruptions in the real world. Tight situations in an hourly model (revealed by high dual variables) can also be further explored by refining some parts of the model when these situations occur. Crucial for exploring variations of a reference strategy, an LP model can easily accommodate policy constraints or targets (here, the degree of electrification or penetration of green gas) at the cost of losing the interpretation of perfect competition, but not the interpretation of co-optimization. Paltsev [16] noted this problem when referring to the MIT LP model (of which Paltsev is a co-author) and explained that policy constraints must be embedded in a CO_2 cost curve, which would then be added to the cost function of the problem. This issue is important; it is further discussed in the methodological Section 6, which explains that this insertion in a CO_2 cost curve is not necessary and the explicit representation of individual policies can be retained in the model.

On the supply side, the model covers electricity, methane, and hydrogen, where it includes generation capacities, storage assets (electro-chemical, hydrogen storage in salt cavern), and transformation processes (electrolyzer, steam methane reforming, methanation). Liquid and solid fuels are also considered in the analysis but with fewer details in their operations (serving energy demands and use in the power sector). The demand side is based on a bottom–up model focusing on the end uses of the different sectors, i.e., transport (with seven different modes), residential and tertiary (space/water heating, lighting, and appliances), and industry (specific consumption of eight different branches). Useful demand is increasing, its evolution being driven by macroeconomic assumptions, population evolution, and living patterns, such as square meters of dwellings and offices. Assumptions on energy efficiency (which embed conservation) are provided both at aggregate yearly levels and for specific sectors. Table 1 lists the reductions due to conservation measures (common for all the scenarios).

Table 1. Reduction in useful consumption due to conservation measures (compared to 2020).

Space Heating	Road Transport	Industry	Appliances
−37%	−18%	−16%	−71%

The analysis is based on two levers of the energy transition: the decarbonization of energy vectors (electricity and other fuels) and the electrification rate of energy end uses. For other levers, it is simply assumed that without further elaboration, energy conservation is fixed (and taken at ambitious levels [8,24]). Carbon dioxide removal from energy related emissions is also excluded. This assumption is important as this technology can influence the results considerably. The assumption does not reflect any ex ante judgment on this technology—It derives from the object of this study, which was to analyze the 100% renewable transition. Substitution at the level of useful energy demand is driven by the scenario storylines and completed by a competitiveness analysis based on commodity prices and measures, e.g., in terms of total cost of ownership for vehicles (representing the total discounted cost for the vehicle owner, including purchase price, tax, insurance, maintenance, fuel operating cost and resale value) or levelized cost of heating for space heating technologies (measuring the average cost of heating in €/MWh_{th} over the lifetime of the heating installation). Deployment and stock replacement constraints set bounds on these substitutions. Additional efficiency gains depending on used energy carriers are also introduced, notably for heating technologies (electric heat pumps, condensing boilers, etc.) and transport (combustion engine, fuel cell, electric motors).

To correctly characterize the impacts of end uses of electrification, it is essential to capture the modification [26] and variability of hourly load profiles of the different end uses. This is especially important for assessing the electricity system balance and the challenges related to renewables integration. This requires a good understanding of demand; dedicated hourly consumption profiles that depend on calendar effect and weather variables (mainly temperature) were thus constructed for each end use. These were obtained by statistically disaggregating the historical electricity load into different profiles depending on calendar patterns (hour, weekdays, season, holidays) and sensitivity to temperature. The work was conducted using linear regression, controlling for the high dimensionality of the model by using regularization method, elastic net, and calibrating the model to also fit the yearly consumption level of each end use. This led to the estimation of four hourly demand profiles: space heating, water heating and cooking, space cooling, and a generic pattern representing the aggregation of industry, appliances, and public transport end uses (i.e., a profile mainly varying with calendar effects linked to weekdays and hours type and without seasonal pattern). The modeling was further detailed for space heating, and two profiles were constructed to appropriately account for technology specificities between heat pump and electric heater. Following Ruhnau et al. [27], the degradation of air-source heat-pump performance with outside air temperature was integrated with the non-linear effects of the coefficient of performance in the end use profile. It turned out to be impossible to directly retrieve a charging profile from historical loads for passenger; profile from the French transmission system operator RTE [28] was then used. Finally, all demands include flexibility and demand-side management potential. Most of the potential can be found in electro-mobility and vehicle-to-grid applications, where it is assumed that 10% of the electric fleet can be used. This corresponds in the MEC and ME scenario to an equivalent battery of 135 GW/810 GWh in 2050 available to manage the electrical load.

3.2. Technological Progress

The energy transition requires a considerable technological shift to renewable and carbon-neutral technologies. As emphasized by the Energy Modeling Forum 24 and associated model comparison exercise [29], technology cost, performance, and availability can substantially impact the macroeconomic costs, and more generally the challenge of meeting long-term global climate goals. This analysis assumes profound technological progress in all the technologies, as depicted in Table 2, which describes the evolution of key technologies.

Table 2. Cost and performances of key technology used in the quantification.

Technology	Item	2020	2030	2040	2050
Photovoltaics	CAPEX (€$_{2018}$/kW)	886	601	469	390
Wind-onshore	CAPEX (€$_{2018}$/kW)	2905	2255	2040	1900
Wind-offshore	CAPEX (€$_{2018}$/kW)	1341	1115	970	950
Electro storage	CAPEX (€$_{2018}$/MWh)	343	221	171	121
Combined cycle gas turbine	CAPEX (€$_{2018}$/kW)	600	600	600	600
Nuclear EPR	CAPEX €$_{2018}$/kW)	6725	6725	6725	6725
Electrolyzer	CAPEX (€$_{2018}$/kW $_{H2\ LHV}$)	1124	600	511	422
	Efficiency (% $_{LHV}$)	66%	69%	69%	69%
Biomethane	(€$_{2018}$/MWh $_{LHV}$)	108	68	68	68
Methanation	CAPEX (€$_{2018}$/kW $_{CH4\ LHV}$)	761	666	587	509
Heat pumps	Nominal coefficient of performance (COP)	2.61	2.93	3.46	4.2

3.3. Resource Evaluation

The full decarbonization objective can be achieved due to the development of renewable energies, i.e., biomass, solar, and wind energy. The potential of these resources was assessed on the basis of information from Ruis et al. [30], leading to a total biomass potential of 1790 TWh, a solar potential of 2200 TWh, and a wind potential of 3200 TWh (including offshore and onshore) for the geographical scope. For biomass, this potential was further divided into solid biomass, bioliquid, and biogas. Given this restricted potential of domestic resources (and the limited yields of wind and solar technologies in Europe), importing decarbonized fuels in Europe could reduce the cost of its decarbonization. Some scenarios therefore allow the possibility of importing synthetic gas from North Africa using existing infrastructure at a price of 75 €$_{2018}$/MWh. Even in the scenarios allowing those imports, the energy independence of the countries would be far higher in 2050 than it is today (82% of domestic energy in 2050, compared to 17% today).

4. Scenario Construction and Global Results

4.1. Two Contrasted Scenarios

France's Low Carbon National Strategy [8] and the global scenario of the EnergyWatch Group [31] provided the first sources of inspiration for constructing contrasted target policies. Those scenarios reflect an extensive use of electricity, which largely relies on domestic resources of wind, solar, and biomass. Drawing on this idea, one constructs a massive electricity development (ME) policy scenario for Central–Western Europe (CW–E) that aims at an end use electrification level of 57% (compared to 24% in 2018). Taking inspiration from national policies is reasonable in an EU organization where energy strategies are decided at the national level. To simplify the discussion, it is assumed that the same policy is implemented in the eight countries; the approach is easily verified to be extendable to the case where different national objectives are applied in each country. Treating this case would, however, require more extensive discussion than a single paper could provide.

ME immediately suggests an alternative policy where the electrification target is reduced by half and compensated by an equivalent (in terms of the capability to satisfy useful energy demand) use of decarbonized fuels. This policy is referred to as lesser electrification (LE). ME and LE are boundary policies [14]; these scenarios are not truly extreme but still strongly differentiated. It is costly, at least on the basis of the results of the model, to increase decarbonization by reinforcing electricity in ME. Similarly, the recourse to green gas reaches its assumed (reasonable) import limits in LE. The embedded conservation effort and technical progress are (plausible) enabling assumptions; they cannot be lowered without jeopardizing the feasibility of the decarbonization goal. These soft limiting assumptions translate into high dual variables, which are standard signs of scarcity and hence justifies them being taken as a boundary. Each ME or LE policy produces a portfolio of assets. As explained in the introduction, portfolios of policies appear attractive in the hope that they efficiently stabilize the portfolio of assets, which is a key requirement given the current uncertainties.

4.2. An Intermediate Scenario

The principle of the variational scenarios is to explore the neighborhood of the reference ME policy. This is achieved by moving on the path from ME to LE by relaxing the limits on the use of green gas and the targets on electrification, and correspondingly reducing the penetration of electricity. ME only relies on domestic resources of green gas whereas the path from ME to LE allows for imports. The approach in principle enables an analysis of a continuum of portfolios of policies for which different tradeoffs can be observed, therefore providing some insight on the flexibility of the system (operations should vary smoothly with the policy parameters). The discussion of the paper is limited to the middle scenario on this path and the analysis conducted on these three cases.

The third policy scenario, multiple energy carriers (MEC), was constructed to create the middle policy portfolio. The electrification rate of MEC was selected to lie between the two electrification targets of ME and LE. Similarly, the availability of biomass is intermediate to those of ME and LE. The decarbonization and conservation objectives remained unchanged. An intuitive interpretation of MEC is that it constructs a portfolio of policies as one would construct a portfolio of sectoral funds.

Decarbonization, in this model, is driven by CO_2 curves constructed by a short iterative process to satisfy a linear pattern of decarbonization through the horizon with full decarbonization in 2050. This indirect procedure is preferred to the one that directly imposes a sequence of decarbonization profiles as it avoids perturbations of dual variables due to overlapping constraints, some of them reflecting EU policies. Because the evolution of the system is constrained by the electrification and green gas policy, the CO_2 curves are expected to be different. This is the case close to the horizon given the need to arrive at full decarbonization (Table 3). This arises when natural gas is displaced from the system, which is in the last period of the horizon (an outcome of the intertemporal optimization) and is eliminated using different technologies in the ME, MEC, and LE scenarios. Domestic bio and synthetic methane in ME implies a CO_2 price of 354 €$_{2018}$/t. Import of synthetic methane until saturation of infrastructure and local synthetic methane makes the CO_2 price, at 399 €$_{2018}$/t in LE higher than in ME because of less favorable integration of electrolyzers in the power sector. MEC can import synthetic gas at 227 €$_{2018}$/t through unsaturated infrastructures. Fixing the decarbonization objective through a 2050 CO_2 price ensures a total conversion to renewable in that year, making this a normative scenario analysis.

Table 3. CO_2 pricing leading to full decarbonization in the various scenarios (€$_{2018}$/tCO_2).

Scenario	2020	2030	2035	2040	2045	2050
Lesser electrification (LE)	15.8	31.6	42.2	79.1	244	399
Multi-energy carriers (MEC)	15.8	31.6	42.2	79.1	193	227
Massive electrification (ME)	15.8	31.6	42.2	79.1	242	354

Policy supports also help with decarbonization, including subsidies for variable renewables until 2030 to drive out coal and subsidies for biogas to build up the industry. Help is also provided for financing energy efficiency and large retrofitting of buildings. Some assumptions on conservation and efficiency are ambitious, which require testing the robustness of the obtained results. The model considers all emissions resulting from combustion except aviation, but leaves those originating in production processes for future work. Overall, the coverage amounts to 81% of total greenhouse gases emissions. In summary, electrification rate and green gas penetration define the scenario domain. The analysis extends over the 2020–2050 horizon with the energy transition supposed to be achieved in 2050.

4.3. Global Results

Figure 1 summarizes the main characteristics of the three scenarios, including a short storyline, the possibility of importing green gas, and the electrification rate. The direct electrification figures reflect the story lines of the scenarios and were derived on the basis of the bottom-up demand model. The ME value reflects that anything that does not require synthetic gas (57% of the demand) switches to electricity. MEC allows for a supply of green gas to be optimized, as stated in the story line, which decreases direct electrification to 46%. LE assigns an equal share of green gas and renewable electricity that saturates the import capacity, and prices the balance at the cost of the synthetic gas at 38%. Although this equivalence between the story line and the electrification rates holds in principle, implementing it numerically requires some trial and error due to the structure of the bottom-up demand model and the staircase form of the implied demand function. Each block of this model switches between fuels on the basis of their competition in total cost. More specifically, for each end use (transport, residential and tertiary, and industrial sectors), the demand trajectory is derived with its different shares of each energy vector to reflect the scenario's storylines regarding electrification and the compatibility with the relative competitiveness of each energy to supply its derived end uses. Some divergences remain in the residential and tertiary sector for heating needs and in the transport sector for light-duty trucks.

	Lesser Electrification	Multi Energy Carriers	Massive Electrification
Scenario Storylines	*Green Gas is as important as decarbonized electricity to achieve full decarbonization.* *Natural Gas enables cost reduction in the first stages of the transition.*	*Biomass constraints are mitigated by green gas imports.* *Electrification rate decided as to optimize the decarbonization cost and limit the risks of failure.*	*Autarkic energy system: strong constraints on biomass ressources require a high electrification rate of energy end uses.*
Direct Electrification in 2050	*38%*	*46%*	*57%*

Figure 1. Scenario storylines and associated electrification rate.

Figure 2 depicts the obtained sectoral composition of electricity, methane, and hydrogen demand in 2050 for each scenario; the main differences in market shares of final energy are found in the building and transport sectors. The decarbonization pathway for industry is similar across scenario. It is mainly based on electrification of the demand (37% with market share gains in low temperature processes). There is a transition of solid or liquid uses toward their biomass equivalent (32%), and gas shares in medium and high temperature processes are unchanged (31% of industry demand).

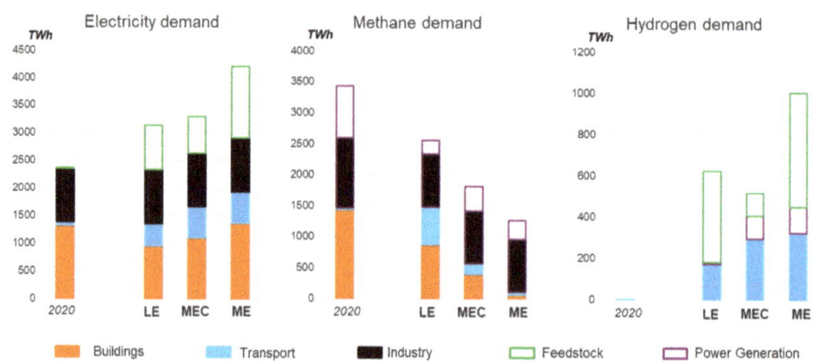

Figure 2. CW–E electricity, methane, and hydrogen demand by final and secondary uses in 2050 (and 2020 as reference) (TWh).

In LE, electricity mainly increases its market share by substituting fuel oil in space heating; only 70% of passenger mobility is electrified, while methane increases its market share in transport. In 2050, the demand of electricity for the production of synthetic fuels increases sharply to cope with the limits on green gas imports. In ME, space heating is almost fully electrified, and methane is almost only used in the industry and power sectors. Final demand in the MEC scenario lies in between those two contrasted scenarios. Regarding secondary uses, the ME scenario requires an important production of hydrogen in 2050, which is used as feedstock for synthetic methane. This scenario only relies on domestic resources to achieve its decarbonization, and biomass resources are insufficient to fulfill non-electric demands. This is related to what occurs in the LE scenario, but hydrogen production is required in this scenario because imports of green gas are reaching their maximum potential in 2050.

The penetration of electricity observed in ME remains in the range of Eurelectric scenarios [1], albeit obtained with different hypotheses of decarbonization and energy efficiency. Similarly, French low-carbon strategy [8] and the global scenario of the Energy Watch Group [31] are close to ME, even if obtained with different assumptions of energy efficiency in transport, industry, and buildings. There was no attempt to replicate these external results as scenarios studies use different data and model assumptions. Still, checking that figures remain in the same range is important for the plausibility of the analysis. A portfolio of policies such as MEC diversifies the portfolio of physical assets compared to a single policy. This mitigates the risk but may result in possible losses of economies of scale. This effect was not studied here.

Figure 3 provides the economic evaluation of the scenario in terms of net present value (NPV) of the energy system cost compared to a current policy scenario. The computation is based on a 5% discount rate and the 2050 values were taken as final steady state situation. It compares the increase in energy expenditures per carrier in cost for network infrastructure (taken from Agora [32] regarding integration of renewables) and in technologies for space heating.

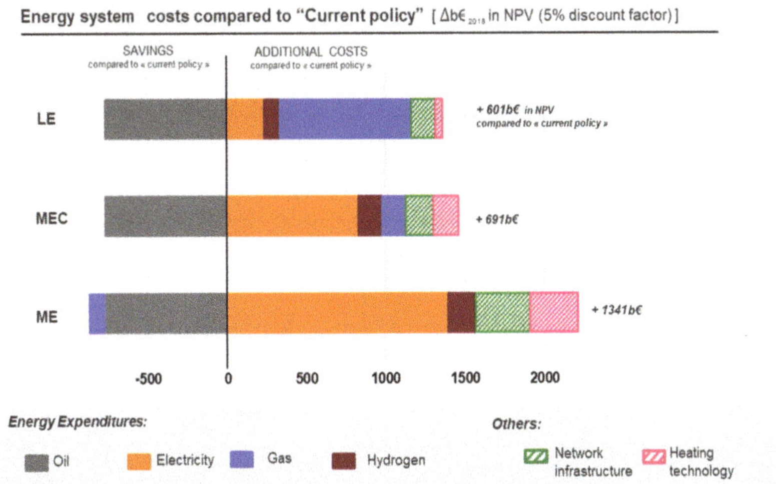

Figure 3. Energy system cost in the different scenarios (b€$_{2018}$).

The multi-energy carrier (MEC) is the most resilient policy in terms of transition costs. Massive electrification (ME) is the most expensive scenario, both in terms of net present value (NPV) over the horizon and decarbonization cost in 2050. Electricity expenditures experience both a volume and a price effect as electrification leads to increasing needs for firm capacity (notably in winter to cover episodes with low wind production). Lesser electrification (LE) is slightly less expensive in terms of net present value but is at risk of a higher final cost due to the uncertainty of green gas import price in case of high adoption of this scenario by the market (for instance, in reaction to the high upfront capital costs of ME). NPV gains in LE are only driven by the lower infrastructure and heating technology costs. These considerations are summarized in Figure 4, which reports the trajectory of energy system costs measured in annualized values.

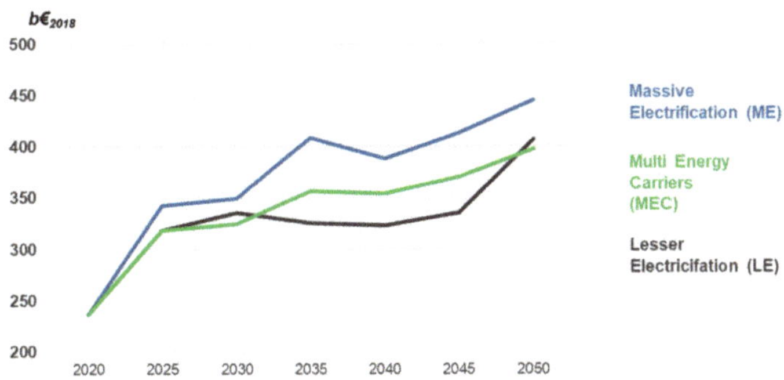

Figure 4. Evolution of energy system cost along the transition in the three scenarios (b€$_{2018}$/year).

5. Decarbonization of the Electricity, Methane and Hydrogen

5.1. Electricity Sector

As expected, electricity demand increases but at different rates in the three scenarios. The main growth drivers are transport, heating, and synthetic gas. Although LE is meant to reduce the penetration of electricity, at least compared to ME, this penetration suddenly increases close to 2050 when the system hits the assumed limitation of green gas import. Conversely, the need to produce synthetic gas, in the absence of green gas imports, justifies a sharp increase in electricity close to 2050 in ME. As expected from its construction, MEC does not show these disruptions and exhibits a much smoother increase in electricity demand throughout the horizon. In all cases, decarbonization requires a major leap in the development pace of renewables (Figure 5), which is monotone when transitioning from LE to ME. Also expected but probably more important given the design of the target scenarios, capital intensive investment in wind and solar are intermediate to those of LE and ME, which increases the adaptability to possible changes in the mood in the market. Note that there is no nuclear EPR investment with the assumed costs.

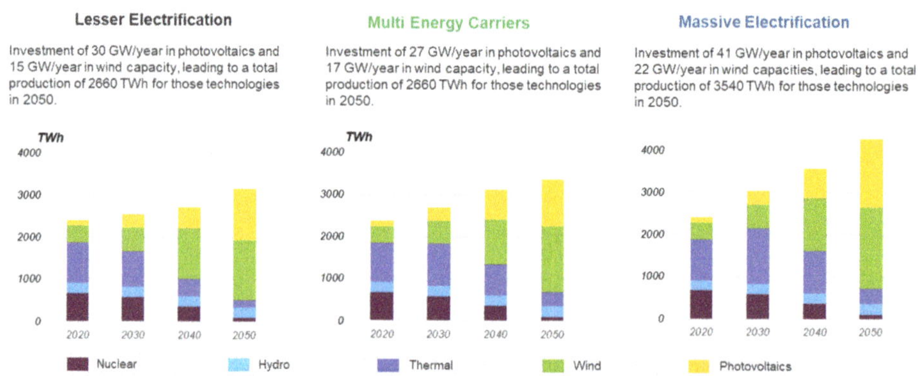

Figure 5. Electricity supply by technology in the different scenarios (TWh).

Figure 5 provides some details on the decarbonization of the power sector. Managing the peak is a standard preoccupation in electricity policy. *Peak* here is defined as the hour where all flexible generations are at their maximum production. As expected, investment in flexibility equipment increases with the penetration of electricity both along the horizon and across the scenarios. The level at which all capacities are fully used (which is the peak level) increases with electrification rate during the period 2030–2040 but later decreases due to progress in energy efficiency and higher demand flexibility. This effect is purely the result of assumptions.

More specifically, balancing the system in 2050 requires thermal assets with low running hours. Defining the *peak moment* as a period of several days with low wind factor in winter, to cope with such events, the system has no other option than to resort to thermal production as batteries and electrolysis are inoperant under these conditions (Figure 6). The low running hours appear in the three scenarios, but the use factor of the thermal plants in the peak of 2050 is significantly smaller in ME and LE compared to MEC.

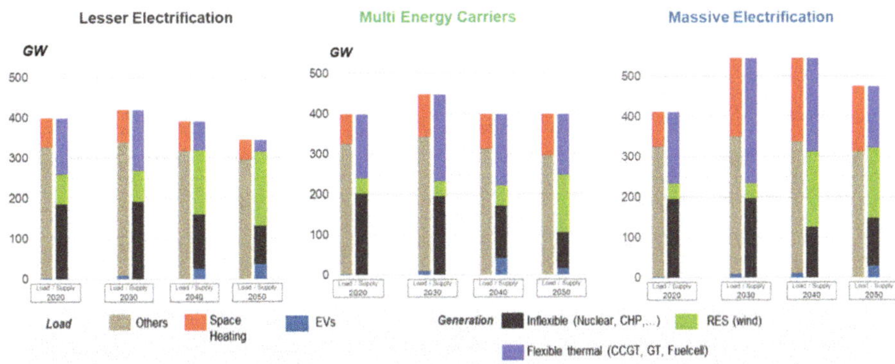

Figure 6. Electricity demand and supply during peak hour (GW).

5.2. Concerns with Investment Signals

Table 4 reports the use rate of combined cycle gas turbines (CCGTs) along the transition in the different scenarios. In all scenarios, the capacity factor of thermal assets increases in 2030 and 2040 to drop in 2050; this is a bad signal for investment in these plants. The low use of CCGT raises the question of the strength of the incentive to invest in these plants in a real market (as explained above, this argument does not apply to batteries and electrolyzers). The cash flow accruing to plant capacities, and hence the justification of the investment, is calculated at opportunity cost in a deterministic LP model that implements co-optimization or represents perfect competition. This is automatic and sufficient in this ideal world. In contrast, the cash flow can become extremely volatile and its forecast complicated for a company operating in the corresponding uncertain market of the real world, which, by design is not based on co-optimization of energy and services (as is the case today in CW–E), and where the definition of services and their remuneration can, and likely will be subject to evolving regulations. The market might recall the vagaries of the history of the EU electricity restructuring and anticipate an insufficient remuneration of services (for example, based on pure accounting cost and not including a capacity and flexibility premium that reflects opportunity cost) and, as a consequence, refrain from investing. This ties in with the problem mentioned in Polzin et al. [13] on the possible difficulties of financing projects that are too risky in the transitions.

Table 4. Average use rate of combined cycle gas turbines (CCGT) in the scenarios (%).

Scenario	2020	2030	2040	2050
Lesser electrification (LE)	16%	30%	21%	4%
Multi-energy carriers (MEC)	16%	31%	29%	9.5%
Massive (ME)	16%	31%	29%	4%

Another view of the same problem is given by the complexity of the management of the peak illustrated in Figure 7, which shows the mix of demand and supply contributed to flexibility by the various agents. Balancing these contributions is automatic in a deterministic model based on co-optimization in perfect forecast and this complexity is not a concern. It is considerably more difficult to anticipate and achieve in a real market system functioning on very short-term granularity and where the market design separates energy and services. This relates to the problem discussed in Blazquez et al. [12], where the authors contemplated the organization of the market that would decentralize these operations, especially when the notion of merit order had disappeared.

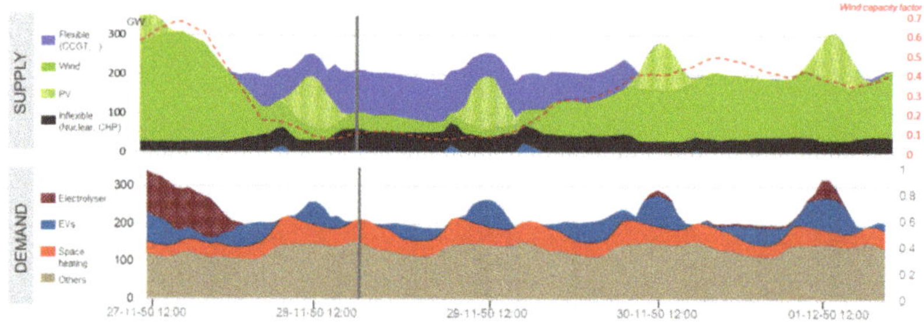

Figure 7. Supply and demand around the peak event in 2050 (gray bar) for the multi-energy carrier (MEC) scenario (geographical scope: AT, BE, CH, DE, FR, GB, NL).

For aggregate flexible thermal capacity, their average use is lower in 2040 for LE than in the two other scenarios, where they are almost equal. A lower use in ME during this period would have been expected, as it relies more extensively on electricity and hence on a higher aggregate and more "peaky" demand. The counterintuitive result was explored in detail with the model; it is unexpected and deep. The higher electricity penetration of ME, even though directly decreasing gas demand through final consumption, indirectly increases it through the need to manage the ME peak through the production and storage of synthetic gas. The lower use of flexible thermal demand in 2050 for LE follows a different logic—it results from the limitation of imported green gas and a return to electricity close to 2050, as discussed before.

As can be expected, a portfolio of policies creates diversification, which in turns offers some hedging (except in the case of full economic meltdown when all correlations turn to one). Simply, what is considered here is the possible turn of the market from an MEC preference to ME or LE. Keeping with Figure 5 and noting that used capacity is equal to the existing available capacity at peak, the pattern of existing capacity in MEC is intermediate to the one of both ME and LE and always increasing or decreasing, depending on the technology. This avoids both stranded cost, whether in investment or retiring, in case of a change of market attitude or policy preference (see [33,34] for a discussion of stranded assets in the decarbonization process).

5.3. *Methane and Hydrogen Decarbonization*

Figure 8 depicts the evolution of methane and hydrogen supply in the different scenarios. Natural gas is a transitional product that is progressively replaced by biogas (from anaerobic digestion and pyrolysis, respectively), synthetic methane (produced locally or imported), and hydrogen. These products can intervene both as substitutes to fossil fuels in demand and as fuel for the electricity supply (for example, to manage critical periods), depending on the electrification scenario (Figure 2).

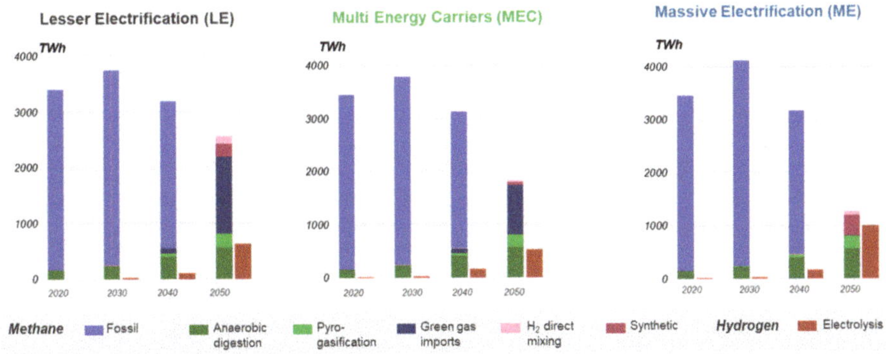

Figure 8. Methane and hydrogen supply by sources in the different scenarios (TWh).

Domestic biomethane is fully exploited in the three scenarios. Hydrogen production by electrolysis is lower in MEC, again avoiding stranded capacity in case policy targets evolve from MEC to LE or ME over time. By construction, the three scenarios differ by the assumptions on allowed green gas export. The immediate result of this assumption is that MEC green gas import lies between those of LE and ME. The possibility of stranded cost due to green gas then depends on the organization of that market. There is no stranded cost in MEC if green gas is spot traded. Stranded assets become a possibility with long-term contracts and a change in target from MEC to ME. LE faces a price risk compared to the other scenarios in the spot market. Except for this point, MEC again is the most flexible arrangement in terms of gas supplies. Notably, even though ME is meant to reduce the need for gas compared to MEC and LE, this is not shown in the graph. This finding is related to preceding remarks on the similar use of thermal plants in ME and MEC. The phenomena have the same cause, which is related to the complex interactions occurring in the power sector where gas is sometimes an input or an output of the power sector. This would justify the joint management of these sectors.

5.4. Sensitivity on Technological Cost

Scenarios usually contain policy, technology, and economic assumptions. The variational scenario approach gives a particular treatment to scenarios reflecting policies, which are the core of the analysis. This is achieved by building upon some initial reference policy to construct new scenarios (only one in this paper), which hopefully improves the initial policy. A relevant question is whether these possible advantages are robust with respect to other parameter scenarios (from other policy clusters). Sustained technological progress is a common assumption in most scenarios [12,18]; it is also key for achieving full decarbonization at moderate economic cost. A sensitivity analysis could be used to verify the robustness of the findings on the relative position of LE, MEC, and ME with respect to different assumptions on technological progress.

The analysis revealed that MEC remains the most resilient scenario as its economic benefits persist over a wide range of assumptions of technological costs, both regarding electric and green gas (synthetic and bio) technologies. It was also observed that the benefit of the LE scenario is very sensitive to optimistic technological assumptions, both on electric renewables and on green gas (Figure 9). As already argued a few times, this robustness was expected as a result of the diversification effect embedded in the construction of MEC.

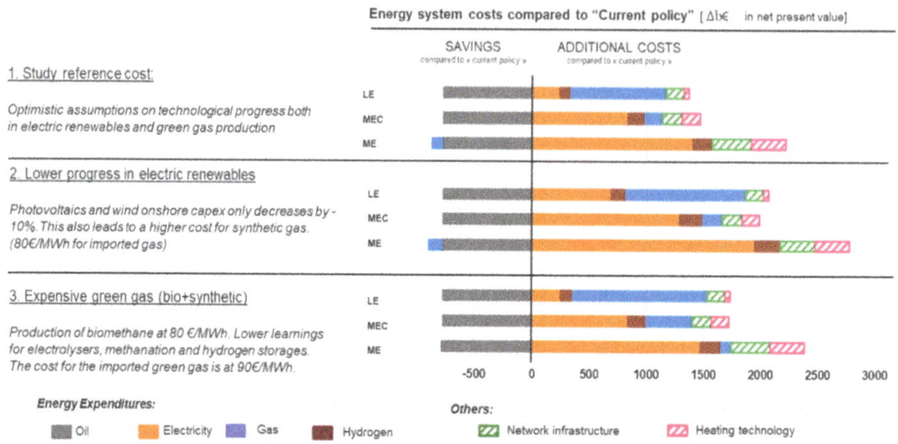

Figure 9. Energy system costs (b€$_{2018}$) sensitivity to technological costs.

6. Generalization: Expanding the Scope of Policies

Energy outlooks are generally based on scenarios that represent different, often contrasted, futures. Their number can be arbitrary but is usually small. This study was limited to three scenarios that were related to a central policy by one parameter that described different mixes of electricity and green gas. The objective was to explore the neighborhood of the ME policy to select a mix of electricity and green gas that improved it. The approach can be extended to include a larger set of policy differentiations. This requires that those policies can be represented by targets that become part of the model; this is generally not difficult with LP formulations. Assuming that this model contains the relevant technology space and an accurate representation of intermittent sources (meant to be achieved with hourly granularity), the methodology can be extended to explore a much wider set of policies represented by quantitative targets. The model can then be solved in terms of asset composition, marginal costs, and other criteria. Various experiments were conducted on the basis of this idea, with special attention paid to more detailed policy objectives such as penetration of hybrid heat pumps (heat pump and gas boiler) in residential and/or tertiary heating, market share of hydrogen and synthetic gas targets in heavy duty transport, and, more specifically, hydrogen and synthetic gas targets in power, decarbonization in large energy industries, or carbon capture and storage. The approach leads to the progressive construction of more complex policies that refine the analysis. This could be seen as an application of the well-known principle of depth first search, where policies are elaborated as analysis proceeds. A relevant question is how this search should be guided.

The treatment of variational scenarios is based on the insertion of quantitative target policies in a single energy model formulated as an LP. These targets destroy the interpretation of the model in terms of perfect competition, but this is not different from policies in the real world that also modify the operation of the market. The interpretation of the dual variables as marginal cost of the policies remains intact. Recalling that energy is priced by dual variables in all existing EU and U.S. restructured electricity markets, dual variables are clearly a useful signal that, even though not frequently used in scenario analysis, should not be neglected. In this particular context, high dual variables provide a scarcity signal and hence stresses in the market that might make policies difficult to implement and thereby suggest searching for another target mix.

Another question is whether it is possible to decipher the implicit costs of policies measured through dual variables of target constraints and derive information on effective market instruments (taxes, subsidies, quotas, etc.) to achieve the targets. Converting the identification of a policy expressed as a constraint and its implicit cost to the instruments that makes it possible to achieve is a difficult problem. Murphy et al. [35] discussed how to solve the reverse problem (computing the Impact of a mix of instruments) through a complementarity problem formulation. Complementarity models extend LP formulations and are now common in energy modeling (see, e.g., Murphy et al. [34], a tutorial emphasizing the advantages of the complementarity formulation compared to the optimization model). The method described in [35] has been extensively applied in looking at various aspects of the highly regulated domestic energy prices in Saudi Arabia. The problem can be stated at different degrees of generality. Specifically, a sector or a particular industry can be submitted to different regulations. Transitioning from policy to instrument is a reverse problem of the one treated by Murphy et al. [36] and this has not yet been studied.

A variant of the above question is completing the model discussed so far by introducing tentative representations of effective policies to directly formulate the model with both targets and some policies intended to contribute to meeting them. Durrmeyer and Samano [37], who compared two effective policies in the automobile sector, provided the inspiration for doing so. The authors formulated the policies through small equilibrium models that included the representation of the economic agents' reaction to the complex policies. The solution of these equilibrium models takes the form of a set of equations and/or complementarity inequalities, similar to what is found in complementarity models. Except for global policy constraints to which market (like CO_2) or additive taxes (like excise taxes) can be associated, introducing these relationships would not be possible with an optimization problem. However, it is, in principle, feasible to introduce them in the complementarity form of the optimization model. A discussion of this question was beyond the scope of this study.

7. Conclusions

Electricity is central to the energy transition; this is particularly true in the context of a 100% renewable policy that aims at full decarbonization at the European level. The question is to assess the content of that statement. Wind and solar are the dominant technologies in this context but biomass can play some role and offer operational advantages in terms of storability and load following. The analysis was conducted through a variational scenario approach that examined a neighborhood of an electricity-intensive policy defined as a combination of electricity and biomass and green gas. The approach differed from the usual studies conducted on a discrete set of contrasted scenarios, which aim to identify promising technological paths. The objective was to explore a variety of options offered by a given technological path to identify a promising variant. Starting from an almost pure 100% intermittent renewable electrification policy inspired by an existing national policy, one considers a partial substitution of intermittent renewable sources by green gas and biomass that remain in a realistic domain in terms of crops surfaces and green gas imports. These scenarios retained the 100% renewable target but varied in a range of electrification objectives. They also relied upon the same assumptions on the other carbon mitigation options (i.e., energy demand reductions, decarbonization trajectory, biomass resources, and potential of carbon dioxide removal). Suggestions for improvements were found: a mixed strategy (MEC) of electricity and biomass/green gas reduces costs, offers more flexibility in terms of investment, and is more robust with respect to uncertainty. Finally, given the controversies on the technical feasibility of a 100% intermittent policy, the strategy improves the operability of the system due to the storage possibilities offered by biomass and green gas. The analysis also illustrated the relevance of some comments on the possible difficulty of financing certain plants given their low number of operating hours and how the mixed strategy mitigates these difficulties. Questions on how to fit zero marginal cost renewable plants in the current market design remain. The analysis can be summarized as demonstrating the tradeoff between the availably of biomass and green gas and the need to produce hydrogen.

The paper only explores a unidimensional variety of policies (replacing electrification by green gas), which was based on the inclusion of a parameterized representation of the policy in the LP model. The approach can be extended to a more elaborate mix of policies. Because these induce more complex asset structures, the results will be more robust to uncertainty than those obtained from less diversified technological paths. Because targets are explicitly introduced in the model, their incremental cost can also be directly observed, enabling the search for an effective portfolio of technological paths centered on a reference strategy. More demanding extensions were also briefly discussed.

Author Contributions: All authors contributed equally to the work. All authors have read and agreed to the published version of the manuscript.

Funding: This research received no specific grant from any funding agency in the public, commercial, or not-for-profit sectors.

Conflicts of Interest: The authors declare no conflict of interest.

References

1. Eurelectric. Decarbonisation Pathways. Available online: https://cdn.eurelectric.org/media/3457/decarbonisation-pathways-h-5A25D8D1.pdf (accessed on 12 October 2019).
2. Pöyry. Fully Decarbonizing Europe's Energy System. 2018. Available online: https://www.poyry.com/sites/default/files/media/related_material/poyrypointofview_fullydecarbonisingeuropesenergysystemby2050.pdf (accessed on 9 September 2019).
3. IPCC. Global Warming of 1.5 °C. An IPCC Special Report on the Impacts of Global Warming of 1.5 °C above Pre-Industrial Levels and Related Global Greenhouse Gas Emission Pathways, in the Context of Strengthening the Global Response to the Threat of Climate Change, Sustainable Development, and Efforts to Eradicate Poverty. 2018. Available online: https://www.ipcc.ch/sr15/ (accessed on 16 January 2020).
4. Keramidas, K.; Diaz Vazquez, A.; Weitzel, M.; Vandyck, T.; Tamba, M.; Tchung-Ming, S.; Soria Ramirez, A.; Krause, J.; Van Dingenen, R.; Chai, Q.; et al. Global energy and climate outlook 2019: Electrification for the low-carbon transition. *JRC Sci. Policy Rep.* **2020**. [CrossRef]
5. Jacobson, M.Z.; Delucchi, M.A.; Bauer, Z.A.F.; Goodman, S.C.; Chapman, W.E.; Cameron, M.A.; Bozonnat, C.; Chobadi, L.; Clonts, H.A.; Enevoldsen, P.; et al. 100% Clean and renewable wind, water, and sunlight all-sector energy roadmaps for 139 countries of the world. *Joule* **2017**, *1*, 14–17. [CrossRef]
6. World Energy Council. World Energy Scenarios. 2019. Available online: https://www.worldenergy.org/publications/entry/world-energy-scenarios-2019-exploring-innovation-pathways-to-2040 (accessed on 24 January 2020).
7. Newell, R.; Raimi, D.; Villanueva, S.; Prest, B. Energy Transition or Energy Addition? Resources for the Future, Report 20-05. Available online: https://media.rff.org/documents/GEO_2020_Report.pdf (accessed on 3 July 2020).
8. Ministère de la Transition Écologique et Solidaire. Projet de Stratégie Nationale Bas-Carbone : La Transition Écologique et Solidaire vers la Neutralité Carbone. Available online: https://www.ecologique-solidaire.gouv.fr/strategie-nationale-bas-carbone-snbc (accessed on 13 November 2019).
9. Heard, B.P.; Brook, B.W.; Wigley, T.M.L.; Bradshaw, C.J.A. Burden of proof: A comprehensive review of the feasibility of 100% renewable-electricity systems. *Renew. Sustain. Energy Rev.* **2017**, *76*, 1122–1133. [CrossRef]
10. Brown, T.W.; Bischof-Niemz, T.; Blok, K.; Breyer, C.; Lunt, H.K.; Mathiesen, B.V. Response to "Burden of Proof" a comprehensive review of the feasibility of 100% renewable-electricity systems. *Renew. Sustain. Energy Rev.* **2019**, *92*, 834–847. [CrossRef]
11. Zappa, W.; Junginger, M.; Van den Broak, M. Is a 100% renewable European power system feasible by 2050? *Appl. Energy* **2019**, *233–234*, 1027–1050. [CrossRef]
12. Blazquez, J.; Fuentes-Bracamontes, R.; Manzano, B. A road map to navigate the energy transition. *Energy Insight* **2019**, *59*. Available online: https://www.oxfordenergy.org/wpcms/wp-content/uploads/2019/10/A-road-map-to-navigate-the-energy-transition-Insight-59.pdf (accessed on 18 September 2020).
13. Polzin, F.; Sanders, M.; Täube, F. A diverse and resilient system for investment in the energy transition. *Curr. Opin. Environ. Sustain.* **2017**, *28*, 24–32. [CrossRef]
14. Ansari, D.; Holz, F.; Al-Kuhlani, H. Energy outlooks compared: Global and regional insights. *Econ. Energy Environ. Policy* **2020**, *9*, 21–42. [CrossRef]
15. Paltsev, S. Energy scenarios: The value and limits of scenario analysis. *Wiley Interdiscip. Rev. Energy Environ.* **2017**, *6*. [CrossRef]

16. Paltsev, S. Projecting energy and climate for the 21st century. *Econ. Energy Environ. Policy* **2020**, *9*, 43–62. [CrossRef]
17. Mohn, K. The gravity of status quo: A review of IEA's world energy outlook. *Econ. Energy Environ. Policy* **2020**, *9*. [CrossRef]
18. Oei, P.-Y.; Burandt, T.; Hainsch, K.; Löffler, K.; Kemfert, C. Lessons from modeling 100% renewable scenario using GENeSYS-MOD. *Econ. Energy Environ. Policy* **2020**, *9*. [CrossRef]
19. del Crespo Granado, P.; Resch, G.; Holz, F.; Welisch, M.; Geipel, J.; Hartner, M.; Forthuber, S.; Sensfuss, F.; Olmos, L.; Bernath, C.; et al. Energy transition pathways to a low-carbon Europe in 2050: The degree of cooperation and the level of decentralization. *Econ. Energy Environ. Policy* **2020**, *9*. [CrossRef]
20. E3MLab. PRIMES Model: Version 2018. Available online: http://www.e3mlab.eu/e3mlab/PRIMES%20Manual/The%20PRIMES%20MODEL%202018.pdf (accessed on 27 March 2020).
21. Directorate-General for Climate Action and Directorate-General for Mobility and Transport. Energy, Transport and GHG emissions: Trends to 2050. Publication Prepared for the European Commission Directorate-General for Energy. 2013. Available online: https://ec.europa.eu/transport/sites/transport/files/media/publications/doc/trends-to-2050-update-2013.pdf (accessed on 6 November 2018).
22. Zampara, M.; Obersteiner, M.; Evangelopoulou, S.; De Vita, A.; Winiwarter, W.; Witzke, H.-P.; Tsani, S.; Kesting, M.; Paroussos, L.; Höglund-Isaksson, L.; et al. EU Reference Scenario 2016. Available online: https://ec.europa.eu/energy/data-analysis/energy-modelling/eu-reference-scenario-2016_en (accessed on 21 November 2018).
23. Capros, P.; Kannavou, M.; Evangelopoulou, S.; Petropoulos, A.; Siskos, P.; Tasios, N.; Zazias, G.; De Vita, A. Outlook of the EU energy system up to 2050: The case of scenarios prepared for European Commission's "clean energy for all Europeans" package using the PRIMES model. *Energy Strategy Rev.* **2018**, *22*, 255–263. [CrossRef]
24. European Commission. A Clean Planet for All: A European Long-Term Strategic Vision for a Prosperous, Modern, Competitive and Climate Neutral Economy: In-Depth Analysis in Support of the Communication COM (2018) 773. 2018. Available online: https://ec.europa.eu/knowledge4policy/publication/depth-analysis-support-com2018-773-clean-planet-all-european-strategic-long-term-vision_en (accessed on 22 November 2018).
25. Loulou, R.; Goldstein, G.; Kanudia, A.; Lettila, A.; Remme, U. Documentation for the TIMES Model. Available online: https://iea-etsap.org/docs/Documentation_for_the_TIMES_Model-Part-I_July-2016.pdf (accessed on 7 March 2020).
26. Quiggin, D.; Buswell, R. The implications of heat electrification on national electrical supply-demand balance under published 2050 energy scenarios. *Energy* **2016**, *98*, 253–270. [CrossRef]
27. Ruhnau, O.; Hirth, L.; Praktiknjo, A. Time series of heat demand and heat pump efficiency for energy system modelling. *Sci. Data* **2019**, *6*, 189. [CrossRef]
28. RTE. Bilan Prévisionnel de L'équilibre Offre-Demande D'électricité En France. Edition 2018. Available online: https://www.rte-france.com/sites/default/files/bp2018_variantes.pdf (accessed on 12 November 2019).
29. Clarke, L.E.; Fawcett, A.A.; Weynant, J.P.; McFarland, J.; Chaturvedi, V.; Zhou, Y. Technology and U.S. emissions reductions goals: Results of the EMF 24 modeling exercise. *Energy J.* **2014**, *35*, 9–30. [CrossRef]
30. Ruis, P.; Nijsa, W.; Tarvydas, D.; Sgobbi, A.; Zucker, A.; Pilli, R.; Jonsson, R.; Camia, A.; Thiel, C.; Hoyer-Klick, C.; et al. ENSPRESO—An open, EU-28 wide, transparent and coherent database of wind, solar and biomass energy potentials. *Energy Strategy Rev.* **2019**, *26*, 100379.
31. Lappeenranta University of Technology and Energy Watch Group. Global Energy System Based on 100% Renewable Energy—Power, Heat, Transport and Desalination Sectors. 2019. Available online: http://energywatchgroup.org/wp-content/uploads/EWG_LUT_100RE_All_Sectors_Global_Report_2019.pdf (accessed on 14 November 2019).
32. Agora Energiewende. The Integration Costs of Wind and Solar Power. 2015. Available online: https://www.agora-energiewende.de/fileadmin2/Projekte/2014/integrationskosten-wind-pv/Agora_Integration_Cost_Wind_PV_web.pdf (accessed on 5 September 2018).
33. Bos, K.; Gupta, J. Stranded assets and stranded resources: Implications for climate change mitigation and global sustainable development. *Energy Res. Soc. Sci.* **2019**, *56*, 101215. [CrossRef]
34. Löffler, K.; Burandt, T.; Hainsch, K.; Oei, P.-Y. Modeling the low-carbon transition of the European energy system. A quantitative assessment of the stranded assets problem. *Energy Strategy Rev.* **2019**, *26*, 100422. [CrossRef]

35. Murphy, F.; Pierru, A.; Smeers, Y. Measuring the effects of price control using mixed complementarity models. *Eur. J. Oper. Res.* **2019**, *275*, 666–676. [CrossRef]
36. Murphy, F.; Pierru, A.; Smeers, Y. A tutorial on building policy models as mixed-complementarity problems model. *Interfaces* **2016**, *46*, 465–481. [CrossRef]
37. Durrmeyer, I.; Samano, M. To rebate or not to rebate: Fuel economy standards versus feebates. *Econ. J.* **2018**, *128*, 3076–3116. [CrossRef]

energies

MDPI

Article

Perspectives for Mitigation of CO_2 Emission due to Development of Electromobility in Several Countries

Karol Tucki [1,*], Olga Orynycz [2,*] and Mateusz Mitoraj-Wojtanek [1]

1 Department of Production Engineering, Institute of Mechanical Engineering, Warsaw University of Life Sciences, Nowoursynowska Street 164, 02-787 Warsaw, Poland; mateusz.wojtanek@gmail.com
2 Department of Production Management, Bialystok University of Technology, Wiejska Street 45A, 15-351 Bialystok, Poland
* Correspondence: karol_tucki@sggw.edu.pl (K.T.); o.orynycz@pb.edu.pl (O.O.); Tel.: +48-593-45-78 (K.T.); +48-746-98-40 (O.O.)

Received: 29 June 2020; Accepted: 6 August 2020; Published: 10 August 2020

Abstract: The creep trend method is used for the analysis of the development of electric car production in three regions: The United States, the European Union and Japan. Based on vehicle registration and population growth data for each year the creep trend method using historical data for the years 2007–2017 is applied for forecasting development up to 2030. Moreover, the original method for calculating the primary energy factor (PEF) was applied to the analysis of power engineering systems in the regions investigated. The assessment of the effects of electromobility development on air quality has been performed, reduction values for pollutant and greenhouse gas emissions have been determined, which was the main objective of this manuscript. Mitigation of air pollutant emissions, i.e., carbon dioxide (CO_2), carbon monoxide (CO) and nitrogen oxides (NO_x) was estimated and compared to the eventual expected increase of emissions from power plants due to an increase of the demand for electricity. It can be concluded that electricity powered cars along with appropriate choices of energetic resources as well as electricity distribution management will play the important role to achieve the sustainable energy economy. Based on the emission reduction projections resulting from the projected increase in the number of electric cars, (corrected) emissions will be avoided in 2030 in the amount of over 14,908,000 thousand tonnes CO_2 in European Union, 3,786,000 thousand tonnes CO_2 in United States and 111,683 thousand tonnes CO_2 in Japan.

Keywords: electromobility; CO_2 reduction; energy distribution management; energy generation technology management; biofuels

1. Introduction

Continuous development of the economies of the European Union, North America and Asia has been clearly observed in recent years [1–3]. This development results in the growth of transport, including private road transport [4,5]. In 2010, approximately 58.4 million passenger cars were produced worldwide and in 2018, production reached 70.5 million [6,7].

The highest number of registered passenger cars in the European Union was recorded in Germany (46.5 million—data from 2017), Italy (37.9 million—data from 2016) and France (32 million—data from 2016) [8]. On the first position of the list of EU's most motorized societies is Luxemburg (670 registered passenger cars per 1000 residents), followed by Italy (625 cars per 1000 residents) and Finland (617 cars per 1000 residents) [8]. During the period of five years (from 2013 to 2017), the highest growth in the number of registered passenger cars was recorded in Slovakia (18%), then in the Czech Republic and Portugal (17% each), Estonia (15%), Malta and Hungary (14% each). In 2017, the highest number of cars aged over 20 years within the European Union occurred in Poland (35.2%).

Preferences as to whether a brand-new passenger car should be powered with petrol or Diesel engine differ among the member states of the European Union (EU). The highest share of petrol-powered vehicles among newly registered ones in 2017 was that in The Netherlands (80.0%), Estonia (74.8%), Finland (68.7%), Denmark (64.4%) and Malta (62.8%). On the other hand, the highest share of cars with Diesel engines among brand new passenger cars was recorded in Croatia (76.1%), Lithuania (68.5%), Romania (67.3%), Ireland (65.6%), Portugal (62.1%) and Spain (50.7%). In 2017, the highest share of alternative fuels in new passenger car registrations occurred in Poland (8.7%) and Italy (8.2%) [6].

Continuous growth in the number of cars with combustion engines contributes to a continuous increase in greenhouse gas emissions—carbon dioxide (CO_2) [9,10]. The volume of this substance in car exhaust gases depends on the type of engine and composition of the air-fuel mixture [11–15].

Irrespective of origin, the CO_2 is recognized as a one of the most influential greenhouse gases [16,17]. State authorities worldwide attempt to resolve that problem in many different ways [18,19]. These include supporting the shift to alternative fuels in the sector of private road transport [20–22]. The electric car is an example of alternative drive vehicle [23,24]. While moving, they do not emit any impurities into the environment [25,26]. Thus, they provide a solution allowing local elimination of increased pollutant emissions, e.g., in large cities where the effect of scale could occur [27,28]. An alternative to those are hybrid vehicles, which have an electric engine and a small combustion engine [29–31]. There are also other low-carbon alternative drive vehicles (for example hydrogen powered), but due to their very low popularity and low availability on the market, they were not included in the calculations.

Growing interest in electric cars has been noticeable on the global automotive market in recent years, which is a signal for the development of innovative research [32–34]. More and more automotive concerns are offering vehicles powered with electric engines or hybrid cars.

Poorly developed network of charging stations is one of the main barriers hindering more dynamic development of the electric vehicle (EV) market in many EU states [35,36]. That poses challenges to electricity providers, transmission system operators and electricity distributors. Undoubtedly, development of electromobility impacts operation of transmission and distribution networks [37,38]. Appearance of an increased number of electric vehicles on the market will cause local electricity consumption changes. Systems for power balancing in low voltage networks, management of balance differences and electricity technical losses are necessary [39–41]. As the sector of electric vehicles and vehicle charging stations is only at an initial development stage and taking into account its interdisciplinary character, it is attractive to potential investors, both in areas directly connected with electric cars and in related industries satisfying the needs of the electromobility sector [42–44].

This study will compare three economic regions belonging to different cultural circles, remaining in different economic environments on three different continents: North America, Asia and Europe.

The United States of America is the third country of the world in terms of population and the most densely populated country of North America. The country is inhabited by nearly 330 million people, a great majority of whom (82.5%) living in urbanized areas [45,46]. Relatively low petrol prices and the country's large area played an important role in popularization of passenger cars (in particular those with a large engine capacity) as a means of personal transport [47,48]. Initially, the global ecological trend caused by climate changes was not an important issue in the US. Recently, however, increasing attention has been paid to air quality [49,50]. The trend initiated by the manufacturer of premium electric cars—Tesla—is being continued and developed [51–53].

The European Union is an economic and political union of 28 democratic European states. It is considered as a single region, because it is formed as a result of deep political, economic and social integration [54,55]. In terms of legal and regulatory aspects, the European Union exercises significant influence onto member states, e.g., by imposing emission limits or regulating environmental issues [56–58]. Nearly one fourth of EU's CO_2 atmospheric emissions come from transport (including 72% accountable to road transport) [59–62]. Many countries recognise the significant role of road transport, including the increasing number of personal forms of transport, in the emissions of greenhouse gases and other harmful substances into the atmosphere, which directly translates into deteriorating quality of air in European cities [63,64]. Some of those countries have undertaken actions aimed at development of electromobility and other alternative fuels to enable reduction of the emissions [65,66].

Japan is an island country in eastern Asia, characterised with one of the world's highest social development indexes [67–69]. According to the report published by the United Nations, Japan takes the nineteenth place in the world and the third place in Asia (after Hong Kong and Singapore) [70,71]. The Human Development Index (HDI) is a summary measure for assessing long-term progress in three basic dimensions of human development: a long and healthy life, access to knowledge and a decent standard of living. Japan has paid great attention to climate related issues, renewables sources of energy and environment protection for many years now. It was already in 1997 that the Japanese automotive concern Toyota launched the first serially manufactured hybrid car—Toyota Prius [72,73]. Until this day, the brand has remained the leader of the hybrid vehicle market [74–76].

2. Materials and Methods

The following list contains a collection of the most important quantities used in calculations, together with appropriate symbols and units (Table 1).

Table 1. Symbols and units used in calculations.

Parameter	Description	Unit
k	smoothing constant equal to the number of consecutive expressions of the time series	-
i	segmental equation number ($i = 1, \ldots, N - k + 1$)	-
a_i, b_i	evaluation of the parameters of the i-th segmental equation	
$f_i(t)$	smoothed (theoretical) value for the period t obtained from the i-th segment equation	-
Y_t	the value of the forecast in period t	-
Y_N	the smoothed value of the forecast variable	-
$\overline{u}$	the smoothed value of the trend increment of the forecast variables	-
N	number of words of the time series of the forecast variables	-
V_T	point prediction error	-
W_r	Theil's discrepancy coefficient	-
E_i	quantity of electricity produced from the given source in the year in the country	[MWh]
E_c	total quantity of electricity produced in the year in the country	[MWh]
η_i	efficiency of primary energy conversion into electricity for the respective fuel	[%]

The study is based on an analysis of official reports devoted to the electromobility sector and the power engineering sector in the European Union, US and Japan. Other source materials available on this subject were also used for the calculations and assumptions. Based on the information obtained, an analysis concerning development dynamics of the sector in the European Union, US and Japan was carried out. The official reports and source materials used in the analysis were referred to in individual parts of the manuscript.

Using the creeping trend method [77], a forecast illustrating the condition of the sector in the countries covered was prepared in the perspective of the year 2030. The primary energy factor (PEF) used therein reflects efficiency of the power engineering system of the analysed areas in the context of primary energy conversion into electricity. The factor connects primary and final energy, it is a "conversion factor" that simplifies the calculations and allows to compare different energy sources. The PEF is a good indicator of the condition of the power grid and the efficiency of the whole system, but it is also a simplification. Of course, it does not consider all influencing factors, such as for example low conversion for RES due to losses. In this paper, authors assume that the PEF is a determining indicator.

Based on data obtained from the review of reports and literature, as well as with the use of calculation results, forecasts and the factors mentioned above, the environmental effect accompanying the changes was calculated. Expected impact of electromobility onto air quality in the regions covered was calculated thanks to the analysis of change dynamics and the forecast made; moreover, that enabled determination of the values related to reduction of pollutant and greenhouse effect emission, which is the main objective of this study.

The analysis was conducted using data since year 2010 because that is the breaking point when something serious started to happen in the electromobility sector. Previously, there were no or a negligible number of electric cars in the areas concerned. The parameters of market (prices, car characteristics, infrastructure etc.) change, but in a positive way for electric cars and inducing their development. The simplified mathematical approach has been chosen because it resembles the phase of introduction and growth of a new product (which can be assumed to be electric cars appearing on the market) in the product life cycle.

The manuscript is a continuation of the article published by the authors on the dynamics of development of the electromobility sector in Poland and Europe [78].

2.1. *Development of Motorization in the Analysed Regions*

Continuous economic development directly contributes to development of the road transport sector, in terms of both passenger and cargo transport [79–81]. According to the US transport statistics department, over 272.5 million "lightweight vehicles" were registered in the US in 2017, a great majority of which were passenger cars [82]. In 2016, nearly 383 million vehicles were registered in the European Union, including over 85% of passenger cars [83,84]. According to the Ministry of Infrastructure, Transport and Tourism of Japan, almost 62 million cars were registered in that country in 2017 [85,86].

Still, the total number of registered vehicles is not a reliable measure. Each of the analysed regions is characterised with a completely different population [87–89]. The value illustrating the correlation among these two figures is the so-called motorization index or passenger car saturation coefficient (PCS) [90,91]. It is expressed as the number of passenger cars per 1000 residents and constitutes an important measure indicating the degree of road transport development as well as civilizational development of the society. A positive change indicates increasing mobility of societies and accompanying development of passenger transport and individual motorization. PCS is treated, in particular with regard to less wealthy countries, as one of economic outlook growth indicators.

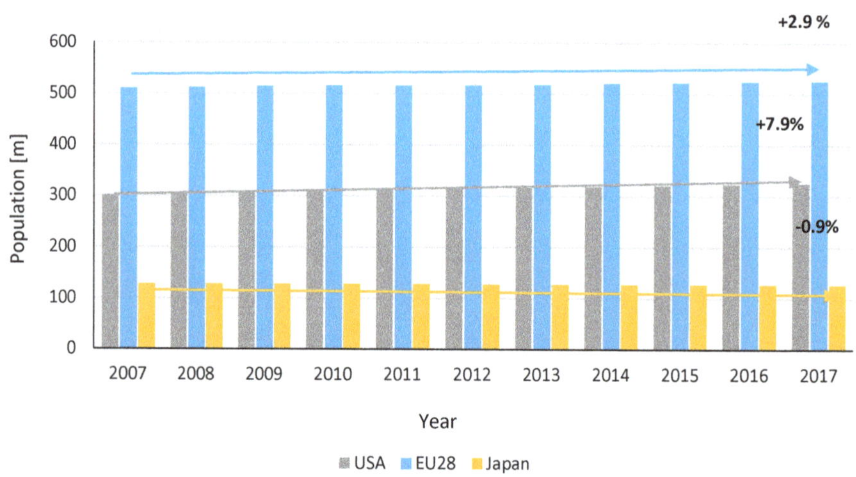

(a)

Figure 1. *Cont.*

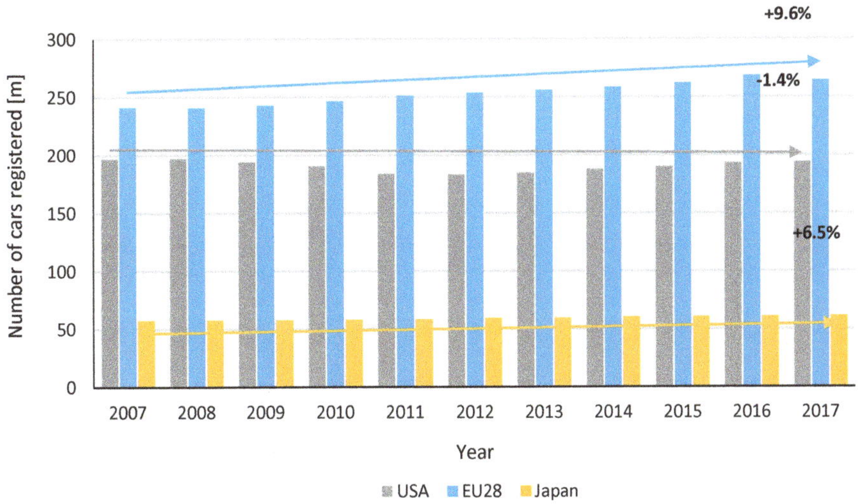

(b)

Figure 1. Motorization development in the analysed regions: (a) Population in the US, EU and Japan in 2007–2017 [m]; (b) Number of cars registered in the US, EU and Japan in 2007–2017 [m vehicles].

Figure 1a,b above present a comparison among the number of registered cars and the number of residents of the analysed regions in years 2007–2017. It can be seen that, over all the years, the European Union had the most numerous population and the highest number of registered passenger cars. Growth in the number of residents in 2007–2017 was the most significant in the US: 7.94%, with 2.85% in the EU and a negative value of −0.95% in Japan. On the other hand, the greatest increase in the number of registered vehicles was observed in the European Union: 9.6%, with 6.51% in Japan, while the United States recorded a decrease of 1.43%.

A more reliable method for evaluating the society's motorization degree is the motorization index mentioned above, whose values in 2007–2017 are illustrated on the graph below (Figure 2). Its definitely highest value is observed in the United States—in 2017, it was equal to 595.7. As the same time, in the European Union it was 503.2, with 483.1 in Japan. That means that per each two residents of the US and EU there falls more than one car, with only slightly less in Japan. During the analysed period, the coefficient showed a growing trend in Japan and in the European Union, with its values increasing year by year. On the other hand, in the United States it reached its maximum value in 2007 and was continuously falling afterwards. That correlates with the trends noticeable in the area of new passenger car registrations and demographic data. Changes of the coefficient over time have been influenced by many different factors. For example, decrease in the number of car registrations in the US since 2008 was strictly connected with high prices of oil which peaked at that time [92]. As far as the European Union is concerned, the increase is related to continuous development of its member states as well as new member states joining the European Union. In this case, even if highly developed countries do not increase the country's PCS coefficient materially, other countries on a lower development stage demonstrate significant year-on-year growth. Consequently, continuous growth of the indicator is maintained over time.

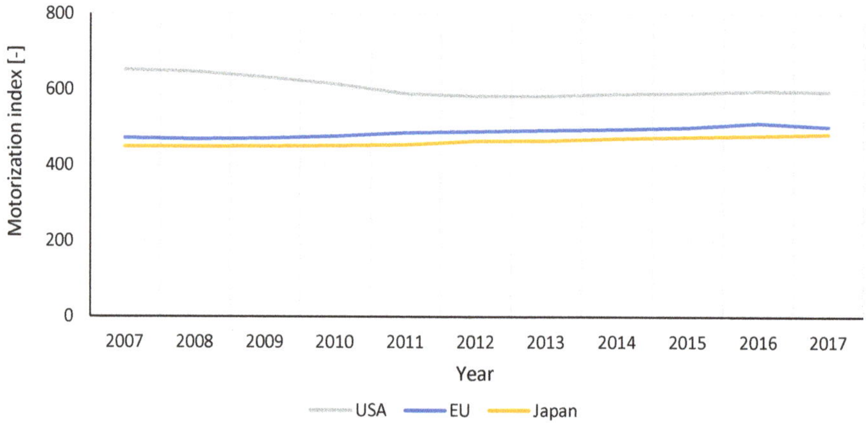

Figure 2. Motorization indexes in the United States, European Union and Japan in 2007–2017.

Regardless of reasons underlying the changes, if the trends observed are maintained, this may cause growth in the absolute number of cars on the roads, increased consumption of oil, followed by increases in the quantities of generated air pollutions, including carbon dioxide.

2.2. Analysis of Electromobility Development Dynamics in the European Union, US and Japan

Based on figures concerning vehicle registrations in European Union member states, contained in the report published by the International Council on Clean Transport (ICCT) [93], data concerning vehicle pools of individual European countries monitored by the European Alternative Fuels Observatory (EAFO), "The Motor Industry of Japan" report published by the Japan Automobile Manufacturers Association [94], as well as sales figures compiled by the US Alliance of Automobile Manufacturers [95], a graph comparing development dynamics of the electric vehicles market in the European Union, United States and Japan was prepared. The graph starts from 2009, as no significant interest in electric vehicles was observed before that year [96].

Historical data concern the period of 2009–2016, while data for the subsequent period result from the creeping trend predictive method forecasting [77].

The mathematical model used for the creep trend method is presented in the literature [77,97–99].

The creeping trend method is one classified as an adaptive model, using harmonic weights. For the predefined $Y_1, Y_2, Y_3, \ldots, Y_N$ time series and the k smoothing constant set on any level (while $k < N$), linear parameters of the trend function are estimated based on consecutive fragments of the series:

$$f_1(t) = a_1 + b_1 t \text{ for } 1 \le t \le k \tag{1}$$

$$f_2(t) = a_2 + b_2 t \text{ for } 2 \le t \le k+1 \tag{2}$$

$$f_{N-k+1}(t) = a_{N-k+1} + b_{N-k+1} t \text{ for } N-k+1 \le t \le N \tag{3}$$

For any t, where $1 \le t \le N$, Y_t values are matched by smoothed theoretical values obtained by means of the $f_i(t) = a_i + b_i t$, functions presented above where $i = 1, 2, 3, \ldots, N-k+1$. This rule concerns the following functions:

$$d(t) \le i \le g(t) \text{ for } d(t) = \begin{cases} 1 & \text{for } t = 1,2,3,\ldots,k \\ t-k+1 & \text{for } t = k+1,\ldots,N \end{cases} \tag{4}$$

and:

$$g(t) = \begin{cases} t & \text{for } t = 1,2,3,\ldots,N-k+1 \\ N-k+1 & \text{for } t = N-k+2,\ldots,N \end{cases} \tag{5}$$

An additional and final smoothening of the function is calculation of average values of all smoothenings:

$$\bar{t} = \frac{1}{1+g(t)-d(t)} \sum_{i=d(t)}^{g(t)} f_i(t) \tag{6}$$

A linear function is constructed from the values calculated as above. Consecutive calculated points are connected by sections, thus creating a graph of the time series' development trend expressed as a segment function. Extrapolation of the model into the future is possible after application of the harmonic weights methods, involving calculation of trend function gains (u), determination of average value of the gains ($\bar{u}$) and determination of standard deviation of the gains. The moment or period T is determined based on the following relationship:

$$Y_t = \overline{Y}_N + (T-N)\overline{u} \tag{7}$$

Values calculated with the use of the above model are applied to construct a graph presenting the aggregated number of electric car registrations in EU member states, the US and Japan. The said graph is presented on the figure below (Figure 3). Continuous lines identify actual data, while dotted lines—forecast gain.

The point prediction error (V_T) is calculated from the relationship:

$$V_T = W_r \cdot Y_t \tag{8}$$

Theil's discrepancy coefficient determined from the relationship:

$$W_r = \sqrt{\frac{\sum_{t=1}^{n}(Y_t - Y_N)^2}{\sum_{t=1}^{n} Y_t^2}} \tag{9}$$

Theil's discrepancy coefficient used in Hellwig's forecasting (in harmonic weighting forecasting) takes into account the sum of the squares of the differences of the empirical values (Y_t) and the theoretical values (Y_N) of the forecast variable. This sum relates (in the sense of a quotient) to squares of empirical values (Y_t). The root from this quotient allows to calculate the relative (percentage) share of the prediction error in the value of forecast point. This contribution (as a percentage) is the same for all point projections, regardless of how far the projections are from the future.

In the analyzed case, the discrepancy between the empirical and theoretical values of the predicted variable was very small (0.95%).

Figure 3 presents as well the relation of the absolute number of electric cars registered during the year to the number of residents of the respective region, together with a forecast. The population related forecast, as well as the one related to the number of electric cars, was compiled with the use of the creeping trend predictive method. The forecast was prepared based on the assumption that the trends observed so far will be maintained in the analysed regions, markets and the automotive industry. Moreover, it is assumed that the European Union will consist of 28 member states until 2020. Due to the procedure related to Great Britain's exits from the European Union, calculations after 2020 do not include Great Britain [100–103].

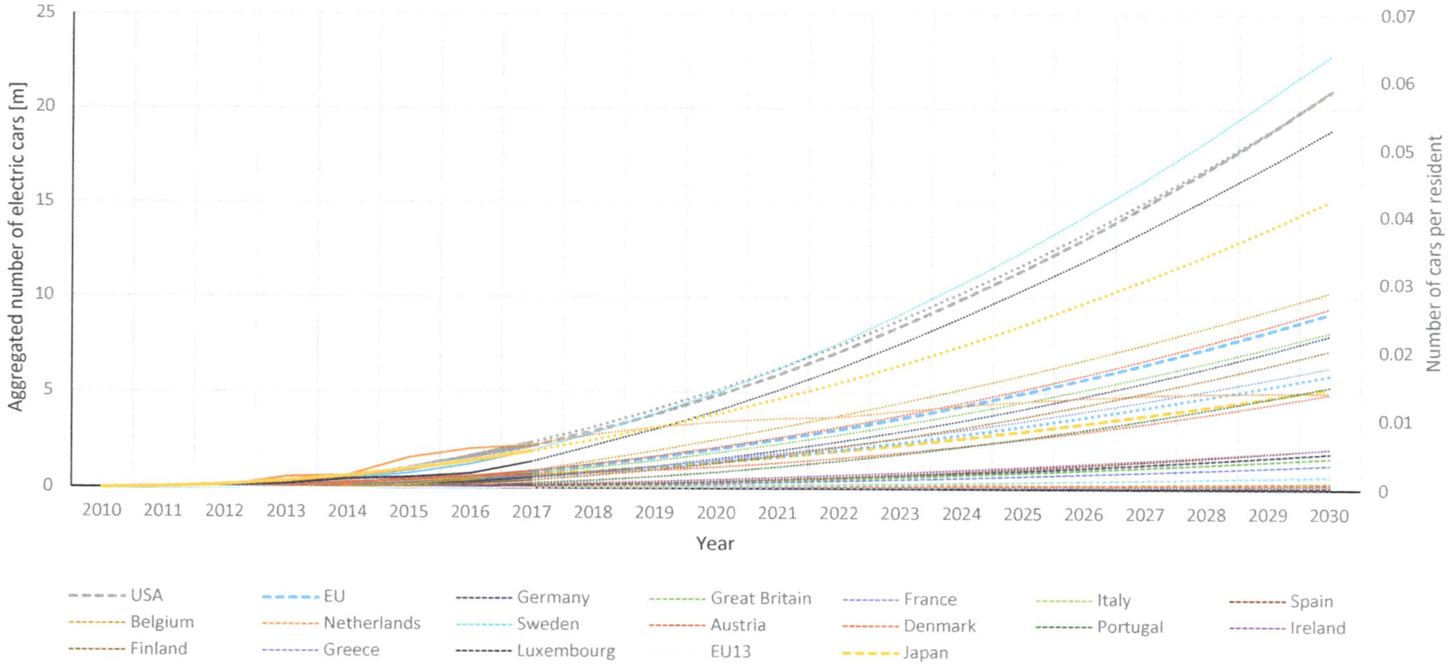

Figure 3. Aggregated number of electric cars in the European Union, United States and Japan, including forecast until 2030 and relation of the absolute number of electric cars to the number of residents, including forecast until 2030 [own calculations].

The graph clearly shows that the United States is the definite leader of the sector, taking into account the aggregated number of vehicles and the relation of car registrations to the population. The aggregated number of electric cars in the US is forecast to exceed 4.8 m in 2020, while in 2030 it is anticipated to be over 20.8 m. In the European Union the values are, respectively, app. 2 m in 2020 and 9 m in 2030. In the EU, countries with the biggest aggregated number of electric cars are: Germany with 348 thousand in 2020 and 1.9 m in Great Britain with 342 thousand in 2020 and 1.6 m in 2030. Japan is characterised with the lowest expected values of app. 1.3 m in 2020 and 5.3 m in 2030.

It ought to be stressed that even though the aggregated number of electric cars anticipated for the coming years is the lowest in Japan, the country clearly outpaces the European Union if the value is recalculated into the number of residents. Depending on the year, the value in Japan is, on average, three times higher than in the EU. In 2017, the relation between the number of electric cars and the number of residents was equal to 0.005 in Japan, while in the European Union it was 0.002.

That number is surprisingly high in Sweden with a value of 0.005. If the current trends and tendencies are maintained, the difference will be growing over time. The relationship is forecast to reach 0.011 in Japan in 2020, with 0.004 in the EU (0.01 in Sweden), while in 2030 it ought to be 0.04 in Japan and 0.02 in the EU (Sweden 0.06), respectively.

2.3. *Energy Sources Mix*

Electric vehicles use electricity accumulated in their batteries in order to drive [104–106]. Batteries are usually charged at charging stations directly connected to the power supply network [107,108]. Electricity supplied through the network comes from the country's production sources. These may be both conventional sources, such as coal power plants or gas ones, but also nuclear power plants and renewable energy sources such as, among others, wind, sun or water power plants. Each of the above mentioned sources of electricity is characterised with different operation and fuel consumption but, above all, they differ among themselves in terms of emission of greenhouse gases and other kinds of pollutant [109–111].

Table 2 below presents a comparison of steam systems using coal and natural gas as fuel and of the gas turbine block in terms of emissions of the following gases: carbon dioxide (CO_2), carbon monoxide (CO), nitrogen oxides (NO_2), sulphur dioxide (SO_2) and dust [112].

Table 2. Comparison of systems using different fuels in terms of emissions of greenhouse gases and air pollutants [112].

Type of System	Fuel	Efficiency [%]	Emission Coefficients [g/kWh]				
			CO_2	CO	NO_x	SO_2	Dust
Steam power plant	Coal	34	1034	0.18	3.13	19.9	1.41
	Natural gas	31	651	0.09	3.04	0	0.05
Gas turbine block	Natural gas	38	532	0.3	0.5	0	0.04

As it can be concluded based on the above table (Table 2), electricity sources using coal or gas as fuel generate a significant amount of air pollutants and greenhouse gases per each kilowatt hour produced. The steam power plant is characterised with the highest level of emissions [113–115]. Gas turbine blocks generate much less emissions [116,117]. The above table (Table 2) also shows that coal causes many times more emissions than natural gas. On the other hand, electricity sources based on nuclear transformations (nuclear power plants) or renewable sources of energy generate virtually no greenhouse gases and other air pollutants during electricity production.

Consequently, considerations concerning use of electric cars as a measure allowing reduction of air pollution and greenhouse gases ought to take into account the crucial element of the energy sources mix used in the respective region. If zero- and low-emission sources have a significant share in electricity generation, then, next to local benefits caused by reduced combustion of fuels in vehicles even in case of electricity prices growth, improvement of air quality and an overall positive environmental effect may be achieved.

Figure 4 below presents a comparison regarding the share of particular energy sources in electricity generation in the United States, European Union and Japan (figures for 2018) [118]. The figure clearly shows that in the case of Japan and the United States, the dominant fuels are natural gas and coal. In both cases, the share of natural gas exceeds 35%, while the share of coal is approximately 30%. Similarly, the share of renewable sources in both cases is 16.8% (US) and 18.4% (Japan). In the case of electricity generated at nuclear power plants, the difference between these countries is significant—whereas nuclear power plants are responsible for production of 19% of electricity in the US, in Japan the ratio is below 5%. The European Union is characterised with an exceptionally high share of renewable sources in electricity production, at 32%. The share of nuclear power plants is 25.2%, coal power plants: 20% and gas power plants at nearly 19%. Liquid fuels in all of the analysed regions account for a marginal share of production, and their share is the highest in Japan at 5.7%. On the figure there is also a scenario on how the energy mix could change in the year 2030. The anticipation is based on the national and international plans and the observed tendencies in energy mix changes.

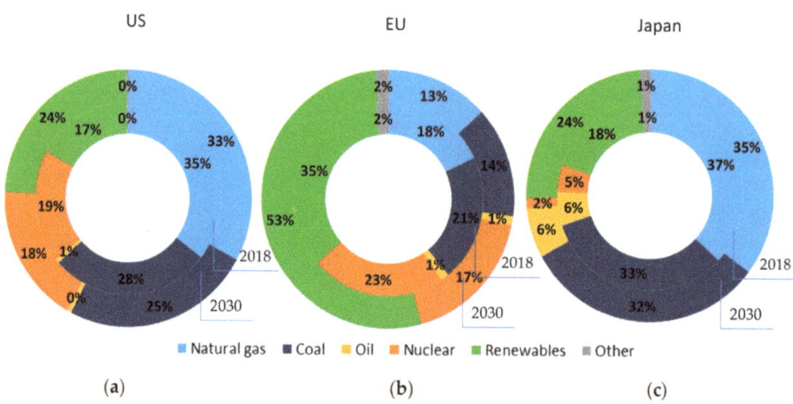

Figure 4. Share of particular energy sources in electricity generation 2018 with a scenario for 2030 in: (**a**) the United States; (**b**) European Union; (**c**) Japan.

The fundamental conclusion which may be drawn from analysis of the data presented hereinabove is that electricity produced in the European Union will involve the lowest level of emissions. That results from the fact that the main sources of electricity are renewable energy sources and nuclear power plants. In total, they account for approximately 57.2% of total production, as opposed to sources based on coal and natural gas, responsible for 38.9%. To compare, a similar comparison in the USE demonstrates that RES together with the nuclear power engineering segment account for 35.8% as opposed to coal and gas at 63.3%. In Japan, the values are, respectively, 23.1% and 69.8%. The share of particular sources of energy in electricity generation is one of several factors impacting environmental friendliness of that kind of electricity and constitutes one of the primary energy factor's components [119–122].

2.4. Primary Energy Factor

Electricity used to drive electric vehicles is the so-called final energy. Its generation and delivery require a certain quantity of primary energy, i.e., energy contained in energy sources and media (e.g., in fuels burned at the power plant) [123–125].

To perform simple recalculation of primary energy needed to produce electricity required to power a vehicle, the primary energy factor (PEF) was used. The factor is a measure of the overall energy efficiency of the power engineering system, and its value depends on many factors, such as the energy sources mix, kind of fuels used in thermal power plants and efficiency of generation units.

In this study, calculations for all analysed regions were performed using an original primary energy factor calculation method. The method constitutes a simplified manner of calculating the factor and enables determination thereof with the use of basic, publicly available data on the power engineering system. It is mainly based on the share of particular fuels in the power generation structure and considers efficiency of electricity generation from different sources. What is more, it considers primary energy contained in particular kinds of fuel.

The most important input data for PEF calculation is the share of particular fuels used at electricity generation sources, expressed as the quotient of electricity produced from the given source and total production. This share is of particular importance due to significant differences regarding efficiency of electricity production from a respective source. The data for 2018 are visualised on the preceding figure (Figure 4).

Efficiency of primary energy conversion in all analysed regions was assumed and estimated pursuant to the Commission Delegated Regulation (EU) 2015/2402 of 12 October 2015. Efficiency in individual years depends on the fuel's calorific value and technological advancement of generation sources [126]. In the case of renewable energy sources, the efficiency of 100% was applied due to the fact that they do not use any fuels for electricity generation. Efficiency values estimated as above are presented in the table below (Table 3).

Table 3. Comparison of energy conversion efficiency for particular sources.

Kind of Primary Energy	Solid Fuels	Gas Fuels	Liquid Fuels	Nuclear Fuel	Renewable Energy Sources
Efficiency	38.9%	43.0%	33.9%	33.0%	100.0%

PEF is calculated pursuant to the following formula:

$$PEF_r = \sum_{i=1}^{n} \frac{E_i}{E_c} \cdot \eta_i^{-1} \tag{10}$$

where, for the number n of sources i:

- E_i [MWh] is the quantity of electricity produced from the given source in the year r in the country,
- E_c [MWh] is the total quantity of electricity produced in the year in the country,
- η_i [%] is the efficiency of primary energy conversion into electricity for the respective fuel.

The lower the value of the primary energy factor, the higher efficiency of primary energy conversion into final energy (electricity) is demonstrated by the power engineering system. Value of the factor strives towards the value of 1, denoting ideal energy conversion without any losses whatsoever.

It is difficult to say clearly how the coefficient will change in the coming years. Although, to make the analysis as realistic as possible, the value of PEF has been forecasted until year 2030. Its final value (in 2030) was estimated on the basis of plans for the share of renewable energy sources in energy production in the areas concerned [127–129]. Values between 2018 and 2030 were interpolated. PEF values calculated as above for the United States, European Union and Japan in 2010–2018 with a forecast until 2030 are presented in Figure 5.

The lowest PEF value is that for the European Union, at 2.18 in 2018. In the case of the United States, the value is 2.31, while for Japan it is equal to 2.26. The United States demonstrated the lowest decrease in the factor over the analysed period (−5.71%) and the lowest average annual decrease (−0.73%). The highest fall is noticeable in the case of Japan (−9.96%), just as the highest average annual decrease (−1.30%). As far as the EU is concerned, the decrease is stable, on average at −0.99% per year and −7.63% throughout the analysed period. In the year 2030 predicted values of the PEF factor are: 2.20 in the US, 1.96 in EU and 2.19 in Japan.

That means that the European Union is characterised with the best efficiency of primary energy conversion into final energy, while the United States is characterised with the worst efficiency in the same area. Change dynamics indicates that the efficiency is being continuously improved in all analysed regions. Improvements are the most noticeable in Japan, and the least noticeable in the US.

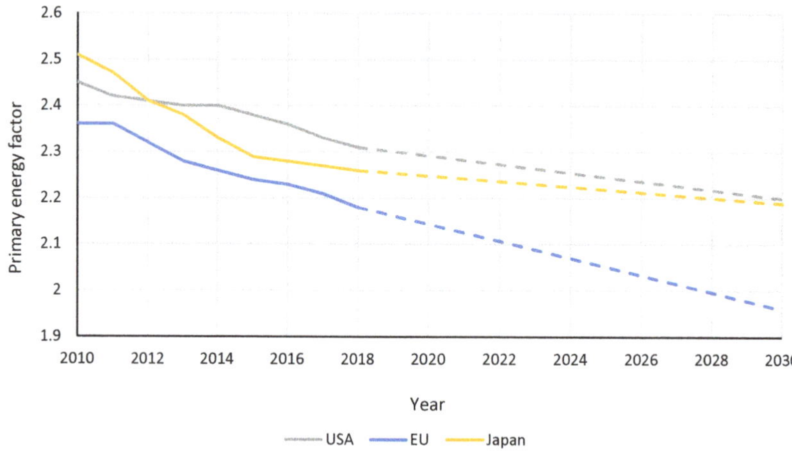

Figure 5. Comparison of the primary energy factor for the United States, European Union and Japan in 2010–2018 with a forecast to year 2030.

Primary energy factor may be considered as an indicator that enables projective evaluation of the environmental effects of electromobility expected in the countries other than discussed in the present paper. The highly emissive power plants, used in particular country, together with an increase of the number of electric automobiles will result only in shift of pollution from the roads to the regions where power stations are located. Moreover, the transmittivity of electric grid has to be taken into account. An increase of the share of biofuels in the electricity mix appears to be a factor improving the situation. It has to be considered that biofuel production, both within the agricultural as in industrial production subsystems, requires consumption of energy what decreases energetic effectiveness of that production [130–132].

As indicated in the papers [133,134] also internal and external transport of crops affects the energetic effectiveness of biofuel's production systems. It is seen, therefore, that the use of biofuels as a remediation for harmful emissions from fossil fuels also requires careful choice of production technology, and appropriate location of industrial conversion facilities with respect to plantations—just in order to optimize the energy consumption and maximize the energy gain. This means, that careful technology management, as well as processes organization management become the important factors in achieving positive environmental and economic gains connected to electro-mobility development.

2.5. *Environmental Effect and Emission Reductions (Tank-To-Wheel)*

Road transport transformations involving replacement of internal combustion vehicles with electric and plug in-hybrid vehicles involve a number of changes. The most important ones are those which have direct influence onto air quality, namely reductions related to emissions of carbon dioxide and other air pollutants [135–139].

The environmental effect of electromobility development in the European Union, United States and Japan is calculated on the basis of data concerning electromobility development dynamics, figures concerning vehicles and drivers' habits as well as emissions from particular fuels. Subsequently, the emission reduction values were adjusted by emissions caused by increased electricity consumption, calculated on the basis of PEF values. Drivers' habits and data concerning vehicles include the average annual distance driven by a passenger car and approximate period of use in the respective country [140–143]. Annual mileage of the vehicle is important for correct estimation of the distance to be driven by an electric car and, consequently, the quantity of electricity it will consume. The period of using the vehicle denotes the period during which the car is used before being replaced with a new one. This datum is important for calculating the environmental effect, as many changes concerning emission limits have been implemented over the last 30 years. As a result, vehicles manufactured over that period differ in terms of quantities of emitted greenhouse gases. The table below (Table 4) presents a comparison of those data in the three analysed regions.

Table 4. Average annual distance covered by passenger car drivers and average period of vehicle use in the European Union, United States and Japan [140–143].

	Distance [km/year]	Period of Use [years]
EU	23,639.69	20.31
US	21,687.56	17.62
Japan	9300.00	12.91

The above figures indicate that the greatest annual distance (over 23,600 km) is covered in the European Union. Only slightly less is driven each year in the United States—nearly 22,000 km. In the case of Japan, the difference is significant, as it is only 9300 km per year. The data look similar regarding the period of use, which is also the shortest in Japan—less than 13 years, and the longest one is in the European Union—over 20 years. This long average period of use in the EU results from the character of the community itself. It is formed by 28 member states, each of them on a slightly different stage of economic development. In highly developed countries with advanced economies, the period of use is much shorter than the average; for example, in Denmark it is 11 years, while in Austria the period is 13 years. In contrast, the average period of use in Greece and Spain is, respectively, 35 and 30 years.

Reduction of CO_2 emissions was calculated using all of the previously mentioned data, and considering the percentage share of cars with Diesel engines and petrol ones, as well as data concerning average carbon dioxide emission for a passenger car in the respective country [144]. The calculations took into account the year of manufacturing and, thus, the emission rate of the car replaced with an electric vehicle. The above was estimated based on the period of vehicle use, typical for the respective country.

Figure 6 presents carbon dioxide emissions avoided on an annual basis. The calculations were based on the data from European vehicle market statistics [9].

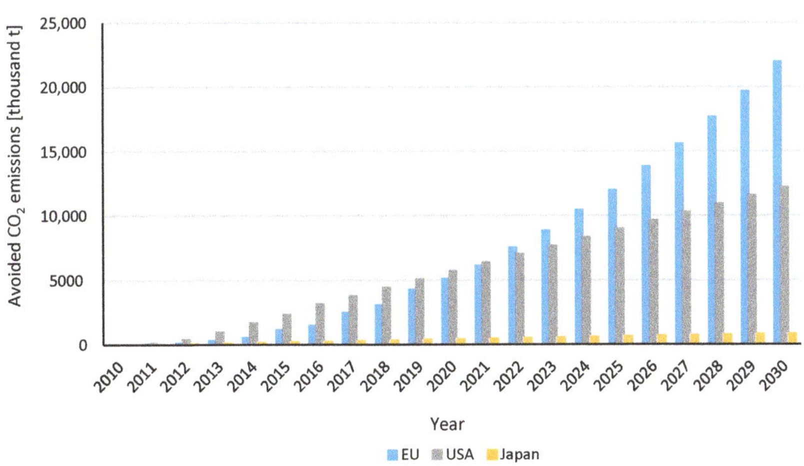

Figure 6. Tank-to-wheel CO_2 emissions avoided on an annual basis in the EU, US and Japan.

The graph shows that the greatest quantity of emissions will be avoided by 2020 in the United States. Still, from 2021 onwards, the European Union will clearly take the first place. Growth of the data bars suggests that development of electromobility in the EU is much more dynamic. Growth regarding avoided emissions is nearly exponential, while in the case of the United States it is more linear. Japan is characterised with such low values that increase in the graph is barely noticeable.

2.6. *Environmental Effect and Emission Reductions—Adjusted Values (Well-To-Wheel)*

The above values concern reduction of air pollutant emissions during use of the vehicle. If they are reduced, that will directly impact air quality, especially in cities. That may allow reducing the effect of smog and improving air quality [145–148]. However, although electric vehicles do not emit any harmful substances during operation, their drive uses electricity. Electricity itself is generated at power plants and, depending on the condition of the power engineering system and the energy sources mix in the respective country, electricity generation causes emissions of various substances into the atmosphere. In connection with the above, the environmental effect from replacement of fleet with electric vehicles was calculated taking account as well emissions occurring in connection with electricity generation.

Adjustment of calculated values was performed on the basis of primary energy factors (PEF) estimated beforehand. This enables determination of primary energy from various media, needed to power an electric vehicle.

Annual electricity consumption was calculated on the basis of data concerning average annual mileage of cars in the respective country as well as the assumed energy consumption by electric cars in road traffic conditions [149]. Next, using PEF values, primary energy consumption was estimated. Based on the energy sources mix typical for the given country and average emission rate values typical for all kinds of fuel (solid, gas and liquid), additional emissions accompanying use of electric cars were determined. Finally, previously calculated emission reductions were adjusted by additional emissions connected with electricity generation. The values of CO_2 emission factors in all the analysed areas come from the data & statistics of the International Energy Agency. The recalculated values are presented in Table 5 below.

The results of the avoided and additional emissions calculations are presented in Table 6 below and in Figures 7 and 8.

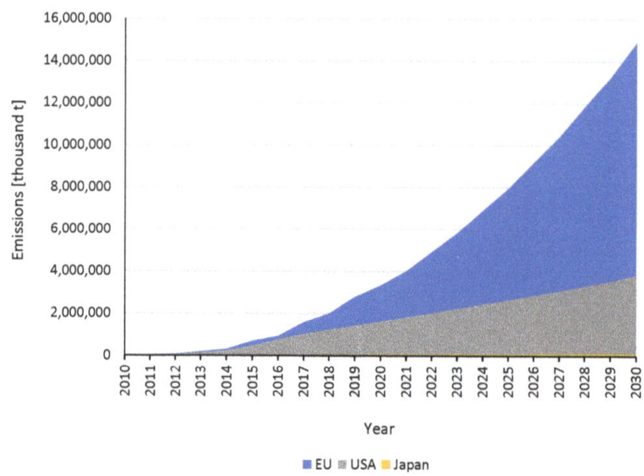

Figure 7. Well-to-wheel avoided CO_2 emissions (adjusted) with forecast for the EU, US and Japan.

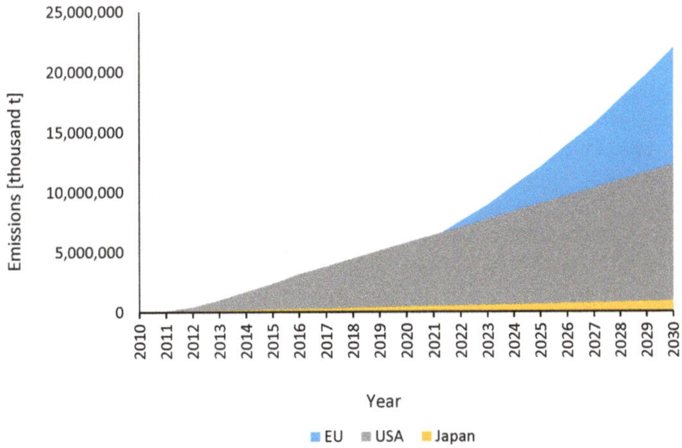

Figure 8. Avoided CO_2 emissions with forecast for the EU, US and Japan.

Table 5. CO_2 emission factor.

	CO_2 Emission Factor [kg/kWh]		
	EU	US	Japan
2010	501.15	529.92	577.88
2011	490.24	594.45	556.56
2012	480.31	620.21	531.61
2013	473.39	625.79	533.42
2014	470.49	628.10	528.44
2015	446.32	592.15	497.14
2016	423.49	571.71	472.19
2017	408.20	558.69	457.67

Table 6. Adjusted CO_2 emission reduction values.

	Avoided Emissions [thousand tonnes]			Additional Emissions [thousand tonnes]		
	EU	US	Japan	EU	US	Japan
2010	29,451	1994	5925	14,206	1935	5897
2011	47,474	105,019	27,172	22,062	96,917	26,105
2012	120,951	412,410	68,307	53,263	362,031	66,671
2013	321,165	970,375	116,917	139,410	851,197	111,647
2014	534,977	1,683,452	171,267	228,747	1,462,883	163,125
2015	1,170,973	2,343,927	212,769	471,331	1,900,218	197,183
2016	1,499,694	3,187,182	254,328	574,613	2,433,557	232,322
2017	2,502,919	3,803,375	296,882	922,341	2,778,997	267,304
2018	3,099,789	4,450,027	339,446	1,106,378	3,223,574	304,281
2019	4,315,720	5,096,686	382,010	1,529,564	3,679,789	341,699
2020	5,125,305	5,743,338	424,574	1,803,432	4,132,652	378,937
2021	6,155,468	6,389,997	467,139	2,149,948	4,582,087	415,993
2022	7,532,924	7,036,649	509,703	2,611,182	5,028,002	452,860
2023	8,855,955	7,683,308	552,267	3,046,024	5,470,321	489,535
2024	10,459,466	8,329,960	594,831	3,569,014	5,908,953	526,013
2025	12,008,579	8,976,619	637,395	4,064,299	6,343,822	562,290
2026	13,838,149	9,623,271	679,959	4,644,509	6,774,837	598,361
2027	15,613,327	10,269,930	722,525	5,195,600	7,201,922	634,223
2028	17,668,980	10,916,582	765,088	5,828,257	7,624,985	669,870
2029	19,670,222	11,563,241	807,652	6,430,280	8,043,950	705,298
2030	21,951,944	12,209,894	850,216	7,044,401	8,423,596	738,533

3. Results

The analysis performed leads to the following conclusions:

- During the period from 2007 to 2017, the population of the European Union increased by 2.9%, in the United States—by 7.9%, while the population of Japan decreased by 0.9%. During the same period, the number of passenger cars in the EU grew by 9.6%, in Japan—by 6.5%, and decreased in the US by −1.4%.
- The motorization index shows a growing trend in Japan and the European Union. In 2000–2017, it increased, respectively, by 33.8 in Japan (from 449.3 to 483.1) and by 31.0 in the EU (from 472.2 to 503.2). It was only in the territory of the US that a decrease by 56.6 was observed (from 652.3 to 595.7).
- The total of 868,320 electric cars were registered in 2010–2017 in the European Union, 2,163,569 in the United States and 685,301 in Japan. Based on the forecast, it was estimated that as many as 2,048,000 electric cars will be registered in the European Union in 2020, and the number will grow to 9,029,000 by 2030. In the US and Japan, the numbers will be respectively 4,808,000 and 1,366,000 cars in 2020, with 20,895,000 and 5,279,000 electric cars in 2030.
- The share of particular sources of energy used in electricity production in the analysed regions varies. The European Union is characterised with the highest share of renewable energy sources (35%) and nuclear power plants (23%). In the United States, RES account for as little as 17%, while in Japan—for 18%. In both those countries, the highest share is that of gas power plants—at 35% in the US and 37% in Japan, respectively. In all of the analysed regions, liquid fuel as a source of energy has a negligible share—under 1% in the US, 2% in the EU and 6% in Japan.
- Calculated pursuant to the authors' original method, the primary energy factor has the lowest value in the European Union at 2.18, while in the United States is takes the value of 2.31, with 2.26 in Japan. As compared with historical data, a declining trend regarding PEF is observed in all the regions. The greatest PEF decrease is recorded in Japan (−9.96% during the analysed period), while the lowest decrease takes place in the US (−5.71%). The European Union is characterised with a stable decrease regarding this factor over the years covered – the average annual decrease is approximately −0.99% per year, while total decrease for the analysed period is −7.63%.
- The environmental effect resulting from car fleet replacement with electric vehicles is a significant reduction of air pollutant emissions, such as carbon monoxide (CO) and nitrogen oxides (NO_X). In the territory of the European Union, emission reductions by 2017 were, respectively: 86,867 thousand tonnes of CO and 20,321 thousand tonnes of NO_X. In the US, emissions of 122,453 thousand tonnes of CO and 17,594 thousand tonnes of NO_X were avoided, while in Japan the reductions concerned 17,208 thousand tonnes of CO and 1719 thousand tonnes of NO_X.
- The environmental effect also concerns carbon dioxide (CO_2) emission reduction. In this area, enormous emission quantities were reduced. By 2017, it concerned 6,228,000 thousand tonnes in the EU, 12,508,000 thousand tonnes in the US and 1,154,000 thousand tonnes in Japan.
- However, this paper considers increased electricity requirements and resulting increased CO_2 emissions from power plants. Despite that, the environmental effect in all the regions is still positive. After adjustment, the EU avoided the emission of 3,802,000 thousand tonnes of CO_2 by 2017, the US—of 2,620,000 thousand tonnes, and Japan—of 83,312 thousand tonnes.
- It is forecast that during the year 2020 alone, emissions (adjusted) of 3,322,000 thousand tonnes of CO_2 will be avoided in the EU, with 1,611,000 thousand tonnes avoided in the US and 46,000 thousand tonnes in Japan. If current trends are maintained, the reductions in 2030 may concern, respectively, over 14,908,000 thousand tonnes (EU), 3,786,000 thousand tonnes (US) and 111,683 thousand tonnes (Japan).
- Strategic planning of electricity generation sources including alternative resources together with implementation of electricity powered automobiles, and appropriate electric energy generation

and distribution management should cause evident environmental impact also in countries other than analysed in the present paper.

- The existing fleet of vehicles with internal combustion engines must be replaced gradually. Future studies should consider vehicle replacement scenarios and the resulting effects on the climate and the economy. Research should also focus on technical aspects regarding the construction of vehicles, engines and methods of supplying electricity.
- During the development of the manuscript, the authors focused mainly on the topic of electromobility development in the studied areas. The calculation of the expected energy mix is based on historical data concerning changes in the structure of electricity sources, development of power grids and energy infrastructure. The assumptions do not take into account situations in which top-down decisions about giving up or subsidizing individual energy sources are taken at government level (for example, switching off nuclear power plants or subsidizing renewable energy). The assumptions made result from the nature of the method used—the creep trend, which is based on historical data. On the basis of these, a trend of change is determined under the assumption of the invariability of unpredictable external factors, such as the above mentioned top-down findings of national governments.

Author Contributions: Conceptualization, K.T., O.O. and M.M.-W.; Methodology, O.O. and K.T.; Software, M.M.-W.; Validation, K.T. and O.O.; Formal analysis, M.M.-W. and K.T.; Investigation, O.O.; Data curation, M.M.-W. and K.T.; Writing—original draft preparation, K.T.; Writing—review and editing, K.T., O.O., and M.M.-W.; Visualization, M.M.-W., K.T.; Supervision, K.T.; Funding acquisition, K.T. All authors have read and agreed to the published version of the manuscript.

Funding: The research was carried out under financial support obtained from the research subsidy of the Faculty of Engineering Management (WIZ) of Bialystok University of Technology. From the grant No. WI/WIZ-INZ/4/2019 (Olga Orynycz). APC was funded by Institute of Mechanical Engineering money, SGGW.

Conflicts of Interest: The authors declare no conflict of interest. The funders had no role in the design of the study; in the collection, analyses, or interpretation of data; in the writing of the manuscript, and in the decision to publish the results.

References

1. Swain, R.B.; Karimu, A. Renewable electricity and sustainable development goals in the EU. *World Dev.* **2020**, *125*, 104693. [CrossRef]
2. Kristensen, H.S.; Mosgaard, M.A. A review of micro level indicators for a circular economy—Moving away from the three dimensions of sustainability? *J. Clean. Prod.* **2020**, *243*, 118531. [CrossRef]
3. Matuszek, J. Trends and directions of production engineering development. *Prod. Innow. Prod. Innov.* **2005**, *1*, 2–7.
4. Brodny, J.; Tutak, M. Analysis of the diversity in emissions of selected gaseous and particulate pollutants in the European Union countries. *J. Environ. Manag.* **2019**, *231*, 582–595. [CrossRef] [PubMed]
5. Wiegmans, B.; Champagne-Gelinas, A.; Duchesne, S.; Slack, B.; Witte, P. Rail and road freight transport network efficiency of Canada, member states of the EU, and the US. *Res. Trans. Bus. Manag.* **2018**, *28*, 54–65.
6. Production of Passenger Cars Worldwide from 1998 to 2018. Available online: https://www.statista.com/statistics/268739/production-of-passenger-cars-worldwide/ (accessed on 27 January 2020).
7. Pettersson, P.; Johannesson, P.; Jacobson, B.; Bruzelius, F.; Fast, L.; Berglund, S. A statistical operating cycle description for prediction of road vehicles' energy consumption. *Trans. Res. Part D Trans. Environ.* **2019**, *73*, 205–229. [CrossRef]
8. Passenger Cars in the EU. Available online: https://ec.europa.eu/eurostat/statistics-explained/index.php/Passenger_cars_in_the_EU (accessed on 8 August 2020).
9. European Vehicle Market Statistics. Pocketbook 2016/17. Available online: https://theicct.org/sites/default/files/publications/ICCT_Pocketbook_2016.pdf (accessed on 8 August 2020).
10. Tucki, K.; Mruk, R.; Bączyk, A.; Botwińska, K.; Woźniak, K. Analysis of the Exhaust Gas Emission Level from a Diesel Engine with Using Computer Simulation. *Rocz. Ochr. Środowiska* **2018**, *20*, 1095–1112.

11. Tucki, K.; Bączyk, A.; Klimkiewicz, M.; Mączyńska, J.; Sikora, M. Comparison of Energy Performance and Toxicity of Diesel Engine Fuelled with Diesel Oil, Rapeseed Oil and Oil Mixture. Available online: https://iopscience.iop.org/article/10.1088/1755-1315/214/1/012102 (accessed on 23 November 2019).
12. Ghadikolaei, M.A.; Yung, K.F.; Cheung, C.S.; Lau, P.C. Chemical properties and composition of PM emitted from a diesel engine fueled with ternary fuel (diesel-biodiesel-ethanol) in blended and fumigation modes. *Fuel* **2019**, *251*, 368–382. [CrossRef]
13. Nordelöf, A.; Romare, M.; Tivander, J. Life cycle assessment of city buses powered by electricity, hydrogenated vegetable oil or diesel. *Trans. Res. Part D Trans. Environ.* **2019**, *75*, 211–222. [CrossRef]
14. Szymczyk, K.; Kadłubek, M. Challenges in general cargo distribution strategy in urban logistics—Comparative analysis of the biggest logistics operators in EU. *Trans. Res. Procedia* **2019**, *39*, 525–533. [CrossRef]
15. Andrés, L.; Padilla, E. Driving factors of GHG emissions in the EU transport activity. *Trans. Policy* **2018**, *61*, 60–74. [CrossRef]
16. Kondratyev, K.Y.; Varotsos, C. Atmospheric greenhouse effect in the context of global climate change. *Il Nuovo Cimento* **1995**, *18*, 123–151. [CrossRef]
17. Adamenko, V.V.; Kondratyev, K.Y.; Varotsos, C.A. Climate Change in the Arctic and its Empirical Diagnostics. *Energy Environ.* **1999**, *10*, 469–482. [CrossRef]
18. Veum, K.; Bauknecht, D. How to reach the EU renewables target by 2030? An analysis of the governance framework. *Energy Policy* **2019**, *127*, 299–307. [CrossRef]
19. Haasz, T.; Vilchez, J.J.G.; Kunze, R.; Deane, P.; Fraboulet, D.; Fahl, U.; Mulholland, E. Perspectives on decarbonizing the transport sector in the EU-28. *Energy Strategy Rev.* **2018**, *20*, 124–132. [CrossRef]
20. Krzywonos, M.; Skudlarski, J.; Kupczyk, A.; Wojdalski, J.; Tucki, K. Forecast for transport biofuels in Poland in 2020–2030. *Przem. Chem.* **2015**, *94*, 2218–2222.
21. Katinas, V.; Gaigalis, V.; Savickas, J.; Marčiukaitis, M. Analysis of sustainable liquid fuel production and usage in Lithuania in compliance with the National Energy Strategy and EU policy. *Renew. Sustain. Energy Rev.* **2018**, *82*, 271–280. [CrossRef]
22. Capros, P.; Zazias, G.; Evangelopoulou, S.; Kannavou, M.; Fotiou, T.; Siskos, P.; De Vita, A.; Sakellaris, K. Energy-system modelling of the EU strategy towards climate-neutrality. *Energy Policy* **2019**, *134*, 110960. [CrossRef]
23. Santini, D.J. Electric Vehicle Waves of History: Lessons Learned about Market Deployment of Electric Vehicles. Available online: http://Cdn.intechweb.org/pdfs/18663.pdf (accessed on 23 November 2019).
24. Chakraborty, D.; Bunch, D.S.; Lee, J.H.; Tal, G. Demand drivers for charging infrastructure-charging behavior of plug-in electric vehicle commuters. *Trans. Res. Part D Trans. Environ.* **2019**, *76*, 255–272. [CrossRef]
25. Cansino, J.M.; Sánchez-Braza, A.; Sanz-Díaz, T. Policy Instruments to Promote Electro-Mobility in the EU28: A Comprehensive Review. *Sustainability* **2018**, *10*, 2507. [CrossRef]
26. Ahmadi, P. Environmental impacts and behavioral drivers of deep decarbonization for transportation through electric vehicles. *J. Clean. Prod.* **2019**, *225*, 1209–1219. [CrossRef]
27. Mersky, A.C.; Sprei, F.; Samaras, C.; Qian, Z. Effectiveness of incentives on electric vehicle adoption in Norway. *Trans. Res. Part D Trans. Environ.* **2016**, *46*, 56–68. [CrossRef]
28. Hooftman, N.; Oliveira, L.; Messagie, M.; Coosemans, T.; Van Mierlo, J. Environmental Analysis of Petrol, Diesel and Electric Passenger Cars in a Belgian Urban Setting. *Energies* **2016**, *9*, 84. [CrossRef]
29. Arat, H.T. Alternative fuelled hybrid electric vehicle (AF-HEV) with hydrogen enriched internal combustion engine. *Int. J. Hydrogen Energy* **2019**, 1–12, in press. [CrossRef]
30. Shankar, R.; Marco, J. Method for estimating the energy consumption of electric vehicles and plug-in hybrid electric vehicles under real-world driving conditions. *Intell. Trans. Syst. IET* **2013**, *7*, 138–150. [CrossRef]
31. O'Driscoll, R.; Stettler, M.E.J.; Molden, N.; Oxley, T.; ApSimon, H.M. Real world CO_2 and NOx emissions from 149 Euro 5 and 6 diesel, gasoline and hybrid passenger cars. *Sci. Total Environ.* **2018**, *621*, 282–290. [CrossRef]
32. Electric Vehicles from Life Cycle and Circular Economy Perspectives. TERM 2018: Transport and Environment Reporting Mechanism (TERM) Report. Available online: https://www.eea.europa.eu/publications/electric-vehicles-from-life-cycle (accessed on 8 August 2020).
33. Tomczak, J.; Pater, Z.; Bulzak, T. Effect of technological parameters on the rotary compression proces. *Eksploat. Niezawodn.* **2013**, *15*, 279–283.

34. Gajek, J.; Awrejcewicz, J. Mathematical models and non-linear dynamics of a linear electromagnetic motor. *Nonlinear Dyn.* **2018**, *94*, 377–396. [CrossRef]
35. Charge Your Car at One of Energa Network of Electric Vehicle Charging Points. Available online: https://emobility.pl/index.php/en/charging-network-eng/ (accessed on 23 November 2019).
36. Kouchachvili, L.; Yaïci, W.; Entchev, E. Hybrid battery/supercapacitor energy storage system for the electric vehicles. *J. Power Sources* **2018**, *374*, 237–248. [CrossRef]
37. Tucki, K.; Orynycz, O.; Wasiak, A.; Świć, A.; Dybaś, W. Capacity market implementation in Poland: Analysis of a survey on consequences for the electricity market and for energy management. *Energies* **2019**, *12*, 839. [CrossRef]
38. Cao, Y.; Huang, L.; Li, Y.; Jermsittiparsert, K.; Ahmadi-Nezamabad, H.; Nojavan, S. Optimal scheduling of electric vehicles aggregator under market price uncertainty using robust optimization technique. *Int. J. Electr. Power Energy Syst.* **2020**, *117*, 105628.
39. Monteiro, V.; Afonso, J.A.; Ferreira, J.C.; Afonso, J.L. Vehicle Electrification: New Challenges and Opportunities for Smart Grids. *Energies* **2019**, *12*, 118. [CrossRef]
40. Varga, B.O.; Mariasiu, F.; Miclea, C.D.; Szabo, I.; Sirca, A.A.; Nicolae, V. Direct and Indirect Environmental Aspects of an Electric Bus Fleet Under Service. *Energies* **2020**, *13*, 336. [CrossRef]
41. Alface, G.; Ferreira, J.C.; Pereira, R. Electric Vehicle Charging Process and Parking Guidance App. *Energies* **2019**, *12*, 2123. [CrossRef]
42. Kvisle, H.H. The Norwegian Charging Station Database for Electromobility (NOBIL). *World Electr. Veh. J.* **2012**, *5*, 702–707.
43. Dong, G.; Ma, J.; Wei, R.; Haycox, J. Electric vehicle charging point placement optimisation by exploiting spatial statistics and maximal coverage location models. *Trans. Res. Part D Trans. Environ.* **2019**, *67*, 77–88. [CrossRef]
44. Zhao, Q. Electromobility research in Germany and China: Structural differences. *Scientometrics* **2018**, *117*, 473–493.
45. United States Population. 2019. Available online: http://worldpopulationreview.com/countries/united-states-population/ (accessed on 23 November 2019).
46. Rivera-Monroy, V.H.; Danielson, T.M.; Castañeda-Moya, E.; Marx, B.D.; Travieso, R.; Zhao, X.; Gaiser, E.E.; Farfan, L.M. Long-term demography and stem productivity of Everglades mangrove forests (Florida, US): Resistance to hurricane disturbance. *For. Ecol. Manag.* **2019**, *440*, 79–91. [CrossRef]
47. Selected Automakers' U.S. Ytd Market Share in 3rd Quarter 2019, by Key Manufacturer. Available online: https://www.statista.com/statistics/343162/market-share-of-major-car-manufacturers-in-the-united-states/ (accessed on 23 November 2019).
48. How Many Polish Genes in the Polish Automotive Industry? Available online: https://www.arp.pl/__data/assets/pdf_file/0011/76178/Raport.pdf (accessed on 7 August 2020).
49. Hu, L.; Dong, J.; Lin, Z.; Yang, J. Analyzing battery electric vehicle feasibility from taxi travel patterns: The case study of New York City, US. *Trans. Res. Part C Emerg. Technol.* **2018**, *87*, 91–104. [CrossRef]
50. Feng, W.; Figliozzi, M. An economic and technological analysis of the key factors affecting the competitiveness of electric commercial vehicles: A case study from the US market. *Trans. Res. Part C Emerg. Technol.* **2013**, *26*, 135–145. [CrossRef]
51. Thomas, V.J.; Maine, E. Market entry strategies for electric vehicle start-ups in the automotive industry—Lessons from Tesla Motors. *J. Clean. Prod.* **2019**, *235*, 653–663. [CrossRef]
52. Liu, J.H.; Meng, Z. Innovation Model Analysis of New Energy Vehicles: Taking Toyota, Tesla and BYD as an Example. *Procedia Eng.* **2017**, *174*, 965–972. [CrossRef]
53. Guo, Z.; Zhou, Y. Residual value analysis of plug-in vehicles in the United States. *Energy Policy* **2019**, *125*, 445–455. [CrossRef]
54. Population and Population Change Statistics. Available online: https://ec.europa.eu/eurostat/statistics-explained/index.php/Population_and_population_change_statistics (accessed on 23 November 2019).
55. EU: Total Population from 2009 to 2019. Available online: https://www.statista.com/statistics/253372/total-population-of-the-european-union-eu/ (accessed on 23 November 2019).
56. Auvinen, H.; Järvi, T.; Kloetzke, M.; Kugler, U.; Bühne, J.A.; Heinl, F.; Kurte, J.; Esser, K. Electromobility Scenarios: Research Findings to Inform Policy. *Trans. Res. Procedia* **2016**, *14*, 2564–2573. [CrossRef]

57. Habib, K.; Hansdóttir, S.T.; Habib, H. Critical metals for electromobility: Global demand scenarios for passenger vehicles, 2015–2050. *Resour. Conserv. Recycl.* **2020**, *154*, 104603. [CrossRef]
58. Pregger, T.; Nitsch, J.; Naegler, T. Long-term scenarios and strategies for the deployment of renewable energies in Germany. *Energy Policy* **2013**, *59*, 350–360. [CrossRef]
59. Council Directive 91/441/EEC of 26 June 1991 Amending Directive 70/220/EEC on the Approximation of the Laws of the Member States Relating to Measures to Be Taken Against Air Pollution by Emissions from Motor Vehicles. Available online: https://eur-lex.europa.eu/legal-content/en/ALL/?uri=CELEX%3A31991L0441 (accessed on 8 August 2020).
60. Directive 94/12/EC of the European Parliament and the Council of 23 March 1994 Relating to Measures to Be Taken Against Air Pollution by Emissions from Motor Vehicles and Amending Directive 70/220/EEC. Available online: https://eur-lex.europa.eu/legal-content/en/ALL/?uri=CELEX%3A31994L0012 (accessed on 8 August 2020).
61. Commission Directive 2002/80/EC of 3 October 2002 Adapting to Technical Progress Council Directive 70/220/Eec Relating to Measures to Be Taken Against Air Pollution by Emissions from Motor Vehicles (Text with EEA Relevance). Available online: https://eur-lex.europa.eu/legal-content/HR/TXT/?uri=CELEX:32002L0080 (accessed on 8 August 2020).
62. Commission Regulation (EU) 2016/427 of 10 March 2016 Amending Regulation (EC) No 692/2008 as Regards Emissions from Light Passenger and Commercial Vehicles (Euro 6) (Text with EEA Relevance). Available online: https://eur-lex.europa.eu/legal-content/EN/TXT/?uri=CELEX%3A32016R0427 (accessed on 8 August 2020).
63. Progress of EU Transport Sector towards Its Environment and Climate Objectives. Available online: https://www.eea.europa.eu/publications/progress-of-eu-transport-sector-1 (accessed on 8 August 2020).
64. Helmers, E.; Leitão, J.; Tietge, U.; Butler, T. CO_2-equivalent emissions from European passenger vehicles in the years 1995–2015 based on real-world use: Assessing the climate benefit of the European "diesel boom". *Atmos. Environ.* **2019**, *198*, 122–132. [CrossRef]
65. Teixeira, A.C.R.; Sodré, J.R. Impacts of replacement of engine powered vehicles by electric vehicles on energy consumption and CO_2 emissions. *Trans. Res. Part D Trans. Environ.* **2018**, *59*, 375–384. [CrossRef]
66. Hooftman, N.; Messagie, M.; Van Mierlo, J.; Coosemans, T. A review of the European passenger car regulations—Real driving emissions vs local air quality. *Renew. Sustain. Energy Rev.* **2018**, *86*, 1–21. [CrossRef]
67. Japan Population. 2019. Available online: http://worldpopulationreview.com/countries/japan-population/ (accessed on 23 November 2019).
68. Yagi, M.; Managi, S. Demographic determinants of car ownership in Japan. *Transp. Policy* **2016**, *50*, 37–53. [CrossRef]
69. Portugal-Pereira, J.; Esteban, M. Implications of paradigm shift in Japan's electricity security of supply: A multi-dimensional indicator assessment. *Appl. Energy* **2014**, *123*, 424–434. [CrossRef]
70. United Nations Reform: Priority Issues for Japan. Available online: https://www.mofa.go.jp/policy/un/reform/priority.html (accessed on 1 February 2020).
71. Universal Periodic Review Japan. Available online: https://www.ohchr.org/EN/HRBodies/UPR/Pages/JPIndex.aspx (accessed on 1 February 2020).
72. Orecchini, F.; Santiangeli, A.; Zuccari, F. Hybrid-electric system truth test: Energy analysis of Toyota Prius IV in real urban drive conditions. *Sustain. Energy Technol. Assess.* **2020**, *37*, 100573. [CrossRef]
73. Prokhorov, D.V. Toyota Prius HEV neurocontrol and diagnostics. *Neural Netw.* **2008**, *21*, 458–465. [CrossRef] [PubMed]
74. Iwata, K.; Matsumoto, S. Use of hybrid vehicles in Japan: An analysis of used car market data. *Trans. Res. Part D Trans. Environ.* **2016**, *46*, 200–206. [CrossRef]
75. Mishina, Y.; Muromachi, Y. Are potential reductions in CO_2 emissions via hybrid electric vehicles actualized in real traffic? The case of Japan. *Trans. Res. Part D Trans. Environ.* **2017**, *50*, 372–384. [CrossRef]
76. Paladugu, B.S.K.; Grau, D. Toyota Production System—Monitoring Construction Work Progress with Lean Principles. In *Encyclopedia of Renewable and Sustainable Materials*, 1st ed.; Hashmi, S., Choudhury, I.A., Eds.; Elsevier: Amsterdam, The Netherlands, 2020; Volume 5, pp. 560–565.
77. Augustyn, K. Forecasts of Electricity Demand in Poland in the Period to 2015—Optimization Role of Business Intelligence. Available online: https://depot.ceon.pl/handle/123456789/9989 (accessed on 27 January 2020).

78. Tucki, K.; Orynycz, O.; Świć, A.; Mitoraj-Wojtanek, M. The Development of Electromobility in Poland and EU States as a Tool for Management of CO_2 Emissions. *Energies* **2019**, *12*, 2942. [CrossRef]
79. Nævestad, T.O.; Størkersen, K.V.; Laiou, A.; Yannis, G. Framework conditions of occupational safety: Comparing Norwegian maritime cargo and passenger transport. *Int. J. Trans. Sci. Technol.* **2018**, *7*, 291–307. [CrossRef]
80. Popova, O.; Gorev, A.; Shavyraa, C. Principles of modern route systems planning for urban passenger transport. *Trans. Res. Procedia* **2018**, *36*, 603–609. [CrossRef]
81. Türe, Y.; Türe, C. An assessment of using Aluminum and Magnesium on CO_2 emission in European passenger cars. *J. Clean. Prod.* **2020**, *247*, 119120. [CrossRef]
82. Number of U.S. Aircraft, Vehicles, Vessels, and Other Conveyances. Available online: https://www.bts.dot.gov/content/number-us-aircraft-vehicles-vessels-and-other-conveyances (accessed on 23 November 2019).
83. Report: Vehicles in Use—Europe. 2018. Available online: https://www.acea.be/uploads/statistic_documents/ACEA_Report_Vehicles_in_use-Europe_2018.pdf (accessed on 8 August 2020).
84. The Automobile Industry Pocket Guide 2018–2019. Available online: https://www.acea.be/uploads/publications/ACEA_Pocket_Guide_2018-2019.pdf (accessed on 8 August 2020).
85. Automobile Inspection & Registration Information Association. Available online: https://www.airia.or.jp/publish/statistics/number.html (accessed on 23 November 2019).
86. Japan Automobile Manufacturers Association (JAMA). Available online: http://jamaserv.jama.or.jp/newdb/eng/index.html (accessed on 23 November 2019).
87. GMS Regional Information Portal. Available online: https://portal.gms-eoc.org/charts/overview/vehicle-motorization-index (accessed on 1 February 2020).
88. International Organization of Motor Vehicle Manufacturers. Available online: http://www.oica.net/world-vehicles-in-use-all-vehicles-2/ (accessed on 1 February 2020).
89. Transport Statistics at Regional Level. Statistics Explained. Available online: https://ec.europa.eu/eurostat/statistics-explained/pdfscache/14273.pdf (accessed on 1 February 2020).
90. Andrzejczak, K. Changes in the growth rate of passenger cars saturation. *Wiad. Stat. Pol. Stat.* **2012**, *11*, 22–34.
91. Jamroz, K. Modelling motorization index at national level. *Zeszyty Naukowo-Techniczne Stowarzyszenia Inżynierów i Techników Komunikacji w Krakowie* **2012**, *2*, 111–120.
92. Crude Oil Prices—70 Year Historical Chart. Available online: https://www.macrotrends.net/1369/crude-oil-price-history-chart (accessed on 23 November 2019).
93. International Council on Clean Transport. Available online: https://theicct.org/publication-type/reports (accessed on 23 November 2019).
94. The Motor Industry of Japan. 2019. Available online: https://www.jama.org/the-motor-industry-of-japan-2019/ (accessed on 23 November 2019).
95. U.S. Light-Duty Advanced Technology Vehicle (ATV) Sales (2011–2019). Available online: https://autoalliance.org/energy-environment/advanced-technology-vehicle-sales-dashboard/ (accessed on 23 November 2019).
96. International Energy Agency. Nordic EV Outlook. 2018. Available online: https://www.nordicenergy.org/wp-content/uploads/2018/05/NordicEVOutlook2018.pdf (accessed on 1 February 2020).
97. Gnat, S. Prediction of Weekly Magazines Sale—Choosen Aproach. *Studia i Prace Wydziału Nauk Ekonomicznych i Zarządzania / Uniwersytet Szczeciński* **2008**, *2*, 97–108.
98. Iwanejko, R.; Bajer, J. Application of mathematical models for prognosing the failures in water supply systems - the Krakow case study. *Czas. Tech. Śr.* **2012**, *109*, 19–148.
99. Sroka, W.; Musiał, W. Multifarious forecasts concerning the use of mineral fertilizers in Poland in 2014–2025. *J. Agribus. Rural Dev.* **2015**, *2*, 291–302. [CrossRef]
100. Holden, P. Territory, geoeconomics and power politics: The Irish government's framing of Brexit. *Polit. Geogr.* **2020**, *76*, 102063. [CrossRef]
101. Shaw, D.; Smith, C.M.; Scully, J. From Brexit to Article 50: Applying Critical Realism to the design and analysis of a longitudinal causal mapping study. *Eur. J. Oper. Res.* **2019**, *276*, 723–735. [CrossRef]
102. Prescott, C.; Pilato, M.; Bellia, C. Geographical indications in the UK after Brexit: An uncertain future? *Food Policy* **2020**, *90*, 101808. [CrossRef]
103. The Society of Motor Manufacturers and Traders (SMMT). Brexit. Available online: https://www.smmt.co.uk/industry-topics/brexit/ (accessed on 1 February 2020).

104. Alamerew, Y.A.; Brissaud, D. Modelling reverse supply chain through system dynamics for realizing the transition towards the circular economy: A case study on electric vehicle batteries. *J. Clean. Prod.* **2020**, *254*, 120025. [CrossRef]
105. Zhou, Y.; Wen, R.; Wang, H.; Cai, H. Optimal battery electric vehicles range: A study considering heterogeneous travel patterns, charging behaviors, and access to charging infrastructure. *Energy* **2020**, 116945, in press. [CrossRef]
106. Wang, Y.N. Power Battery Performance Detection System for Electric Vehicles. *Procedia Comput. Sci.* **2019**, *154*, 759–763. [CrossRef]
107. Bellocchi, S.; Klöckner, K.; Manno, M.; Noussan, M.; Vellini, M. On the role of electric vehicles towards low-carbon energy systems: Italy and Germany in comparison. *Appl. Energy* **2019**, *255*, 113848. [CrossRef]
108. Palkowski, K. Electric Car and Atmosphere Protection. *Rocz. Ochr. Srodowiska Ann. Set Environ. Prot.* **2016**, *18*, 628–639.
109. Kosowski, K.; Tucki, K.; Piwowarski, M.; Stępień, R.; Orynycz, O.; Włodarski, W.; Bączyk, A. Thermodynamic Cycle Concepts for High-Efficiency Power Plans. Part A: Public Power Plants 60+. *Sustainability* **2019**, *11*, 554. [CrossRef]
110. Mikielewicz, D.; Kosowski, K.; Tucki, K.; Piwowarski, M.; Stepien, R.; Orynycz, O.; Wlodarski, R. Gas Turbine Cycle with External Combustion Chamber for Prosumer and Distributed Energy Systems. *Energies* **2019**, *12*, 3501. [CrossRef]
111. Tucki, K.; Orynycz, O.; Wasiak, A.; Swic, A.; Wichlacz, J. The Impact of Fuel Type on the Output Parameters of a New Biofuel Burner. *Energies* **2019**, *12*, 1383. [CrossRef]
112. Dołęga, W. Ecology in Manufacturing. Available online: https://www.cire.pl/pliki/2/wytwarzanieenergiiaakologia.pdf (accessed on 23 November 2019).
113. Meratizaman, M.; Monadizadeh, S.; Sardasht, M.T.; Amidpour, M. Techno economic and environmental assessment of using gasification process in order to mitigate the emission in the available steam power cycle. *Energy* **2015**, *83*, 1–14. [CrossRef]
114. Kosowski, K.; Domachowski, Z.; Próchnicki, W.; Kosowski, A.; Stępień, R.; Piwowarski, M.; Włodarski, W.; Ghaemi, M.; Tucki, K.; Gardzilewicz, A.; et al. *Steam and Gas Turbines with the Examples of Alstom Technology*, 1st ed.; Alstom: Saint-Quen, France, 2007; ISBN 978-83-925959-3-9.
115. Chmielniak, T. *Technologie Energetyczne*; Wydawnictwo Politechniki Śląskiej: Warrenton, OR, USA, 2008; pp. 230–560. (In Polish)
116. Bejan, A. *Advanced Engineering Thermodynamics*; John Wiley & Sons: Hoboken, NJ, USA, 2016; pp. 394–460.
117. Kotowicz, J.; Brzęczek, M. Analysis of increasing efficiency of modern combined cycle power plant: A case study. *Energy* **2018**, *153*, 90–99. [CrossRef]
118. BP Statistical Review of World Energy. 2019. Available online: https://www.bp.com/content/dam/bp/business-sites/en/global/corporate/pdfs/energy-economics/statistical-review/bp-stats-review-2019-full-report.pdf (accessed on 8 August 2020).
119. Latõšov, E.; Volkova, A.; Siirde, A.; Kurnitski, J.; Thalfeldt, M. Primary energy factor for district heating networks in European Union member states. *Energy Procedia* **2017**, *116*, 69–77. [CrossRef]
120. Wilby, M.R.; González, A.B.R.; Díaz, J.J.V. Empirical and dynamic primary energy factors. *Energy* **2014**, *73*, 771–779. [CrossRef]
121. Schicktanz, M.D.; Wapler, J.; Henning, H.H. Primary energy and economic analysis of combined heating, cooling and power systems. *Energy* **2011**, *36*, 575–585. [CrossRef]
122. Data and Statistics. Explore Energy Data by Category, Indicator, Country or Region. Available online: https://www.iea.org/data-and-statistics (accessed on 1 February 2020).
123. Final Report. Evaluation of Primary Energy Factor Calculation Options for Electricity. Available online: https://ec.europa.eu/energy/sites/ener/files/documents/final_report_pef_eed.pdf (accessed on 8 August 2020).
124. Directive of the European Parliament and of the Council amending Directive 2012/27/EU on energy efficiency. Available online: https://www.actu-environnement.com/media/pdf/news-27990-directive-efficacite-energetique.pdf (accessed on 8 August 2020).
125. Bukrejewski, P.; Skolniak, M.; Kowalski, L. Comparison of the environmental effect of M1 category vehicles fed with traditional and alternative fuels. *Arch. Automot. Eng.* **2017**, *75*, 5–21.

126. Commission Delegated Regulation (EU) 2015/2402 of 12 October 2015 Reviewing Harmonised Efficiency Reference Values for Separate Production of Electricity and Heat in Application of Directive 2012/27/EU of the European Parliament and of the Council and repealing Commission Implementing Decision 2011/877/EU. Available online: https://eur-lex.europa.eu/eli/reg_del/2015/2402/oj (accessed on 8 August 2020).
127. 2030 Climate and Energy Framework. Available online: https://ec.europa.eu/clima/policies/strategies/2030_en (accessed on 8 January 2020).
128. Strategic Energy Plan. Available online: https://www.enecho.meti.go.jp/en/category/others/basic_plan/ (accessed on 8 January 2020).
129. Center for Climate and Energy Solutions. Available online: https://www.c2es.org/content/renewable-energy (accessed on 8 January 2020).
130. Wasiak, A. *Modeling Energetic Efficiency of Biofuels Production. Green Energy and Technology*, 1st ed.; Springer Nature Switzerland: Cham, Switzerland, 2018; pp. 29–47. ISBN 978-3-319-98430-8. [CrossRef]
131. Wasiak, A.; Orynycz, O. Formulation of a model for energetic efficiency of agricultural subsystem of biofuel production. In Proceedings of the 2014 IEEE International Energy Conference, Cavtat, Croatia, 13–16 May 2014; pp. 1333–1337. Available online: https://ieeexplore.ieee.org/document/6850586 (accessed on 29 November 2019).
132. Wasiak, A.; Orynycz, O. The effects of energy contributions into subsidiary processes on energetic efficiency of biomass plantation supplying biofuel production system. *Agric. Agric. Sci. Procedia* **2015**, *7*, 292–300. [CrossRef]
133. Wasiak, A.; Orynycz, O. The Effect of Transportation Choices on Energetic Effectivenes of Rapeseed Plantation. IX International Scientific Symposium Farm Machinery and Processes Management in Sustainable Agriculture 2017, 400–405. Available online: https://depot.ceon.pl/handle/123456789/14763 (accessed on 29 November 2019).
134. Orynycz, O.; Świć, A. The Effects of Material's Transport on Various Steps of Production System on Energetic Efficiency of Biodiesel Production. *Sustinability* **2018**, *10*, 2736. [CrossRef]
135. Climate for Poland—Poland for Climate: 1988–2018–2050. Available online: https://cop24.gov.pl/news/news-details/news/climate-for-poland-poland-for-climate-1988-2018-2050/ (accessed on 8 August 2020).
136. Plötz, P.; Funke, S.A.; Jochem, P. The impact of daily and annual driving on fuel economy and CO_2 emissions of plug-in hybrid electric vehicles. *Trans. Res. Part A Policy Pract.* **2018**, *118*, 331–340. [CrossRef]
137. Lin, W.Y.; Hsiao, M.C.; Wu, P.C.; Fu, J.S.; Lai, L.W.; Lai, H.C. Analysis of air quality and health co-benefits regarding electric vehicle promotion coupled with power plant emissions. *J. Clean. Prod.* **2020**, *247*, 119152. [CrossRef]
138. Pearce, P.; Johnston, V. A new fast lane or just a roadblock? Mitigating road transport GHG emissions under Australia's Emissions Reduction Fund. *Environ. Plan. Law J.* **2016**, *33*, 181–202.
139. Hunter, T.S. Offshore Petroleum Drilling and Risk. A Study of Proposed Deep-Sea Exploration Drilling in Commonwealth Regulated Waters of the Great Australian Bight. Available online: https://www.greenpeace.org.au/wp/wp-content/uploads/2019/05/Professor-Soliman-Hunter-GAB-Report-April-2019.pdf (accessed on 27 June 2020).
140. Federal Highway Administration. Average Annual Miles per Driver by Age Group. Available online: https://www.fhwa.dot.gov/ohim/onh00/bar8.htm (accessed on 23 November 2019).
141. Japan's Used Cars Are Newer with Lower Mileage. Available online: https://integrityexports.com/japan-car-auction-academy/why-cars-from-japan/japans-used-cars-are-newer-with-lower-mileage/ (accessed on 23 November 2019).
142. Average Lifespan for U.S. Vehicles. Available online: https://berla.co/average-us-vehicle-lifespan/ (accessed on 23 November 2019).
143. Japan Automobile Manufacturers Association (JAMA). Motor Vehicle Statistics of Japan. Available online: http://www.jama-english.jp/publications/industry.html (accessed on 23 November 2019).
144. Tucki, K.; Mruk, R.; Orynycz, O.; Wasiak, A.; Botwińska, K.; Gola, A. Simulation of the Operation of a Spark Ignition Engine Fueled with Various Biofuels and Its Contribution to Technology Management. *Sustainability* **2019**, *11*, 2799. [CrossRef]
145. European Vehicle Market Statistics, 2018/2019. Available online: https://theicct.org/publications/european-vehicle-market-statistics-20182019 (accessed on 23 November 2019).
146. Den Tonkelaar, W.A.M. CAR Smog: System for on-line extrapolation from hourly measurements to concentrations along standard roads within cities. *Sci. Total Environ.* **1996**, *189–190*, 423–429. [CrossRef]

147. Frankowski, J. Attention: Smog alert! Citizen engagement for clean air and its consequences for fuel poverty in Poland. *Energy Build.* **2020**, *207*, 109525. [CrossRef]
148. Reducing Pollution with Electric Vehicles. Available online: https://www.energy.gov/eere/electricvehicles/reducing-pollution-electric-vehicles (accessed on 23 November 2019).
149. Moćko, W.; Ornowski, O.; Szymańska, S. Estimation of energy consumption of electric vehicle during road driving test. *Zeszyty Problemowe Maszyny Elektryczne* **2013**, *2*, 31–35. Available online: http://www.komel.katowice.pl/ZRODLA/FULL/99/ref_05.pdf (accessed on 8 January 2020).

energies

MDPI

Article

Energetic, Economic and Environmental (3E) Assessment and Design of Solar-Powered HVAC Systems in Pakistan

Sajid Mehmood [1,2], Serguey A. Maximov [1], Hannah Chalmers [1] and Daniel Friedrich [1,*]

1 School of Engineering, Institute for Energy Systems, The University of Edinburgh, Colin Maclaurin Road, Edinburgh EH93DW, UK; S.Mehmood@ed.ac.uk (S.M.); Serguey.Maximov@ed.ac.uk (S.A.M.); Hannah.Chalmers@ed.ac.uk (H.C.)

2 Department of Mechanical, Mechatronics and Manufacturing Engineering (New Campus), University of Engineering & Technology, Lahore 54000, Pakistan

* Correspondence: D.Friedrich@ed.ac.uk; Tel.: +44-(0)131-6505662

Received: 29 June 2020; Accepted: 19 August 2020; Published: 21 August 2020

Abstract: Rapid urbanization, global warming and enhanced quality of life have significantly increased the demand of indoor thermal comfort and air conditioning systems are not a luxury anymore, but a necessity. In order to fulfil this need, it is imperative to develop affordable and environmentally friendly cooling solutions for buildings. In this work, the 3E performance (energetic, economic and environmental) of electrically driven water-cooled vapour compression systems and thermally (solar) driven vapour absorption cooling systems are evaluated and the parameters affecting the performance of solar-driven vapour absorption systems are investigated. The energy simulation software TRNSYS is used to simulate the performance of both systems in order to fulfil the cooling needs of an industrial manufacturing building for the typical climate conditions for Lahore, Pakistan. Primary energy saving, initial investment, operational cost, and carbon footprint indices are used to analyse the performance of both systems. In addition, a parametric code is written in Python and linked with TRNSYS to perform a parametric study to investigate the effects of various parameters such as solar field size, storage tank volume, optimum annual and monthly collector angles, and flow rate in the solar field on the solar-driven vapour absorption chiller performance. The results reveal that around 5% more energy can be absorbed per collector surface area by changing the solar tilt angle on a monthly basis compared to one fixed angle. The analysis shows that electrically driven vapour compression-based cooling systems have much higher running cost and are potentially hazardous for the environment but have lower capital costs. On the other hand, solar thermal systems have lower running costs and emissions but require further reductions in the capital costs or government subsidies to make them viable.

Keywords: solar thermal driven absorption chiller; optimum collector tilt angle; TRNSYS; parametric study; heating ventilation & air conditioning; solar cooling

---

1. Introduction

Increasing energy demand and environmental pollution are among the major challenges for the world and research is underway to develop more environmentally friendly systems. Currently most of the world energy is provided by burning fossil fuels which results in large CO_2 emissions [1]. For example, Pakistan covered 54% of its total energy requirements by burning imported fossil fuels like natural gas and oil [2], which presents a significant burden on the national finances [3]. In addition, the country is still facing a 4000 MW [2] energy shortfall in the peak summer season due to high demand for electrically driven space cooling. While energy consumption is directly linked to level of development, it is important to decouple them and to develop lower carbon technology alternatives [4]. Among the South Asian countries, Pakistan is strongly affected by global warming and faces severe energy crises in peak summer seasons. It is forecasted that the huge investment in China's "belt and road initiative" will have many positive aspects for country development but will substantially increase the greenhouse gas emissions (GHG) as well [5]. Energy consumption in Pakistan is increasing by around 8% [6] per year and a large part of this increase is due to cooling needs in the peak summer season.

People spent up to 80% of their time indoors [7] and in the industrialized world, energy consumption in the building sector accounts for around 40% of the total world energy demand [8]. This is mainly used to maintain a comfortable temperature inside buildings. Population growth, global warming and enhanced quality of life (high internal gains from equipment inside buildings and lighting, increased occupant comfort demands, architectural characteristics, such as growing popularity of glass buildings and window to wall ratio) have significantly increased the energy demand of air conditioning. The amount of energy spent for space cooling has increased by 4% annually since 1990 and it is forecasted that the energy demand for all cooling applications will increase by 15% by 2050 [9]. Reducing the energy consumption in the building sector without compromising on occupants' comfort while fulfilling heating, ventilation and air conditioning (HVAC) energy demand from renewable resources like solar could have environmental and financial benefits. To achieve these benefits, it is essential to evaluate and improve the feasibility of solar energy utilization for cooling applications, which could tackle the energy crises during peak summer season and utilize local renewable resources to reduce the import and associated costs of fossil fuels.

Based on thermodynamic principles, a cooling effect can be produced either with electrically driven or thermally driven cycles [10]. Vapour Compression Cycle (VCC) systems are mostly used to fulfill cooling demands but rely on an electrically driven compressor and contribute around 10% of global GHG emissions [9]. The market share of VCC-based technology is around 80% [9] despite the large carbon footprint and often hazardous refrigerants. Most of the refrigerants used in these cooling systems exhibit significant global warming potential and also contribute to ozone depletion. On the other hand, thermally driven cooling systems are based on four different technologies: vapour absorption cycle, vapour adsorption cycle, desiccant cooling system and ejector cooling cycle [10]. Of these, vapour Absorption Cooling Systems (ACS) are most widely used due to high performance and technical maturity [11]. The source of heat to operate ACS can be solar energy, industrial exhaust gases or fossil fuels. Among the different heat sources, solar thermal driven ACS are under rapid development and are a promising technology to provide sustainable peak summer cooling due to the coincidence of cooling demand and solar irradiance. According to the International Energy Agency (IEA), it is expected that solar cooling will have a potential market of 417 TWh per year in 2050 [12]. Pakistan has a high annual global solar irradiance of 2071 kWh/m^2 [13] and around 10–12 sunshine hours per day, which makes it attractive for solar-driven cooling applications.

Most of the studies found in the literature are based on energetic performance of vapour absorption chillers, while the evaluation of economic and environmental performance of solar-driven ACS are relatively unexplored, particularly for the climate and conditions of Pakistan. A few studies related to solar-driven absorption chillers and different types of solar collectors are summarized here. Different types of collectors are available to capture solar energy and to convert it into thermal energy at different temperatures. Mostly stationary (non-concentrating) collectors (evacuated tube collector or flat plate collector) are used to harness solar energy to operate single effect chillers [14], while concentrating collectors (e.g., parabolic trough collector) were used by [15,16] to achieve the higher temperature required to operate multi-effect absorption chillers. Photovoltaic thermal collectors [17] combine photovoltaic and thermal collectors and provide both heat and electricity. Solar tracking systems (single axis or double axis tracking) are used to enhance the annual efficiency of the collector. The performance of all these collectors depends on solar irradiance and local climate conditions such as humidity and temperature. A review of different types of solar collectors and cooling technologies was conducted by Shirazi et al. [8]. Buonomano et al. [18] used stationary flat plate collectors to drive double effect chillers and compared this technology with concentrating collectors. Xu et al. [19] analysed single, double and variable effect chillers with Lithium bromide(LiBr)-water as a working pair. They found that the variable effect chiller achieved higher solar fraction and low auxiliary heat input. Gomri et al. [20] simulated solar ACS with natural gas as an auxiliary heat source in times of low solar irradiance. They discussed coefficient of performance (COP) of the refrigeration unit, exergy efficiency with varying generator inlet temperature and reducing condenser temperature. A configuration-based model of solar assisted ACS was carried out by [21] to fulfill cooling load requirement for an education building.

The performance and economic benefits of solar-driven ACS vary based on many factors such as weather, location, cost of components and system design [22]. The climate of Pakistan is hot and humid with a prolonged summer season adversely affected by global warming and people using VCC cooling systems which contribute to GHG emissions. The government of Pakistan is encouraging people through tax incentives to use solar energy to reduce GHG emissions [23]. While thermally driven cooling systems have significant potential to reduce the use of fossil fuels and consequently associated GHG emissions, they need to be carefully designed to provide both environmental and economic benefits. Considering this fact, there are only a few studies that compare electrically and solar thermal driven cooling systems simultaneously on their energy, economic and environmental (3E) performance. The performance depends critically on local weather conditions and costs of ACS and VCC system components, and to the authors' knowledge, there is not a single study available for the hot climate of Pakistan based on 3E assessment. Studies on evacuated tube solar collector tilt angle adjustment for Pakistan are also unexplored, although the country has huge potential for solar energy.

In this work, the comparative performance of vapour compression and solar assisted vapour absorption cooling systems were compared based on energetic, economic and environmental performance. The objective of this work was to compare two cooling technologies for the same chosen cooling capacity based on energy, economic and environmental indices; an electrically driven water-cooled vapour compression cycle chiller and a solar thermal driven vapour absorption cooling system. In addition, this study discusses the parameters affecting the performance of solar-driven vapour absorption chillers.

2. Working Principle and Description of Two Cooling Systems

This section presents the working principles of the solar-driven vapour absorption cycle and conventional vapour compression cycle.

2.1. Conventional Vapour Compression Cycle

Figure 1 shows the schematic of the conventional VCC when a mechanical compressor is used. The refrigerant is passed as low-pressure gas to an electrically driven compressor. The electrically driven compressor turns it into a high-pressure gas before it goes to the condenser. In the condenser, the refrigerant exchanges heat with water and condenses into a liquid state. The condensed refrigerant is then passed through an expansion valve which reduces its pressure. Finally, the low-pressure refrigerant is evaporated in the evaporator to produce the cooling effect and closes the cycle. To remove heat from the condenser, either air or water can be used. In this study, we used a water-cooled condenser due to its higher specific heat capacity as compared to air, which results in higher chiller performance.

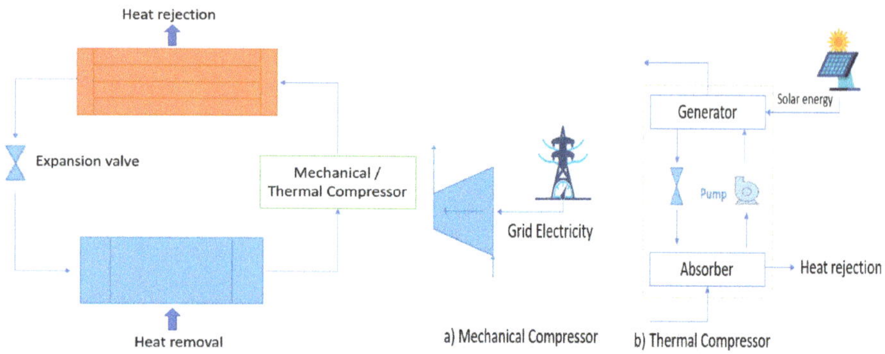

Figure 1. Schematic diagram of the conventional vapour compression cycle and the solar-driven vapour absorption cycle.

2.2. Solar-Driven Vapour Absorption Cycle

The schematic of the vapour absorption cycle is shown in Figure 1 in which the electrically driven compressor of the VCC is replaced by a thermal compressor which consists of a generator, solution pump and an absorber unit. The refrigerant (typically water) returns from the evaporator to the absorber section where it is absorbed in Lithium bromide (LiBr) which has a strong affinity to absorb water. Once the LiBr mixes with water, a pump moves this LiBr-water mixture to the generator section where heat from solar energy is used to desorb water from this solution. Due to the different boiling points of water and LiBr, water evaporates earlier and flows to the condenser. The liquid LiBr is returned back to the absorber section to absorb water for the next cycle. The evaporated refrigerant (water) then follows the standard refrigeration cycle (condenser, expansion valve and evaporator). To dissipate heat from the absorber and condenser, air or water can be used.

3. Simulation Methodology

The methodology employed in this work is shown in Figure 2. We used the energy simulation software TRNSYS which is a well known, flexible graphical-based simulation software widely used by researchers in the energy simulation community to model and simulate the transient behavior of energy systems [24]. This software solves differential equations generated from the system configuration with the modified Euler method and uses a successive iteration method to solve the non-linear equations for each component. The building geometry is defined in SketchUp which is widely used in creating 3D models of architectural drawings, interior design and in other civil and mechanical engineering applications [25]. We used the TRNSYS 3D plugin for SketchUp to create a dynamic 3D-building model. TRNSYS 3D zones are different from SketchUp zones, particularly because TRNSYS 3D zones are used to simulate the dynamic flow of energy. The geometrical characteristics of the building zones of the SketchUp model were imported into the TRNSYS TRNBuild software. The important building envelope (walls, windows, roof) characteristics, internal gains (light, equipment and occupants) and their variability over time were considered and incorporated in the building model. Building characteristics design parameters for indoor comfort are tabulated in Table 1. Finally, TRNSYS simulation studio was used to calculate the cooling load profile for a typical metrological year weather data for Lahore, Pakistan. To fulfil daily cooling load demand and to make a 3E assessment, two models were developed, i.e., solar assisted vapour absorption cooling system (ACS) and water-cooled electrically driven VCC, for an industrial building located in Lahore, Pakistan as shown in Figures 3 and 4, respectively. To analyse the parameters affecting the performance of solar-driven absorption chillers, a parametric study was performed over the collector area, flow rate in the collector loop, collector tilt angle, collector azimuthal angle and storage tank volume to determine optimum values. We linked the TRNSYS simulation to Python to perform the parametric study.

The Python script modifies the TRNSYS deck file, runs the simulation and compiles the results of each simulation as shown in Figure 2.

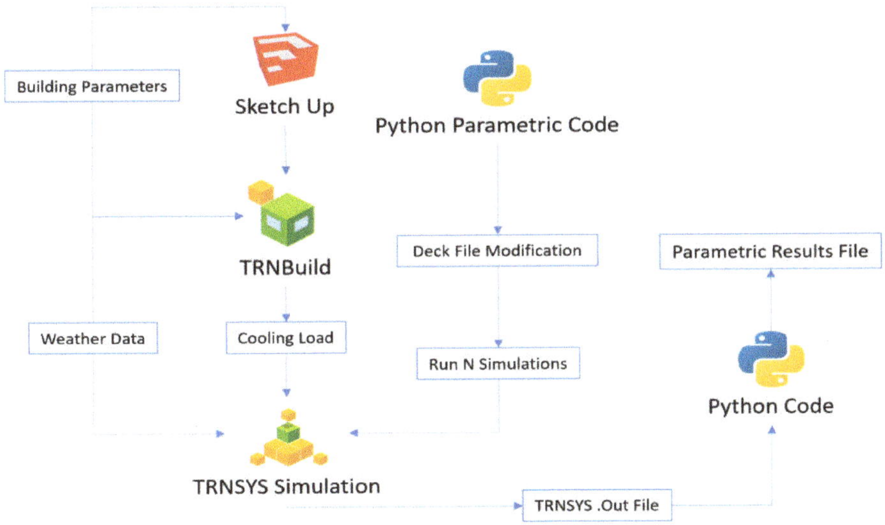

Figure 2. Simulation and parametric study methodology.

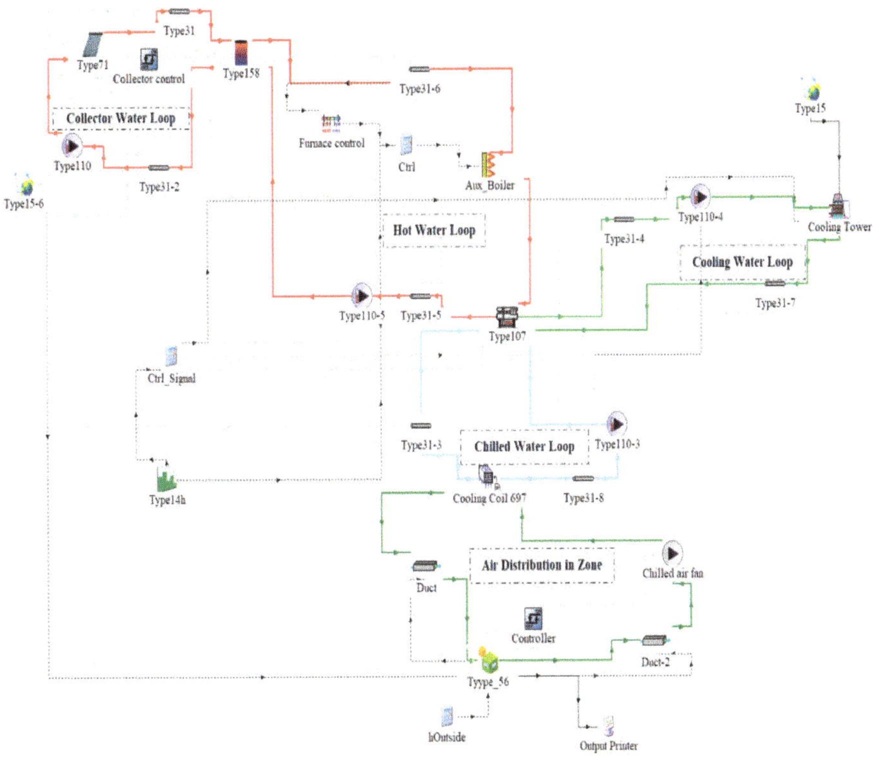

Figure 3. Solar absorption cooling system simulated in TRNSYS.

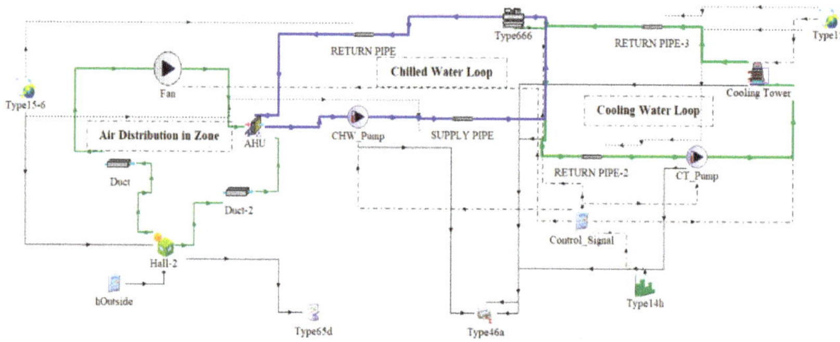

Figure 4. Water-cooled vapour compression cooling system simulated in TRNSYS.

Table 1. Building envelop characteristics and design cooling load conditions.

Parameter	Comfort & Gains Unit	Amount	Parameter	Wall Layers & Windows Unit	Amount
Temperature set point	°C	26	U_{wall}	W/m^2 K	2.003
Relative Humidity set point	%	50	Wall thickness	m	0.268
Air change rate	vol./h	2	U_{roof}	W/m^2 K	0.535
Infiltration	vol./h	0.6	Roof thickness	m	0.052
Lighting	W/m^2	1.1	U value, window	W/m^2 K	1.1
Equipment	kW	340	G value, window	%/100	0.62
Occupants	no.	100	Roof solar absorptance	%/100	0.1
U_{floor}	W/m^2 K	0.336	$U_{skylights}$	W/m^2 K	5.69

3.1. General Approach

As shown in Figure 3, the solar-driven ACS consists of four water loops, i.e., solar collector loop, hot water (HW) loop, cooling water (CW) loop and chilled water (CHW) loop. Water passes through the collectors (Type 71) and is heated before being passed to a storage tank (Type 158). For smooth operation of the chiller, the hot water is stored in a thermally stratified storage tank. To control the outlet temperature of the collector, a feedback controller was used to adjust the fluid flow rate of a variable speed pump. The pump was controlled by a differential controller to operate only when there is at least an 8 °C temperature rise from collector; this saves electrical energy during the night and in times of low solar irradiance. An auxiliary gas boiler was installed in the hot water loop to increase the temperature of hot water to the absorption chiller (Type 107) set point if solar irradiance was insufficient to achieve that temperature. To produce the cooling effect as described in Section 2.2, this hot water acts as a heat source for the desorption of water and LiBr. The chilled water is then passed through an air handling unit (cooling coil 697) to fulfil the cooling load requirements of the building. A fan coil air handling unit was used to exchange the heat of chilled water with zone air. A cooling tower was used to reject the heat of condenser and absorber of the absorption chiller. A control strategy based on working hours was used to control pumps, fans of the air handling unit and cooling tower to save electricity. Among different available collector options, we selected an evacuated tube collector for this study because it is more efficient than flat plate collectors [26], and it can provide the required temperature to drive a single effect absorption chiller. In addition, the authors are not aware of a study that evaluates the performance of evacuated tube collectors with optimum annual and monthly tilt adjustments for the climate of Lahore, Pakistan. Figure 4 shows the TRNSYS model of the water-cooled VCC. Electrically driven water-cooled VCC produced chilled water as explained in Section 2.1, which was then passed to the air handling unit (AHU) and exchanged heat with air to meet the cooling load requirements of the building. A cooling tower is used to reject heat of the condenser. The characteristics of the main components are given in Tables 2 and 3. The mathematical descriptions of the main TRNSYS types (components) are described below.

Table 2. Evacuated tube collector and air handling unit specifications.

Parameter	Solar Collector Unit	Amount	Parameter	Air Handling Unit Unit	Amount
Collector area	m^2	2100	Capacity	CFM	87,272
Optical efficiency	-	0.73	Water flow rate	kg/h	119,764
a_1	$W/(m^2\ K)$	1.21	Air flow rate	kg/h	249,657
a_2	$W/(m^2\ K^2)$	0.0075			

Table 3. The absorption chiller specifications.

Parameter	ACS Chiller Temperature Unit	Amount	Parameter	ACS Chiller Flow Rates Unit	Amount
HW	°C	110	HW	kg/h	112,984
CHW	°C	6.67	CHW	kg/h	119,764
CW	°C	29.44	CW	kg/h	314,300

3.2. The Case Study Building

In terms of sector-wise energy consumption of Pakistan, the industrial sector consumes the most (37.7%) followed by the transport sector (32.2%) and households sector (22.2%) [2]. We selected a tyre manufacturing building hall located in Lahore, Pakistan as a case study. This is an actual existing building for which the building materials and dimensions are available: it is a single-floor building with total floor area of 3065.8 m^2 and total volume of 30,369.8 m^3. The building is exposed to the sun from all sides and has a glazing fraction (window to wall ratio) of 29.26% and 41.81% on the east and west sides, respectively. There are no windows on the south and north sides of the building, but 36 skylights (total area 86.96 m^2) in the building roof provide daylight. To calculate building hourly cooling load over a one-year period, the SketchUp model was imported into TRNBuild and connected with the in-built typical meteorological year (TMY2) of Lahore. To model the thermal behavior of this building, we used a multi-zone building model (Type 56) which considers the variability of internal gains due to lights, equipment and occupants. The set point temperature (26 °C) and relative humidity (50%) were considered as design conditions according to the comfortable range of temperature and relative humidity specified by ASHRAE (American Society of Heating, Refrigerating and Air-Conditioning Engineers) standards [27]. The building is an industrial building and its operating hours are from 09:00 to 18:00. The three-dimensional archetype of the selected building is shown in Figure 5 and thermal characteristics of the building envelop, gains and other design conditions considered for simulation are tabulated in Table 1.

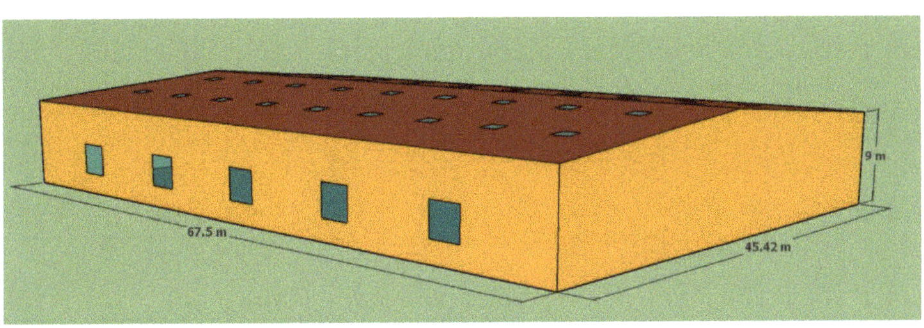

Figure 5. SketchUp model of the case study building.

3.3. Evacuated Tube Collector

TRNSYS Type 71 was used to model this collector using a standard quadratic efficiency curve. The design parameters used in this type are representative of the Viessmann Vitosol 200 T SPE solar collector [28]. At steady state, we can write the energy balance of the solar collector as

$$A_A G = Q_u + Q_{losses} \tag{1}$$

A_A represents collector aperture area, G solar irradiance, Q_u useful energy gain from collector. Q_{losses} indicates energy losses due to optical properties of collector absorber surface ($Q_{loss,opt}$) and by conduction, convection and radiation from collector surface as written in Equation (2).

$$Q_{losses} = Q_{loss,opt} + Q_{loss,convective} + Q_{loss,conductive} + Q_{loss,radiative} \tag{2}$$

The rate of useful energy gain and efficiency of solar collector can be expressed by Equations (3) and (4), respectively [29].

$$Q_u = mC_p(Tout - Tin) \tag{3}$$

where m and C_p represent the water mass flow rate and specific heat capacity of fluid, respectively. Tout and Tin represent the outlet and inlet temperatures of the collector.

$$\eta_{th} = a_0 - a_1\left(\frac{\Delta T}{G}\right) - a_2\frac{(\Delta T)^2}{G} \tag{4}$$

where a_0, a_1, a_2 represents optical collector efficiency and first and second order heat loss coefficients, respectively. The parameters are given in Table 2 and were selected according to ASHRAE standards [30] rated by Solar Rating & Certification Corporation (SRCC) from manufacturer catalogue. ΔT indicates difference of average collector outlet temperature and ambient temperature.

3.4. Hot Water Storage Tank

In solar-driven ACS, an insulated water tank is usually used to store hot water for smooth operation of the absorption chiller and to operate the chiller in times of less or no solar irradiance. A large size tank can be expensive and is a source of heat loss, so size of tank and its aspect ratio should be chosen carefully [10]. We modelled the hot water vertical storage tank using TRNSYS Type 158. The storage tank dissipates heat while interacting with the environment from the top, bottom and edges. The tank was divided into 10 iso-thermal temperature nodes for thermal stratification and the aspect ratio was chosen as 3.5 to minimize thermal losses as recommended by [31]. Heat can be transferred into and out of the storage tank depending on the temperature difference. Rate of change of temperature and heat loss from the top, edges and bottom of the storage for tank node j is expressed by Equations (5)–(8).

$$\frac{dT_{tan\,k,j}}{dt} = (Q_{in,tan\,k,j} - Q_{out,tan\,k,j})/C_{tan\,k,j} \tag{5}$$

where Q_{in}, Q_{out} represent heat transfer in and out of the tank and C_{tank} shows thermal capacitance of the tank.

$$Q_{loss,top,j} = A_{top}U_{top}\left(T_{tan\,k,j} - T_{env,top}\right) \tag{6}$$

$Q_{loss,top}$ shows amount of energy lost from top surface of tank, A is area, and U is heat transfer coefficient. T_{tank} is tank temperature and T_{env} is surrounding or environmental temperature.

$$Q_{loss,bottom,j} = A_{bottom}U_{bottom}\left(T_{tan\,k,j} - T_{env,bottom}\right) \tag{7}$$

$$Q_{loss,edge,j} = A_{edge}U_{edge}\left(T_{tan\,k,j} - T_{env,edge}\right) \tag{8}$$

3.5. Absorption Chiller

To model a hot water operated absorption chiller, Type 107 (available in standard TRNSYS library) was chosen, which required a data file. The data file contains the fraction of design energy and cooling capacity of the machine, which are functions of HW, CW, CHW and part load conditions. Rated or design capacity is defined as the capacity of the chiller at design or rated conditions. It is also called nameplate capacity. Actual capacity is capacity of chiller at current conditions of hot water, chilled water and cooling water. The main design parameters of this chiller are given in Table 3.

The fraction of design load (ratio of current load w.r.t. rated capacity) at which the machine is required to operate can be calculated using the following equations.

$$f_{\text{Design Load}} = \frac{\dot{Q}_{\text{removed}}}{\text{Cap}_{\text{Rated}}} \tag{9}$$

where $\dot{Q}_{\text{removed}}$ is the amount of energy removed from the CHW stream to achieve the desired set point temperature and $\text{Cap}_{\text{Rated}}$ is the rated capacity of the absorption chiller.

The actual capacity of the machine at any time step is calculated by using Equation (10).

$$\text{Capacity} = f_{\text{FullLoadCapacity}} * f_{\text{NominalCapacity}} * \text{Cap}_{\text{Rated}} \tag{10}$$

where, $f_{\text{FullLoadCapacity}}$ is ratio of load divided by actual capacity and $f_{\text{NominalCapacity}}$ capacity at current conditions to capacity at design conditions.

The model used the following relations to find the outlet temperature of the hot water ($T_{\text{hw,out}}$), chilled water ($T_{\text{chw,out}}$) and cooled water ($T_{\text{cw,out}}$) streams,

$$T_{\text{hw,out}} = T_{\text{hw,in}} - \frac{\dot{Q}_{\text{hw}}}{\dot{m}_{\text{hw}} C_{\text{phw}}} \tag{11}$$

$$T_{\text{chw,out}} = T_{\text{chw,in}} - \frac{\dot{Q}_{\text{chw}}}{\dot{m}_{\text{hw}} C_{\text{pchw}}} \tag{12}$$

$$T_{\text{cw,out}} = T_{\text{cw,in}} + \frac{\dot{Q}_{\text{cw}}}{\dot{m}_{\text{cw}} C_{\text{pcw}}} \tag{13}$$

where $\dot{Q}_{\text{hw}}$, $\dot{Q}_{\text{cw}}$ and $\dot{Q}_{\text{chw}}$ are HW, CW and CHW energy streams, respectively. The parameters m and C_p give the mass flow rate and specific heat capacity of water in the loop indicated by the subscript. The amount of energy dissipated from the cooling water can be written by Equation (14),

$$\dot{Q}_{\text{cw}} = \dot{Q}_{\text{chw}} + \dot{Q}_{\text{hw}} + \dot{P}_{\text{el,aux}} \tag{14}$$

where $\dot{P}_{\text{el,aux}}$ accounts for the energy consumed by solution pumps and controllers.

The Coefficient of Performance (COP) of the chiller is defined as

$$\text{COP} = \frac{\dot{Q}_{\text{chw}}}{\dot{Q}_{\text{aux}} + \dot{Q}_{\text{hw}}} \tag{15}$$

Solar fraction (SF) is the fraction of load covered by solar energy. It can be expressed as

$$\text{SF} = \frac{Q_{\text{solar}}}{Q_{\text{solar}} + Q_{\text{aux}}} \tag{16}$$

Performance maps for a single-stage absorption chiller are shown in Figure 6a,b. Figure 6a shows the fraction of nominal capacity as a function of cooling water inlet temperature for different chilled water temperatures, which illustrate that cooling capacity increased by reducing the cooling water temperature and setting chilled water temperature to a higher value. Figure 6b illustrates part load performance of the machine at different cooling water temperatures and it shows that the machine requires more energy at higher part load ratios. For a fixed part load ratio, the machine efficiency can be improved by reducing cooling water temperature.

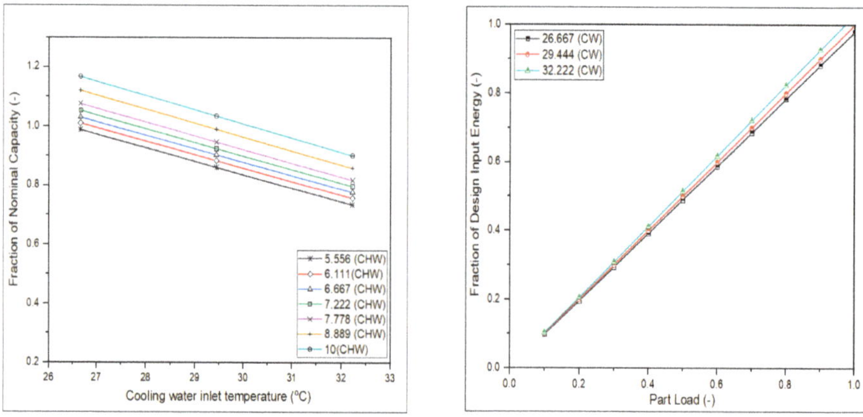

(**a**) (**b**)

Figure 6. (**a**) Fraction of nominal capacity with varying cooling water inlet temperature at different chilled water temperatures for absorption cooling system (ACS); (**b**) fraction of design input energy at different part load ratio for ACS.

3.6. Vapour Compression Chiller

To compare the conventional cooling system with the solar-driven vapour absorption cycle, a water-cooled VCC chiller was modelled by TRNSYS Type 666.

The chiller load and part load ratio (PLR) for each time step are calculated as

$$\dot{Q}_{load} = \dot{m}_{chw} * C_p(T_{chw,in} - T_{chw,set}) \tag{17}$$

$$PLR = \frac{\dot{Q}_{load}}{Capacity} \tag{18}$$

3.7. Cooling Tower

A cooling tower which is modelled with TRNSYS Type 510 is used to remove heat from the condenser and absorber of the ACS. A variable flow rate fan was used to achieve the desired fluid temperature with control signals. The performance map (see Figure 6a,b) shows that cooling water temperature (CW) strongly influences ACS performance. Most of the time, this low-grade heat is dissipated to the environment but it could be used for some industrial processes as well [32]. The most commonly used heat rejection device is a cooling tower, which achieves cooling through the evaporation of water. One of the limitations of an air cooled cooling tower is that the lowest temperature that can be achieved is the wet bulb temperature of the air [10].

3.8. Pumps

Pumps are used to circulate fluid in the different loops, i.e., collector, HW, CW and CHW loops. These pumps consume energy to perform mechanical work and electricity is the most common source to operate these pumps.

The pump shaft power (P_{shaft}) and amount of energy transferred to fluid stream (Q_{fluid}) is calculated by Equations (19) and (20), respectively.

$$P_{shaft} = P * \eta_m \tag{19}$$

where P is power drawn by the pump in the current time step of the simulation and η_m is the motor efficiency.

$$Q_{fluid} = P_{shaft}\left(1 - \eta_p\right) + \left(P - P_{shaft}\right) f_{motorloss} \tag{20}$$

where η_p represents pump efficiency and $f_{motorloss}$ shows fraction of pump motor efficiencies that contribute to temperature rise in the fluid stream passing through the pump.

3.9. Pipes

TRNSYS Type 31 was used to model the thermal behavior of fluid flow in pipes. The equal friction method was used to determine the size of pipes in each loop based on flow rate.

3.10. System Performance Indices

The performance of the chillers was evaluated based on energetic, economic and environmental (3E) performance metrics. Total primary energy consumption, levelized annual capital investment taking into account both initial investment and running cost, and carbon footprint indices were used to compare both systems. Annualized cost of both system were calculated considering cost function tabulated in Table 4 and in order to carry out energetic assessment, total primary energy consumption of the system was calculated. For the operation of solar-driven ACS, it utilizes electricity (to drive different pumps and fans) and natural gas (used in an auxiliary boiler to raise the temperature of hot water in the case of less solar irradiance), but water-cooled VCC is electrically driven and only utilizes electricity (to drive the compressor, different pumps and fans). Therefore, the primary energy consumption for a solar-driven ACS system (PEC_{ACS}) and water-cooled VCC system (PEC_{VCC}) can be expressed by Equations (21) and (22), respectively.

$$PEC_{ACS} = \left(PEF_E \times E_E + PEF_{NG} \times E_{NG}\right) \tag{21}$$

$$PEC_{VCC} = \left(PEF_E \times E_E\right) \tag{22}$$

where E_E and E_{NG} are total annual consumption of electricity and natural gas, respectively. PEF_E and PEF_{NG} represents primary energy factor for electricity and natural gas, respectively. The values of these factor are given in Table 5. The primary energy saving (PES) in the case of using solar-driven ACS can be expressed as:

$$PES = \left(PEC_{VCC} - PEC_{ACS}\right) \tag{23}$$

Table 4. Capital cost functions of main components obtained from market analysis (Pakistani Rupees) [33].

Parameter	Cost Function	Parameter	Cost Function
ETC collector	25,000 per m^2	Hot water storage tank	40,000 per m^3
Auxiliary boiler	110,000 per kW	Single-effect absorption chiller	31,250 per kW
Vapour compression chiller	25,568 per kW	Cooling tower	2840 per kW
Pump	6000 per kW		

To calculate annualized capital cost of both systems, the cost functions tabulated in Table 4 were considered. Total capital investment is the sum of cost of individual components and installation cost. Levelized capital cost (CI_L) can be calculated as

$$CI_L = \left(\sum Z_k + C_{INSTL}\right) * CRF \tag{24}$$

where $\sum Z_k$ represents sum of capital cost of all components and C_{INSTL} shows installation cost.

To take into account the project lifetime (n) and interest rate (i), we considered capital recovery factor (CRF), which can be expressed as:

$$CRF = \frac{i(1+i)^n}{(1+i)^n - 1} \tag{25}$$

The running cost in the case of the conventional VCC is given by the electricity required to run pumps and the compressor, while for the thermal driven ACS, electricity or a combination of electricity and natural gas is used. Electricity is used to drive pumps and fans while natural gas or electricity (electric resistance heater) is used to provide auxiliary energy to the hot water loop to achieve the required driving temperature for the absorption chiller in times of insufficient solar irradiance.

Annual cost of electricity and natural gas consumption can be calculated by using Equations (26) and (27).

$$FC_E = E_E \times c_E \tag{26}$$

$$FC_{NG} = E_{NG} \times c_{NG} \tag{27}$$

where FC_E and FC_{NG} are annual cost of electricity and natural gas consumption, while E_E, E_{NG}, c_E and c_{NG} are total electricity consumed, total natural gas consumed, unit cost of electricity and unit cost of natural gas, respectively. The total annual CO_2 emissions (CDE) are the sum of emissions produced from electricity (CDE_E) and emissions from burning natural gas (CDE_{NG}) in the auxiliary boiler. These are given by

$$CDE = CDE_E + CDE_{NG} \tag{28}$$

$$CDE_E = E_E * EF_{CO_2,E} \tag{29}$$

$$CDE_{NG} = E_{NG} * EF_{CO_2,NG} \tag{30}$$

where E_E and E_{NG} are total electricity and natural gas consumed. $EF_{CO_2,E}$ and $EF_{CO_2,NG}$ are CO_2 emission factors for grid electricity and natural gas, which are listed in Table 5.

3.11. Model Validation

TRNSYS, which was developed by the University of Wisconsin, validates each type (component) before providing it to the TRNSYS users [34], which indicates that all components available in TRNSYS are themselves validated. However, it is still important to check that the types and their parameters are correctly implemented in the model. Typically, building energy modelling tools are benchmarked using a standard called ASHRAE 140 [35] and the building modelling tool used for this study (TRNSYS) has been through that test protocol a number of times.

To validate the building model, we compared the peak calculated cooling load (712 kW) with installed cooling equipment capacity (774.4 kW) which shows the correct implementation of the building model. The installed cooling capacity is slightly higher (a general practice) to incorporate future needs (more installation of machinery inside the building, which will result in higher internal gains) and global warming issues (increase in outdoor temperature). In this study, the VIESMANN Vitsol 200 T collectors (SRCC certified) were selected for the ACS simulation model. Solar thermal collectors are tested according to Solar Rating and Certificate Corporation (SRCC) standards. There are different categories of tests like high temperature resistance (to access that a collector can withstand stagnation under high irradiance level), thermal performance test (to determine how energy can be gained from a collector at different temperatures and irradiance), exposure and rain test, internal and external thermal shock test, and mechanical load and impact resistance test. Thermal efficiency of the collector at different temperatures (Figure 7a) and optical efficiency, and first and second order heat loss coefficients of this collector (tabulated in Table 2) were extracted from the manufacturer's catalogue [28]. Then, to verify the correct implementation of this collector in TRNSYS, it was tested against the SRCC thermal performance test under similar conditions. Results indicate that a maximum discrepancy of less than 2% between the simulated and reported efficiency was found, which indicates acceptable accuracy of the collector's model in TRNSYS as well as correct implementation in the model. The absorption chiller was modelled by the characteristic equation method [36,37] which utilizes two correlations to determine cooling capacity and design heat input. This cooling capacity and fraction of design energy input further depend on cooling water temperature, chilled water temperature, hot water temperature and part load performance (see Figure 6). In this study, data provided by the manufacturer [38] was compared with the performance data file of TRNSYS Type 107 and discrepancy was found to be less than 5%, which shows acceptable accuracy of the simulation models.

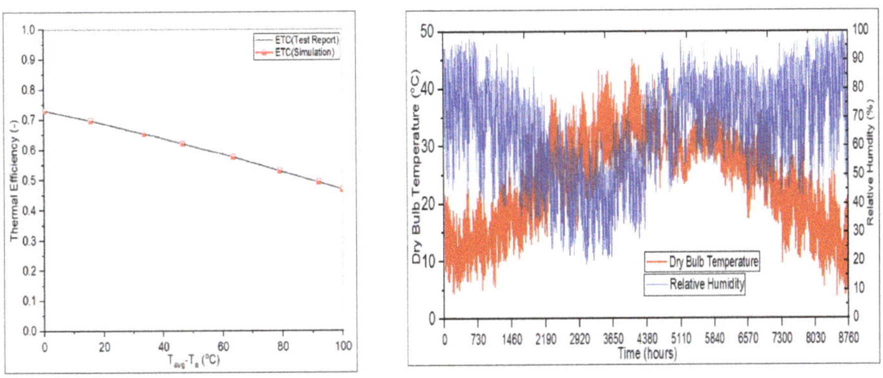

(a) (b)

Figure 7. (**a**) Comparison of collector efficiency curve (simulation vs. experimental data reported by the collector's manufacturer at G = 1000 W/m^2); (**b**) daily temperature and humidity variation for Lahore, Pakistan.

Table 5. Parameters considered for energetic, economic and environmental (3E) analyses in this study.

Parameter	Unit	Amount	Parameter	Unit	Amount
Interest rate or discount rate (i) [39]	%	13.25	Installation cost of ABS	% of CI	150
Average electricity price [40]	Rs./kWh	20	Installation cost of VCS	% of CI	130
Average natural gas price [41]	Rs./m^3/h	15	CO_2 emission factor for electricity [42]	kg per MWh	566
Lifetime of solar plant	Years	23	CO_2 emission factor for natural gas [42]	kg per MWh	202
Primary energy factor for natural gas [43]	kWh_{PE}/kWh_{NG}	1.22	Primary energy factor for electricity [43]	kWh_{PE}/kWh_E	3.05

4. Results and Discussion

This section presents the TRNSYS simulation results. First, the weather data, monthly cooling load and breakdown analysis of cooling load results are discussed. Then water-cooled compression chiller and solar-driven absorption chiller performance results are presented, and in the last section parametric results are presented.

4.1. Weather Data

Figure 7b shows daily variation in temperature and relative humidity for the case study building located in Lahore, Pakistan, which has a warm and humid climate. Ambient temperature of Lahore varies from 3 °C in winter to 45 °C in peak summer season and relative humidity remains around 50% at the start of summer and reaches up to 70%–80% in the rainy season (July–September). The summer season of Lahore lasts from March to the mid of November. The natural atmospheric temperature for climate of Lahore remains above the ASHRAE defined comfort zone for most of the year. Therefore, some type of HVAC equipment is required to displace room heat gains to maintain comfortable indoor temperatures.

4.2. Cooling Load

The first step in sizing any HVAC system is to determine the cooling load, which is the amount of energy required to offset heat gains so that the room temperature stays within the comfort zone. The designed conditions, such as temperature and relative humidity, used for calculation of cooling load are listed in Table 1. Building zone air receives heat from internal and external gains. External heat gains are given by the energy that enters the building through its envelop, such as walls, windows, doors, roof and floor. Internal gains are due to thermal energy released by lights, equipment and occupants. Gains that increase the temperature of air are called sensible gains, while gains that add moisture to the space are named latent heat gains.

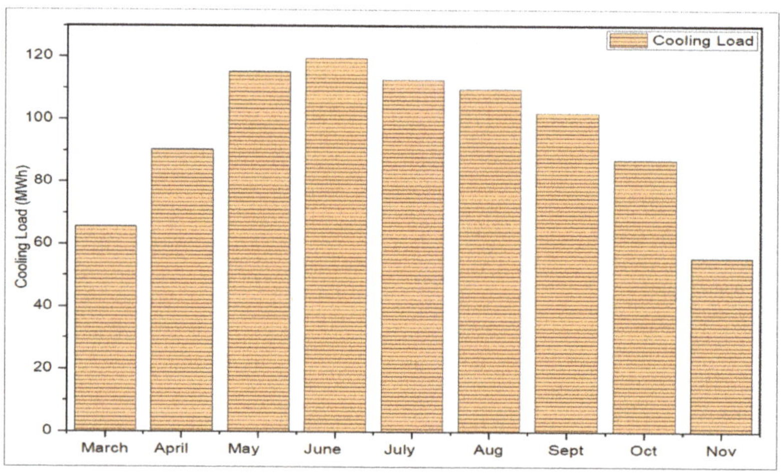

Figure 8. Monthly building cooling load.

In the developed model, both sensible and latent heat gains and variability of these heat gains at different times of the day were considered. The share of internal and external gains in the total cooling load was calculated as 47% and 53%, respectively. The share of high internal gains is due to equipment load (machinery) and number of occupants working inside the building. The room cooling load is the amount of energy required to displace all internal and external heat gains to achieve design set point temperature and relative humidity. The largest heat gains are from building envelop (walls, roof and floor) followed by heat released by building's equipment. The other major gain is heat from infiltration, i.e., influx of outside air through windows or door usage. The amount of heat gains from people and windows through conduction were calculated as 2% and 6%, respectively. The lowest heat gains (<1%)

are from lights because the selected building operates only during the daytime (9 a.m. to 6 p.m.) and to utilize solar irradiance for day lighting, sky lights of 87 m^2 were used. It is noted from the simulation results that maximal amount of solar irradiance was absorbed by the building roof and large number of sky lights (which on the one side provide daylight but on the other hand this irradiance becomes a heat load). It was also noted that solar irradiance from the east and west walls were higher compared to the north and south side of the building. Therefore, some retrofitting techniques (insulation) can be applied to the roof and east and west walls to reduce the external heat gains to the building, thereby reducing overall cooling load. The bar chart (Figure 8) shows the monthly cooling load variation over the summer season. In the months of March and November, outside air temperature lies in the range of 20 °C to 30 °C so there is less need of cooling in these months. The maximum cooling load demand is in the month of May, June and July when ambient temperature normally goes above 40 °C.

4.3. Water-Cooled Vapour Compression Chiller Results

This section presents the performance of the VCC chiller to fulfil the peak cooling load of 712 kW. The TRNSYS model of the water-cooled VCC chiller is shown in Figure 4.

The simulation of the water-cooled VCC model was carried out over a summer season with a time step of 1 h. Since the peak cooling load requirements are in June, Figure 9a illustrates the variation of temperature (ambient and zone) and solar flux over two sunny and one cloudy day in June. Looking at ambient temperature and solar flux of peak hours of the sunny days shows that ambient temperature goes above 40 °C with a corresponding solar flux of around 1000 W/m^2. The operating hours for the building were only from 9 a.m. to 6 p.m. and the thermostat was set at 26 °C. The building zone temperature (Tair_Hall) is maintained by water-cooled VCC at the required design set point temperature by offsetting all external and internal gains. The monthly energy consumption of the compressor to fulfil the cooling load and the monthly heat rejection from the condenser are shown in Figure 9b. It is interesting to note that compressor monthly energy consumption in the months of July and August is higher compared to June. This is due to the effect that the VCC system performs poorly in a humid environment. It can be seen from Figure 7b that July and August are humid months in Lahore, Pakistan, which results in high latent load and lower performance of the VCC system.

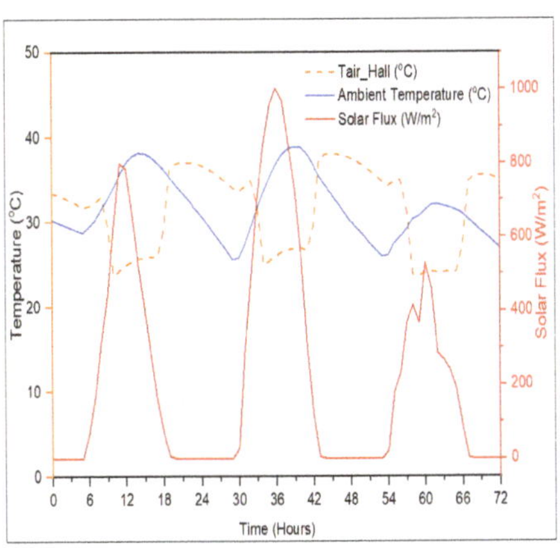

(a)

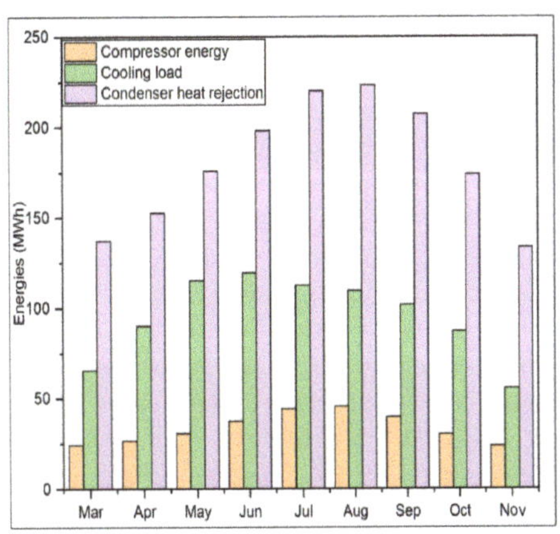

(b)

Figure 9. (**a**) Simulation results of water-cooled vapour compression cycle (VCC) chiller for three days (3rd to 5th June of the typical metrological year). (**b**) Water-cooled VCC chiller monthly results.

4.4. Solar-Driven Vapour Absorption Chiller Results

This section presents the performance (temperature and energies of different loops) of hot water solar operated absorption chillers. Figure 10a shows the temperature and energies of hot water, cooling water and chilled water streams. The chiller starts working according to the defined scheduled (9 a.m.) and operates continuously for 9 h. During the chiller's operation, solar collectors are used to harness solar energy in the form of hot water.

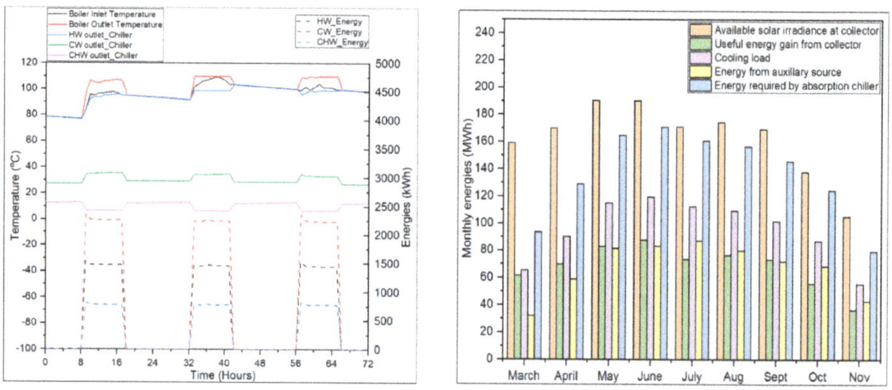

(**a**) (**b**)

Figure 10. (**a**) Chiller outlet temperature and energies of hot water, cooling water and chilled water loop for three days from 03.06. to 05.06. (**b**) evacuated tube collector (ETC) + absorption chiller results at collector area (A_c) = 1000 m^2.

This HW is stored in a thermally stratified vertical configuration tank. The pump of the hot water loop circulates this HW from the tank to the boiler. To raise the temperature of hot water in the case of insufficient solar irradiance, an auxiliary boiler is installed in series. One benefit of solar-driven chillers is the coincidence of the higher solar flux with the peak cooling load. In this model, chilled water outlet temperature from chiller was set to 6.7 °C during operating hours as evident from Figure 10a, and the cooling water outlet temperature rose to 34.9 °C. The cooling tower in this model was designed for 5.5 °C temperature difference. The right *y*-axis of Figure 10a shows the amount of energy required in different fluid streams. The amount of energy rejected by the cooling tower is the sum of energy entered to the generator and chilled water energy. Cooling water dissipates heat from the condenser and absorber of the ACS. Monthly yield of solar assisted absorption chiller is shown in Figure 10b, which represents monthly energy data, i.e., energy available at solar collector, useful energy gain from collector, monthly auxiliary energy requirement, monthly cooling load and amount of monthly energy absorption chiller needed to produce the required cooling effect. The efficiency of the solar collector can be calculated by dividing the useful energy gain from the collector by the total solar irradiance available at the collector according to Equation (4). In this simulation, we got the maximum monthly collector efficiency in the month of May and June, which is in line with maximum cooling load demand. As the solar irradiance and ambient temperature varies throughout the year, so does the collector efficiency as it depends on solar irradiance and difference of average fluid and ambient temperature. To produce the cooling effect and to offset heat gains of the building zone to maintain design temperature, the absorption chiller needed heat energy that comes from two sources, i.e., energy from solar collector and energy from auxiliary source. The COP of the absorption chiller can be calculated by dividing the provided cooling load by the total amount of energy the absorption chiller needs to produce this cooling effect according to Equation (15). We found a maximal COP of 0.68, which agrees with the manufacturer data [38]. It is interesting to note that peak cooling load requirements in the months of May, June and July are in line with maximal solar irradiance available at the solar collector, thereby reducing the energy consumption requirement from the auxiliary source.

4.5. Energetic, Economic and Environmental (3E) Results

This section discusses comparative performance of a water-cooled vapour compression system with a solar-driven vapour absorption cooling system on energetic, economic and environmental indices as tabulated in Table 6. Energetic analysis compares the annual energy consumed by water-cooled VCC and solar-driven ACS to meet the cooling demand. Economic analysis takes into account both the capital investment and running costs. Total energy CO_2 emissions factor for electricity and natural gas were considered to carry out the environmental analysis of both systems. Primary energy consumption of both systems was calculated using the TRNSYS model using Equations (20) and (21), and results show that the solar-driven cooling system utilized 5.74×10^5 kWh of annual primary energy. On the other hand, annual primary energy consumption of water-cooled VCC systems was calculated as 2.23×10^6, which is 3.88 times higher than the solar-driven cooling system. The calculation of primary energy saving shows that 1.65×10^6 kWh of primary energy can be saved annually in the case of using the solar-driven ACS system to meet the cooling demand.

Table 6. 3E results of VCC and ACS system.

Sr.No	Parameter	VCC System	ACS System
1	Annual primary energy consumption (kWh)	2.23×10^6	5.74×10^5
2	Annualized capital investment cost (Rs.)	1.07×10^7	3.31×10^7
3	Monthly running cost (Rs.)	1,612,186	393,269
4	Annual CO_2 emissions (tons)	346.78	108.58

The calculation of annualized capital investment results shows that the solar-driven ACS cooling system is 3.1 times more expensive than the water-cooled VCC system based on the current discount rate of Pakistan in 2020. It was found that the annualized capital investment cost of solar-driven ACS system is 3.31×10^7 Pakistani rupees (with collector area of 2100 m^2), while cost of the water-cooled VCC system was 1.07×10^7 Pakistani rupees. For the calculation of annualized cost, only discount rate was considered for initial investment cost of components. Fuel escalation rate of electricity and natural gas was not taken into account while calculating annualized costs of both systems. The high cost of the solar-driven ACS system is due to high discount rates. Further sensitivity analysis was performed by considering a subsidized discount factor of 10% and 8% for solar-driven ACS systems only. Results show that with 10% and 8% subsidized discount factors, the solar-driven cooling system is 2.4 and two times more expensive than a conventional VCC system for the climate of Pakistan. To calculate running cost of both systems, total annual energy consumption was calculated. It was calculated that monthly bills in the case of solar-driven ACS (with collector area 2100 m^2) will be 393,269 Pakistan rupees, while for water-cooled VCC, this will cost 1,612,186 Pakistani rupees, which is 4.1-times higher than for the solar system.

Finally, the environmental impact of both systems was studied and the results show that annual CO_2 emissions in the case of using solar-driven ACS (with collector area of 2100 m^2) for the selected building will be 3.19-times lower than water-cooled VCC. The total emissions were calculated using Equation (28) considering emission factors for electricity and natural gas and annual CO_2 emissions from solar-driven ACS and water-cooled VCC were calculated as 108.58 and 346.78 tons, respectively.

4.6. Parametric Study

In this section, a detailed parametric study is performed to investigate the effect of the critical parameters like solar field size, storage tank volume, flow rate in solar collector loop, solar tilt and azimuthal angle with annual and monthly optimized adjustment on the performance of the solar-driven ACS.

Figure 11a shows the relation of annual useful energy gain and annual auxiliary energy requirement to operate the absorption chiller by varying the collector area from 500 m^2 to 3000 m^2. This shows that annual useful energy gain from the solar field increases by increasing the solar field size and the auxiliary energy requirement decreases. By increasing the solar field, the solar fraction increases almost linearly up to an optimum collector area which depends on the cooling load. After this optimum point, solar fraction does not increase linearly, and it is not beneficial to increase the collector area further. It is evident from Figure 11a that solar fraction is not increasing significantly after the 2100 m^2 solar collector area. By increasing the collector area from 2100 m^2 to 3000 m^2, solar fraction increased only by 8%. So, rather than increasing the solar fraction to higher values, it is better to use an auxiliary heat source to achieve absorption chiller driving temperatures. The effect of variation of storage tank size on solar fraction was also investigated and it was found that storage tank volume is less sensitive to solar fraction. The solar fraction does increase with an increase in solar field area, but it does not show any significant increment with the storage tank volume because the system operates only during the day and no energy is needed for nighttime operation. The parametric study results indicate that a storage tank volume of 5 m^3 for the current cooling load is sufficient as solar fraction has no significant improvement by increasing the tank volume beyond this size. Actually, a relatively small storage tank is sufficient because we are operating the chiller only during the daytime, and in fact, a larger storage tank will have higher heat loss coefficients.

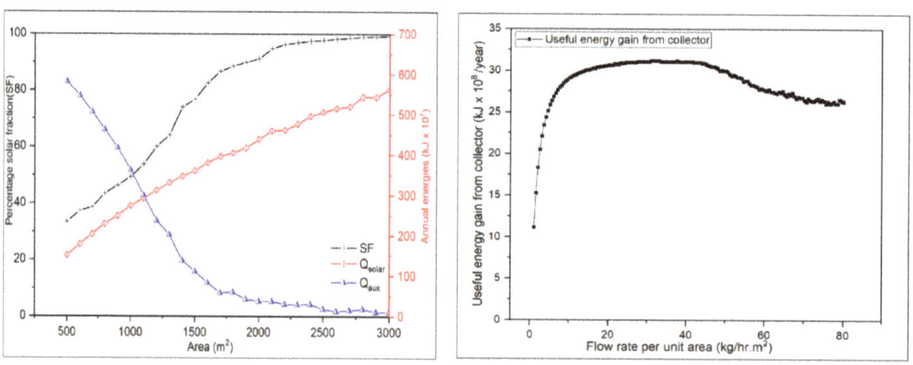

(**a**) (**b**)

Figure 11. (**a**) Variation of solar fraction, solar and auxiliary energy by varying collector area; (**b**) the effect of mass flow rate on energy gain from collector (Ac = 1000 m^2).

Another critical parameter for the performance of absorption chillers is the flow rate in the solar field. It is important to select a flow rate which on the one side consumes low electrical energy by the pump, and on the other hand maximizes the useful energy gain from the collector. Figure 11b illustrates the useful energy gain from the collector with respect to varying flow rates. The graph shows that increasing the flow rate in the solar loop increases the useful energy gain from the collectors. However, for flow rates greater than 50 kg/h m^2, useful energy gain from the collectors starts decreasing. For larger flow rates we need a larger pump which will consume more energy. We found that collector flow rate in the range of 16–50 kg/h m^2 is good for overall collector performance and these results are in good agreement with the literature [44].

Figure 11a shows that solar fraction increases with an increase in solar collector area. Figure 12a shows the effect of the solar collector area on annualized cost of ACS, and Figure 12b shows annualized running cost and associated emissions at different solar field sizes. It was found that at a lower solar fraction, more energy comes from the auxiliary source, which means large size of boiler and high consumption of fuel. The more consumption of fuel results in high running costs and more emissions. The selection of a suitable collector size and auxiliary boiler affects the annualized cost. For example, at a collector area of 500 m^2 when solar fraction is only 33%, the annualized cost of the system is 45.10 × 10^6 Pakistani rupees due to the large size of auxiliary boiler, and annualized running cost and associated emissions are 8.42 × 10^6 rupees and 575.4 tons of CO_2 emissions, respectively. By increasing the solar collector area, solar fraction increases, which results in low auxiliary energy requirements (small size of boiler), lower annualized costs and associated emissions. At a collector area of 1700 to 2100 m^2, the annualized cost of the system changes slightly because at this point the high cost of the solar field is compensated by a reduction in the capital cost of the auxiliary boiler (small size boiler is needed). However, selecting the collector area from 2100 to 3000 m^2 results only in an 8% increase in solar fraction which shows that, though the size of boiler reduces marginally, the high cost of solar field results in an increase in annualized cost. Similarly, at a collector area of 2100 m^2, annualized running cost is 3.53 × 10^6 rupees, which is 2.4 times less than for a collector area of 500 m^2 and produces a similarly large reduction in CO_2 emissions compared to a system with a collector area of 500 m^2. Further, an increase in collector area does not significantly reduce the annualized running cost and emissions. Therefore, 2100 m^2 is a suitable collector area in terms of reduction in annualized costs and CO_2 emissions.

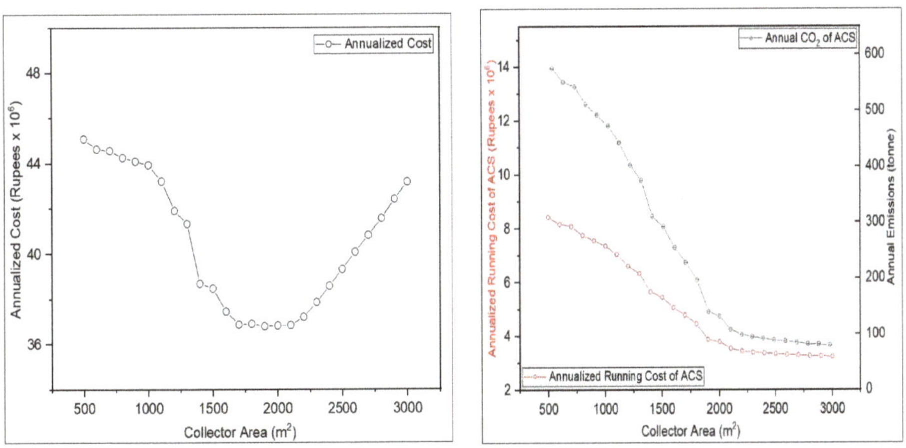

(**a**) (**b**)

Figure 12. (**a**) Variation of annualized cost of ACS by varying collector area; (**b**) variation of annualized running cost and annual CO_2 emissions of ACS by varying collector area.

To maximize the amount of solar irradiance on the collector surface, the tilt angle of the collector should be set carefully. Parametric results for the investigation of optimum collector tilt angles are shown in Figure 13a. The right axis of Figure 13a shows the variation of annual solar irradiance available at the collector surface by varying the tilt angle. The maximum annual solar irradiance is reached by setting the collector tilt angle to 31.5°. Setting the tilt angle greater than this reduced annual solar irradiance at the collector surface. Similarly, the variation of monthly solar irradiance as a function of tilt angle is shown for a few months on the left axis in Figure 13a. Parametric results for the monthly optimized angles are shown in Figure 13b. Results shows that monthly tilt angle varies from 61° to −7° towards the south (negative sign means towards the north) throughout the year. Maximum solar irradiance is received at the collector in the winter season by setting the collector at higher tilt angles and at lower angles for the summer season. For peak summer season (May, June and July) when the sun rises in the northeast and sets in the northwest, the optimized tilt angle should be set as negative (towards the north). For the months of March and October, the optimized angle is almost equal to the latitude of the location. The performance of the collectors is also affected by orientation (azimuthal angle) of the collector throughout the year as this changes the amount of available solar energy at the collector surface. The variation of annual solar irradiance by changing collector position towards east (0 to 90 degree) or towards west (0 to −90 degree) is shown in Figure 14a. Parametric results for the optimum annual azimuthal position show that facing the equator (0 degrees) is the best collector orientation. Monthly variation of solar irradiance for a few months of the summer and winter seasons as a function of collector azimuthal angle is shown in Figure 14a as well. It is interesting to note that in the summer season, variation of monthly irradiance at collector surface by varying collector orientation is not significant, but this is more prominent in winter months.

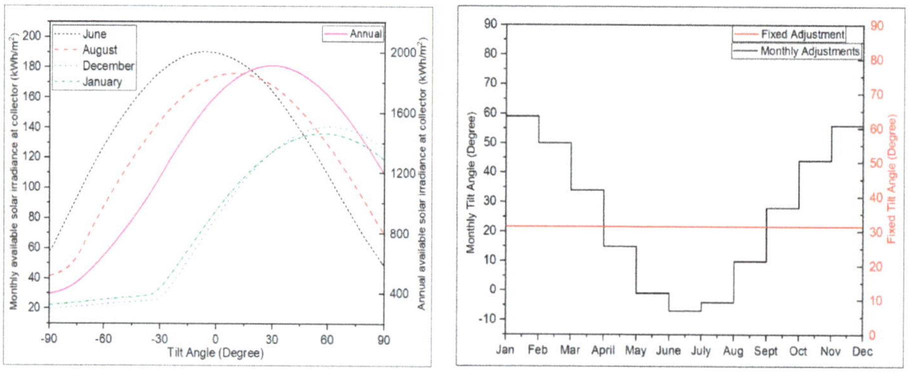

(**a**) (**b**)

Figure 13. (**a**) Variation of solar irradiance as a function of collector tilt angle; (**b**) monthly optimized collector tilt angles.

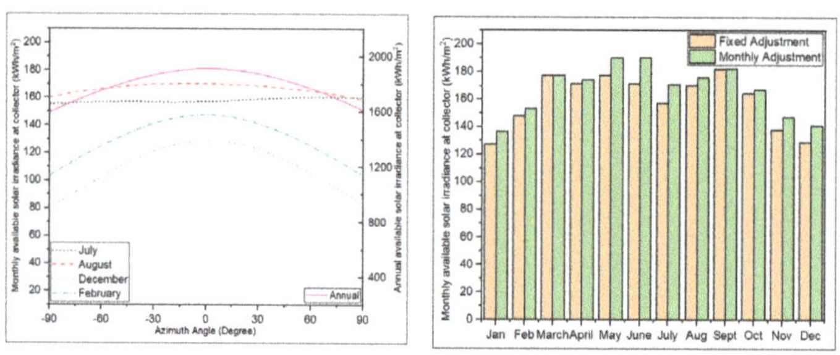

(**a**) (**b**)

Figure 14. (**a**) Variation of solar irradiance as a function of collector orientation; (**b**) solar irradiance available at collector surface with fixed and monthly adjustment.

The benefits of setting the collector angle at the optimized annual and optimized monthly angles are shown in Figure 14b. The bar chart shows the amount of monthly available solar irradiance at the collector surface by setting it at the annual optimized angle and monthly optimized angles. It can be seen from the bar chart that minimum difference of energy gain with both adjustments is achieved in the months of March and September (because optimized angles for these months are close to the annual optimized angle) and the maximum difference lies in May, June and July. Results indicate that around 5% more solar irradiance is available per m^2 of collector surface by adjusting the collector angles on a monthly basis compared to the annual optimized angle. By increasing the frequency of tilt adjustment of the collector on a fortnightly and daily basis improves the solar irradiance by 5.1% and 5.35%, respectively. The improvement of solar irradiance for fortnightly and daily adjustment is marginal compared with monthly adjustments. So, instead of using an expensive tracking system, manual adjustment of the collector at optimized monthly angle is a good alternative option for thermal collectors.

5. Conclusions

In this study, two different cooling systems, vapour compression-cycle based (water-cooled chiller) and thermal-driven based (solar-driven vapour absorption chiller) were studied. Electricity from a national grid was used to drive the compressor of the water-cooled vapour compression chiller while the evacuated tube collector and/or auxiliary gas boiler were used to supply hot water to drive the absorption chiller. A large industrial building was modelled in TRNSYS TRNBuild as a reference case to calculate hourly and peak cooling load for the design of both systems. Based on peak cooling load, both systems were modelled in TRNSYS and were compared energetically, economically and environmentally for the climate of Lahore, Pakistan. In addition, to perform a parametric study a parametric code was written in Python and coupled to the TRNSYS simulation model of the solar-driven absorption chiller. Main parameters affecting the performance of absorption chillers such as collector area, storage tank volume, flow rate in solar field, solar collector annual and monthly optimized tilt and azimuthal angles were investigated. The following conclusions can be drawn:

(1) At the current discount rate for Pakistan, the comparison of solar thermal and electric cooling technologies showed that the levelized capital cost of solar-absorption chillers is about three times higher than for conventional water-cooled chillers. The cost of the ACS system depends on selection of collector area and boiler for auxiliary energy requirements. On the other hand, the running cost of the solar-driven ACS is 4.1-times lower than for the water-cooled VCC. It was found that 3.19-times less CO_2 emissions will be produced annually by using a solar-driven vapour absorption chiller for the selected building instead of an electrically driven water-cooled vapour compression chiller. In the case of using a solar-driven vapour absorption chiller to fulfil cooling needs of the selected building archetype, 1.65×10^6 kWh of primary energy can be saved annually.

(2) The parametric study results indicate that solar fraction increases with increasing collector area, but the rate of increase of solar fraction decreases at higher collector areas. The rate of increase of solar fraction above an optimum collector area is not significant. There is no strong dependence between solar fraction and storage tank volume for the tested system. This is most likely due to the daytime only operation of the system so that a small storage tank is sufficient to balance solar resource and demand differences.

(3) Another critical parameter that affects the performance of solar collectors is flow rate in the solar collector loop. It is found that increasing the flow rate results in higher useful energy gain and higher collector efficiency. The parametric study shows that the mass flow per unit collector area for optimal collector performance lies in the range of 16–50 kg/h m^2. For higher flowrates, collector yield decreases and it also results in higher energy consumption of pumps.

(4) To maximize the amount of solar irradiance on the collector surface, collector tilt (slope) angle and azimuthal angle with annual optimum fixed adjustment and monthly adjustments were investigated. It was found that adjusting the collector tilt on a monthly basis results in 5% more solar irradiance per m^2 of collector surface with respect to annual optimum fixed adjustment. Further, increasing the collector tilt angle adjustment frequency on a fortnightly and daily basis results in 5.1% and 5.35% more solar irradiance per unit collector area compared to annual fixed adjustment, as increasing the frequency of slope adjustment from monthly to fortnightly or daily marginally increased the solar irradiance. It is beneficial to adjust collector slope on a monthly basis. Maximum solar irradiance is received at the collector surface in the winter season by setting the collector at higher tilt angles and at lower angles for the summer season.

Overall, in this paper, energetic, economic and environmental performance of two different cooling technologies, i.e., water-cooled vapour compression system and solar-driven vapour absorption system, were compared. Critical parameters affecting the performance of the solar-driven ACS system such as solar field size, flow rate in solar collector loop, storage tank volume, annual optimum fixed angle and monthly optimized angles for solar thermal collectors are presented, which will be useful for readers who work in solar-driven cooling systems.

Author Contributions: Conceptualization, S.M. and D.F.; methodology, S.M. and D.F.; software, S.M. and S.A.M.; validation, S.M. and S.A.M.; formal analysis, S.M. and D.F.; data curation, S.M.; writing—original draft preparation, S.M.; writing—review and editing, S.M., S.A.M., H.C. and D.F.; visualization, S.M.; supervision, H.C. and D.F.; project administration, D.F.; funding acquisition, S.M., S.A.M. and D.F. All authors have read and agreed to the published version of the manuscript.

Funding: This research received no external funding.

Acknowledgments: Sajid Mehmood acknowledges the support from Higher Education Commission (HEC), Pakistan in the form of PhD scholarship and University of Engineering and Technology, Lahore, Pakistan for study leave. Serguey A. Maximov acknowledges the support from National Agency for Research and Development (ANID), Chile in terms of PhD scholarship.

Conflicts of Interest: The authors declare no conflict of interest.

Nomenclature

A_A	aperture area (m^2)
a_0	optical collector efficiency
a_1	first order collector heat loss coefficient (W/m^2 K)
a_2	second order heat loss coefficient (W/m^2 K^2)
ACS	absorption cooling system
C_p	specific heat capacity of water (kJ/kg K)
chw	chilled water
c_E	unit cost of electricity (Rs./kWh)
cw	cooling water
Cap_{Rated}	rated capacity of chiller (kW)
CI_L	levelized capital investment (Rs.)
c_{NG}	unit cost of natural gas (Rs./m^3/h)
E_E	total annual consumption of electricity (kWh)
CDE_E	annual CO_2 emissions from electricity (kg)
E_{NG}	total annual consumption of natural gas (kWh)
CDE_{NG}	annual CO_2 emissions from natural gas (kg)
FC_E	annual cost of electricity (Rs.)
ETC	evacuated tube collector
G	solar irradiance (kJ/h m^2)
FC_{NG}	annual cost of natural gas (Rs.)
i	interest rate (%)
hw	hot water
n	project lifetime (years)
m	mass flow rate of water (kg/h)
P_{shaft}	shaft power (kJ/h)
Rs.	Pakistani Rupees
Q_u	useful energy gain from collector (kJ/h)
Q_{losses}	energy losses in collector (kJ/h)
$Q_{in,tank}$	heat transfer into the tank (kJ/h)
$Q_{in,tank}$	heat transfer out of tank (kJ/h)
$\dot{Q}_{chw}$	chilled water energy stream (kJ/h)
$\dot{Q}_{hw}$	hot water energy stream (kJ/h)

Q_{fluid}	rate of energy transferred to fluid stream (kJ/h)
Q_{aux}	rate of energy delivered by auxiliary source (kJ/h)
$Q'_{removed}$	energy removed from chilled water stream (kJ/h)
Q_{solar}	rate of energy delivered by solar (kJ/h)
SF	solar fraction
$T_{chw,in}$	chilled water inlet temperature (K)
$T_{chw,set}$	chilled water set point temperature (K)
U	heat transfer coefficient (W/m^2 K)
η_{th}	collector thermal efficiency
VCC	vapour compression cycle
η_m	motor efficiency
η_p	pump efficiency

References

1. Kibria, A.; Akhundjanov, S.B.; Oladi, R. Fossil fuel share in the energy mix and economic growth. *Int. Rev. Econ. Financ.* **2019**, *59*, 253–264. [CrossRef]
2. Baz, K.; Xu, D.; Ampofo, G.M.K.; Ali, I.; Khan, I.; Cheng, J.; Ali, H. Energy consumption and economic growth nexus: New evidence from Pakistan using asymmetric analysis. *Energy* **2019**, *189*, 116254. [CrossRef]
3. Shahid, M.; Ullah, K.; Imran, K.; Mahmood, I.; Mahmood, A. Electricity supply pathways based on renewable resources: A sustainable energy future for Pakistan. *J. Clean. Prod.* **2020**, *263*, 121511. [CrossRef]
4. Flores-Chamba, J.; López-Sánchez, M.; Ponce, P.; Guerrero-Riofrío, P.; Álvarez-García, J. Economic and spatial determinants of energy consumption in the European union. *Energies* **2019**, *12*, 4118. [CrossRef]
5. Akbar, U.; Popp, J.; Khan, H.; Khan, M.A.; Oláh, J. Energy Efficiency in Transportation along with the Belt and Road Countries. *Energies* **2020**, *13*, 2067. [CrossRef]
6. Talbi, B.; Nguyen, D.K. An empirical analysis of energy demand in Tunisia. *Econ. Bull.* **2014**, *34*, 452–458.
7. O'Connor, D.; Calautit, J.K.; Hughes, B.R. A novel design of a desiccant rotary wheel for passive ventilation applications. *Appl. Energy* **2016**, *179*, 99–109. [CrossRef]
8. Shirazi, A.; Taylor, R.A.; Morrison, G.L.; White, S.D. Solar-powered absorption chillers: A comprehensive and critical review. *Energy Convers. Manag.* **2018**, *171*, 59–81. [CrossRef]
9. She, X.; Cong, L.; Nie, B.; Leng, G.; Peng, H.; Chen, Y.; Zhang, X.; Wen, T.; Yang, H.; Luo, Y. Energy-efficient and -economic technologies for air conditioning with vapor compression refrigeration: A comprehensive review. *Appl. Energy* **2018**, *232*, 157–186. [CrossRef]
10. Henning, H.; Motta, M.; Mugnier, D. *Solar Cooling Handbook*, 3rd ed.; TEcSol, F., Ed.; AMBRA|V: Vienna, Austria, 2013; ISBN 9783990434383.
11. Henning, H.M. Solar assisted air conditioning of buildings—An overview. *Appl. Therm. Eng.* **2007**, *27*, 1734–1749. [CrossRef]
12. International Energy Agency (IEA). Technology Roadmap Solar Heating and Cooling. Available online: https://www.iea.org/publications/freepublications/%0Apublication/Solar_Heating_Cooling_Roadmap_2012_WEB.pdf (accessed on 21 July 2020).
13. Alternative Energy Development Board. Available online: http://www.aedb.org/ae-technologies/solar-power/solar-resources (accessed on 6 April 2019).
14. Chen, J.F.; Dai, Y.J.; Wang, R.Z. Experimental and analytical study on an air-cooled single effect LiBr-H2O absorption chiller driven by evacuated glass tube solar collector for cooling application in residential buildings. *Sol. Energy* **2017**, *151*, 110–118. [CrossRef]
15. Bi, Y.; Qin, L.; Guo, J.; Li, H.; Zang, G. Performance analysis of solar air conditioning system based on the independent-developed solar parabolic trough collector. *Energy* **2020**, *196*, 117075. [CrossRef]
16. Zheng, X.; Shi, R.; Wang, Y.; You, S.; Zhang, H.; Xia, J.; Wei, S. Mathematical modeling and performance analysis of an integrated solar heating and cooling system driven by parabolic trough collector and double-effect absorption chiller. *Energy Build.* **2019**, *202*, 109400. [CrossRef]
17. Al-Alili, A.; Hwang, Y.; Radermacher, R.; Kubo, I. A high efficiency solar air conditioner using concentrating photovoltaic/thermal collectors. *Appl. Energy* **2012**, *93*, 138–147. [CrossRef]

18. Buonomano, A.; Calise, F.; D'Accadia, M.D.; Ferruzzi, G.; Frascogna, S.; Palombo, A.; Russo, R.; Scarpellino, M. Experimental analysis and dynamic simulation of a novel high-temperature solar cooling system. *Energy Convers. Manag.* **2016**, *109*, 19–39. [CrossRef]
19. Xu, Z.Y.; Wang, R.Z. Comparison of CPC driven solar absorption cooling systems with single, double and variable effect absorption chillers. *Sol. Energy* **2017**, *158*, 511–519. [CrossRef]
20. Gomri, R. Simulation study on the performance of solar/natural gas absorption cooling chillers. *Energy Convers. Manag.* **2013**, *65*, 675–681. [CrossRef]
21. Khan, M.S.A.; Badar, A.W.; Talha, T.; Khan, M.W.; Butt, F.S. Configuration based modeling and performance analysis of single effect solar absorption cooling system in TRNSYS. *Energy Convers. Manag.* **2018**, *157*, 351–363. [CrossRef]
22. Al-Ugla, A.A.; El-Shaarawi, M.A.I.; Said, S.A.M.; Al-Qutub, A.M. Techno-economic analysis of solar-assisted air-conditioning systems for commercial buildings in Saudi Arabia. *Renew. Sustain. Energy Rev.* **2016**, *54*, 1301–1310. [CrossRef]
23. THOMSON REUTERS FOUNDATION NEWS. Available online: https://news.trust.org/item/20130625141050-sw1nq/ (accessed on 14 September 2019).
24. TRNSYS. Available online: http://www.trnsys.com/ (accessed on 9 February 2019).
25. SketchUp. Available online: https://www.sketchup.com/industries/architecture (accessed on 24 March 2019).
26. Herrando, M.; Pantaleo, A.M.; Wang, K.; Markides, C.N. Solar combined cooling, heating and power systems based on hybrid PVT, PV or solar-thermal collectors for building applications. *Renew. Energy* **2019**, *143*, 637–647. [CrossRef]
27. ASHRAE. Available online: https://ashrae.iwrapper.com/ViewOnline/Standard_90.2-2018 (accessed on 5 June 2019).
28. VIESSMANN. Available online: http://www.viessmann-us.com/content/dam/internet-ca/pdfs/solar/vitosol_200-t_spe_tdm.pdf (accessed on 15 June 2019).
29. Trnsys 18: A TRaNsient SYstem Simulation program. *Sol. Energy Lab. Univ. Wis. Madison Math. Ref.* **2018**, *4*, 705.
30. ASHRAE. *Solar Collectors and Photovoltaic in EnergyPRO*; ASHRAE: Peachtree Corners, GA, USA, 2013; pp. 1–23.
31. Han, Y.M.; Wang, R.Z.; Dai, Y.J. Thermal stratification within the water tank. *Renew. Sustain. Energy Rev.* **2009**, *13*, 1014–1026. [CrossRef]
32. Khankari, G.; Munda, J.; Karmakar, S. Power Generation from Condenser Waste Heat in Coal-fired Thermal Power Plant Using Kalina Cycle. *Energy Procedia* **2016**, *90*, 613–624. [CrossRef]
33. Munwir, A.; Damson Consulting Engineers, Lahore, Pakistan. Personal communication, 2020.
34. The University of WISCONSIN. Available online: https://sel.me.wisc.edu/trnsys/validation/index.html (accessed on 2 February 2020).
35. Energy Efficiency & Renewable Energy. Available online: https://www.energy.gov/eere/buildings/ashrae-standard-140-maintenance-and-development (accessed on 21 July 2020).
36. Puig-Arnavat, M.; López-Villada, J.; Bruno, J.C.; Coronas, A. Analysis and parameter identification for characteristic equations of single- and double-effect absorption chillers by means of multivariable regression. *Int. J. Refrig.* **2010**, *33*, 70–78. [CrossRef]
37. Kühn, A.; Ziegler, F. Operational results of a 10 kW absorption chiller and adaptation of the characteristic equation. *Proc. First Int. Conf. Sol. Air Cond.* **2005**, *10*, 5–9.
38. Broad X Non-Electric Chiller: Model Selection & Design Manual. Available online: https://www.broadusa.net/en/wp-content/uploads/2015/03/Broad-X-chiller-Model-selection-design-manual-C.pdf (accessed on 6 January 2020).
39. State Bank of Pakistan. Economic Data. Available online: http://www.sbp.org.pk/ecodata/index2.asp (accessed on 24 December 2019).
40. National Electric Power Regulatory Authority. LESCO. Available online: https://nepra.org.pk/tariff/DistributionLESCO.php (accessed on 24 December 2019).
41. Sui Northern Gas Pipelines Ltd. Consumer Gas Prices. Available online: https://www.sngpl.com.pk/ (accessed on 24 December 2019).
42. Yousuf, I.; Ghumman, A.R.; Hashmi, H.N.; Kamal, M.A. Carbon emissions from power sector in Pakistan and opportunities to mitigate those. *Renew. Sustain. Energy Rev.* **2014**, *34*, 71–77. [CrossRef]

43. Central Power Purchasing Agency. Available online: http://www.cppa.gov.pk/Home/DownloadDetails?Type=AnnualReports (accessed on 2 February 2020).
44. Ko, M.J. A novel design method for optimizing an indirect forced circulation solar water heating system based on life cycle cost using a genetic algorithm. *Energies* **2015**, *8*, 11592–11617. [CrossRef]

energies

MDPI

Article

Environmental Sustainability of the Vehicle Fleet Change in Public City Transport of Selected City in Central Europe

Vladimír Konečný [1], Jozef Gnap [1,*], Tomáš Settey [1], František Petro [1], Tomáš Skrúcaný [1] and Tomasz Figlus [2]

[1] Faculty of Operation and Economics of Transport and Communications, University of Žilina, SK-010-26 Žilina, Slovakia; vladimir.konecny@fpedas.uniza.sk (V.K.); tomas.settey@fpedas.uniza.sk (T.S.); frantisek.petro@fpedas.uniza.sk (F.P.); tomas.skrucany@fpedas.uniza.sk (T.S.)

[2] Faculty of Transport, Silesian University of Technology, 40-019 Katowice, Poland; tomasz.figlus@polsl.pl

* Correspondence: jozef.gnap@fpedas.uniza.sk

Received: 22 June 2020; Accepted: 23 July 2020; Published: 28 July 2020

Abstract: Diesel is the most used fuel for buses and other urban transport vehicles in European countries. This paper deals with impacts on emissions production from the operation of the urban public transport fleet after its renewal. To what extent can the renewal of the urban public transport fleet in the city of Žilina contribute to increasing environmental sustainability in the way of reducing air pollution? The vehicle fleet change has partially consisted of vehicle traction system transition-diesel buses were substituted by hybrid driven (HEV) and electric driven buses (BEV). How can the direct and indirect emissions from the operation of vehicles be calculated? These were the posed research questions. The research aimed to propose a methodology for the calculation of direct and indirect emissions. Indirect emissions values (WtT—Well-to-Tank) for different types of fuels and tractions were obtained based on regression functions. These WtT emission factors together with the existing TtW (Tank-to-Wheels) emission factors (direct emissions) can be used for the assessment of environmental impacts of specific types of vehicles concerning energy source, fuel, or powertrain and type of operation. Direct pollutants such as CO, NO_x and PM were calculated with the use of simulation methodology of HBEFA (Handbook of Emission Factors for Road Transport) software. The calculated CO_2 savings for the period 2019–2023 about fleet renewal in absolute terms are EUR 1.3 million tons compared to the operation of the original fleet while maintaining the same driving performance. The renewal of the vehicle fleet secured by vehicle traction transition can be a way to reduce the energy intensity and environmental impacts of public transport in Žilina.

Keywords: public transport; vehicle fleet change; vehicle traction transition; electric vehicles; direct and indirect emissions; environmental sustainability

---

1. Introduction and Literature Review

The Europe 2020 strategy for smart, sustainable and inclusive growth sets out five headline targets, which determine the position where Europe should be in 2020. One of these targets is related to climate and energy. Member States have committed to reduce their greenhouse gas emissions by 20% by 2020, to increase the share of renewable energy sources in the EU energy mix to 20% and to achieve the 20% energy efficiency improvement. The EU is currently on the right track to achieve two of the above targets, but it will not be able to meet the energy efficiency target without further action [1].

Air protection is one of the areas where the EU is highly active because of the need to ensure cleaner air. Air pollution can significantly harm human health and the environment. Annually, up to 400,000 premature deaths in the EU are caused by poor air quality [2]. Action is being taken at EU and national level as well as through active cooperation at international conventions level. The aim is to improve air quality by controlling emissions of pollutants into the atmosphere improving fuel quality and integrating environmental protection requirements into other sectors e.g., transport, energy [2]. The transition to a competitive low carbon economy means that the EU should be prepared to ensure that by 2050 it will reduce its internal emissions by 80% compared to 1990. The European Commission has carried out an extensive model analysis with various possible scenarios showing how these objectives could be met. This analysis of different scenarios shows that in terms of cost-effectiveness it would be optimal to achieve an internal reduction of emissions of 40% by 2030 compared to the level in 1990 and of 60% by 2040. The reduction of emissions are of 25%. In this way, there would be an annual reduction of approx. 1% in the first decade by 2020, compared to the 1990 level. In the second decade from 2020 to 2030, there would be a reduction of 1.5% and in the last two decades by 2050, there would be a reduction of 2%. It is envisaged that with the greater availability of more cost-effective technologies also in the transport sector, efforts will be intensified.

One of the objectives of the White Paper Roadmap to a Single European Transport Area: Creation of a competitive resource-efficient transport system to make a significant contribution to achieving the 60% greenhouse gas emission reduction target; halve the use of "conventionally fueled" cars in urban transport by 2030; phase them out in cities by 2050; achieve the introduction of essentially CO_2-free city logistics in major urban centers by 2030 [3].

Urban public transport with extensive fleets of city buses as well as taxis and lorries used in urban logistics are particularly well suited for the introduction of alternative propulsion systems and fuels. This could make a significant contribution to reducing the intensity of carbon oxides in urban transport and at the same prepare the conditions for testing new technologies and the opportunity for their timely introduction to the market.

The Slovak Republic, as an EU Member State, has also joined these targets in the area of reducing greenhouse gas emissions.

The objective described in the Public Transport Development Strategy is to increase the attractiveness of public passenger transport through the modernization and reconstruction of public transport infrastructure, including the provision of ecological and low-floor vehicle fleet [4,5]. The specific objective "Increasing the attractiveness and accessibility of public passenger transport through the renewal of public transport vehicles (urban public transport)" has been set. The deployment of low-floor and energy-efficient vehicles in urban public transport will not only increase the accessibility of urban public transport for disabled passengers, as well as passenger comfort and time savings, but it will also reduce energy consumption and the related costs. The condition for supporting the renewal of vehicles in urban public transport is the existence of a comprehensive strategic plan for sustainable development of transport in individual cities (transport master plan, plans of sustainable urban mobility) and implementation of measures to ensure the preference of urban public transport on the routes for which they will be designed and the building of Integrated Transport Systems [6–8].

Results to be achieved by this:

- increasing the attractiveness of public passenger transport,
- improving the quality of services provided by urban public transport in large agglomerations (travel time savings, expanding the range of services, increasing comfort and reliability, etc.) [9],
- increasing accessibility of urban public transport vehicles,
- reduction of negative impacts on the environment (reduction of noise, emissions of CO_2, NO_2 and PM_{10}, vibrations, etc.),
- reducing the morbidity of the population and increasing the standard of living of the population's life expectancy,
- increasing the share of public passenger transport in the division of transport work.

The reduction of the costs of operation of urban public transport vehicles and energy is also expected [10,11]. The other member states of the European Union have also adopted strategies to reduce emissions in passenger transport [12].

Air pollution from transport and its impacts can be monitored by measurement, modelled on the basis of suitable simulation models based on historical data and calculated on the basis of suitable emission calculators and emission factors.

Suna, S. et al. [13] deal with the analysis of past and future trends in the area of emissions from transport in a selected Chinese city for the period 2000 to 2030.

There are studies focusing on research into the reduction of emissions by the management of traffic flow and optimization of the operation of vehicles in order to limit the stopping of vehicles in the traffic flow. Such research in the field of bus transport (Arti Choudhary, Sharad Gokhale, 2019) confirmed a significant reduction of emissions [14].

Changes in passenger behavior can have a major impact on reducing energy consumption and the associated reduction of greenhouse gas emissions. Research into energy consumption trends and greenhouse gas emissions up to 2050 at the national level in China was published by Li, P. et al., 2018 [15].

Impacts of transport on the environment, especially on air pollution, are greatest in large agglomerations. They are connected with economic development, increasing incomes of the population and the associated increase in the degree of automobilization. Impacts on air pollution are often multiplied by insufficient road infrastructure and the associated rise of congestions. Traffic congestions, deteriorating air conditions, and a negative impact on the population are also becoming a problem for smaller cities such as Žilina, if city bypasses are not completed. Research on the development of the number of passenger cars and pollutant emissions in the conditions of Romania in the Lasi metropolitan area was conducted and published by Rosu Lucian, Istrate Marinela and Banica Alexandru, 2018 [16]. They carried out the research with a use of a questionnaire survey, statistical data and a simulation model. The results indicate that the metropolitan area is confronted with a significant expansion that leads to high levels of various emissions of air pollutants from the massive use of passenger cars in the peri-urban area. The number of premature deaths due to environmental pollution in cities can also be reduced by promoting public transport which can substitute some part of the driving passenger cars and decrease the emissions production [17]. Within the frame of local impacts on the environment, scientists investigate a mutual relationship or more precisely a dependence between emissions of individual pollutants and parameters of traffic flow such as its structure, age of vehicles in it, traffic flow intensity (Catalano, M. et al., 2016) [18], taking into account a maximum peak load and considering options for reducing pollution at local (street) level, e.g., in Turkey, Istanbul (Elbir, T. et al., 2010) [19].

The methods of reducing pollutant emissions also involve the operation of more environmentally acceptable types of fuels and renewal of the vehicle fleet with more modern vehicles. For example, Kuranc, A. et al., 2017 [20] deal with the issue of fleet renewal in the agricultural sector.

A specific area of production of emissions from transport and transport services are greenhouse gas emissions. Transport is one of the largest greenhouse gas (GHG) producers. The amount of emissions produced can be expressed as the equivalent of carbon dioxide (CO_2) emissions, which is the amount of CO_2 emissions that represents the same global warming potential as the actual greenhouse gas mix-carbon dioxide (CO_2), methane (CH4) and nitrous oxide (N_2O) (Stojanovic et al., 2012) [21].

Lizbetin et al. (2018) also address the issue of GHG emissions in the road haulage area [22]. GHG emissions influence the ozone layer and share to the greenhouse effect that causes global warming problems that are closely related to weather changes and extreme weather events. It needs to be pointed out discrepancies associated with FAME biofuels (Fatty Acid Methyl Esters) in particular the fact that, although their use produces nearly zero GHG emissions, their production is highly energy intensive. Article by Ivkovic et al. (2018) is concentrating on the production of GHG emissions in the field of long distance transport of persons, especially in road and air transport [23]. The aim of authors research was to develop and select a suitable method for modeling the estimation of GHG costs in the road and air transport sector in Serbia, as well as to apply a method aimed at special calculation by type of transport.

In addition to monitoring and measurement, the amount of emissions generated from transport can also be determined by the application of suitable energy and emission calculators, most of which are based on the use of emission factors of individual types of pollutants for specific groups of vehicles and the operational fuel consumption is used as input for the calculation. In particular, for the calculation of greenhouse gas emissions, emission factors and methodology according to European Standard EN 16258:2012 methodology for calculation and declaration of energy consumption and GHG emissions of transport services (freight and passengers) can be used. In this way, it is possible to compare the amount of emissions produced by the different modes of transport according to the approach (the well-to-wheels and tank-to-wheels principles), for example, Petro and Konečný (2017), Skrúcaný et al. [24,25]. Using the emission factors prescribed by the standard, for example, the comparisons of greenhouse gas emissions or energy consumption of several modes of transport are becoming more objective [26]. L'upták, V. et al., (2019) published a case study focused on a comparative analysis of environmental impact assessment (greenhouse gases according to EN 16258: 2012 standard) of two modes of transport-railway and bus transport in the field of passenger transport [27]. Hlatká, M. et al. (2018) also used that standard and published a study that compares the production of greenhouse gas emissions on a particular transport route in passenger transport with the use of bus transport and a combination of air transport with bus transport [28].

The problem of emission calculators is the fact that they consider emission factors for a period of several years without updating them whereby with the use of electrical energy there is a year-on-year change in the energy mix from its production and thus also a change in the amount of indirect emissions (well-to-tank). Petro et al. (2019) point to this problem and propose a structure of a dynamic calculator that would update the amount of emissions from electrical energy production on a year-on-year basis. Several scientific studies and articles refer to the use of EN 16258 standard, which was adopted in 2012 [29,30].

Keskisaari, V. et al. (2017) assessed the links between the urban structure and socio-economic and demographic variables in a published study, and also considered the lifestyle of the population, in relation to the production of greenhouse gas emissions from land transport in Helsinki, Finland. The aim of this study was to identify and improve our understanding of the latent ways of modality that guide the possibilities of daily travel of people and the resulting greenhouse gas production [31]. In their study in Finland, Ottelin, J. et al. (2014) found, inter alia, that in the metropolitan region there is a relation between ownership of cars and the use of air transport in the middle-income group of the population. The main political implication of their study was that air transport needs to be included in the assessment of greenhouse gas emissions (as confirmed by other studies) and strategies focused on the reduction of greenhouse gas emissions related to the transport behavior of the population [32].

Emissions from transport are also affected by the system of regular emission controls [33]. Milosavljevic, B. et al. (2015) [34] dealt with a dispersion of pollutants from transport in urban space and emission factors.

Blaž, J et al. (2019) [35] focused on the issue of the use of hybrid-drive buses in public passenger transport and Napoli, G. et al. (2017) [36] studied the development of a fuel cell hybrid electric powertrain. Lebkowski, A. (2019) analyzed various configurations of hybrid power systems, consisting only of batteries, combinations of batteries and supercapacitors, and only supercapacitors. For these configurations, mathematical models were developed. These models were used in the research on energy consumption and carbon dioxide emissions using a city bus with a length of 12 m [37]. These are procedures that can be used to refine the presented methodology for calculating emissions from the operation of hybrid buses (16 buses in urban public transport of Žilina) and to optimize their deployment according to the characteristics of public transport lines.

Kivekas, K. et al. (2018) [38] dealt with the issue of deployment of electric buses in urban public transport. The environmental effects of electromobility in urban public transport in Gdansk, Poland were addressed by Pietrzak, K. et al. (2020) [39]. The methodology applied in this paper is focused on the evaluation of the gradual replacement of diesel buses by electric buses. In terms of the evaluation electricity production, the methodology is based on the current state of the energy mix of electricity production, which is mainly based on production in coal-fired thermal power plants. The consumption of fuel and electricity is estimated, so it can distort the results.

Csiszár, C. et al. (2019) proposed a method of locating of charging stations for electric vehicles, where they pointed out that in terms of effectiveness the most suitable locations are P + R (park and ride) car parks where transport by passenger car is combined with public passenger transport [40]. After modifications, their location method can also be used for the selection of places for partial charging of electric buses during their operation on public transport lines in order to increase their range and thus the efficiency of operation and benefit for the city's air.

In the paper focused on the use of hydrogen as a renewable energy source, Ozawa et al. (2017) in their methodology also considered the WtT greenhouse gas emissions in the supply chain [41]. Khan (2017) dealt with the same issue of the calculation of indirect emissions, but only greenhouse gases. For WtT emissions, he also considered greenhouse gas emissions in the transport and distribution of fuels [42].

Many scientific works deal with problems of vehicle operation emissions and ways to decrease environmental impacts of transport on air pollution. The gap is that scientific works give just comprehensive and general results of vehicle operation and its impact on air pollution. These results are unusable for a real specific region with taking in to account all conditions affecting the final environmental impact.

Each vehicle fleet is different, the depth of renewal is a very varying factor, each region is different, electricity production is different—these and many other factors are influencing the real environmental impact of the vehicle fleet renewal in real conditions of chosen region.

These facts are reasons why the primary research questions were stated:

- To what extent can a change of vehicle fleet of urban public transport in a small regional city of Žilina with a population of 82,931 (as of 31 December 2019) contribute to meeting the objectives of reducing the impact of transport on energy intensity and air pollution?
- How can the direct and indirect emissions from the operation of vehicles be calculated?

This paper continues by sections. "Materials and Methods" described materials and methods used in the research and some results are firstly represented. Deeper results interpretation and following discussion can be found in the chapter "Results and Discussion". Concluded remarks to the results and their discussion are stated in the section "Conclusion".

2. Materials and Methods

2.1. Methodology of Calculation and Declaration of Direct and Indirect Emissions from Transport Services of Public Transport

The members of the author team have been working on the proposal of the calculation methodology and the calculation of WtT and TtW emissions of pollutants from transport and transport services since 2006 (the results in the field of CO_2 emissions production were published, for example [43,44]. The published studies were also focused on the comparison of environmental impacts from the operation of bus and trolleybus transport in a particular city. For the calculation of TtW emissions, the emission limits of buses were used. The efficiency of transforming the energy of fuel (diesel) into the required power was considered using the Sankey diagram. The energy value of used diesel and its consumption by public transport buses was taken into account.

Gradually, the authors modified and applied the methodology. It is improving with regard to the changes of values of emission factors, the adoption of the standard EN 16258 in 2012, the availability of results of certified measurements of direct emissions from road transport vehicles (HBEFA—Handbook of Emission Factors for Road Transport), the availability of statistical data on production and pollutant emissions from the power industry and the petrochemical industry.

The method of calculation regards both direct and indirect emissions of harmful substances produced by the operation of a vehicle in public passenger transport. This approach is more objective compared to considering only direct emissions of exhaust gases of a vehicle.

Emissions from transport services represent the sum of direct and indirect emissions. The indirect emissions from transport operation taken into account come from:

- the production of electricity necessary for the production of fuel in a refinery,
- the production of fuel in a refinery.

The direct emissions are related to the fuel consumption of a vehicle during providing transport services.

The structure of considered direct and indirect emissions is demonstrated in Figure 1. An (in block diagram) version of this picture can be found in the Appendix **??** in Figure A1. The extended version in Figure A1 includes contains the sequence of steps for the calculation of total emissions, which the inputs and outputs of these steps are identified.

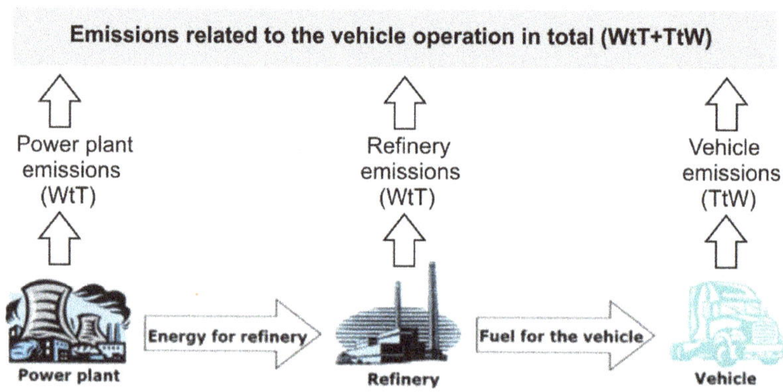

Figure 1. The chain of emissions production in transport with the use of fossil fuels (direct TtW and indirect WtT emissions).

An extended (in block scheme format) version of this picture can be found in the Appendix A The extended version in annex include contains the sequence of steps for the calculation of total emissions, which the inputs and outputs of these steps are identified.

To assess the impact of the operation of vehicles of a particular transport system on the environment, it is necessary to set values of emission factors for the used type of fuel or for the electricity. In the conditions of the Slovak Republic, we consider Slovenské Elektrárne as a producer and the historical development of the amount of electricity and the related harmful substances produced by this producer. For diesel, it is necessary to identify emission factors related to its production (indirect emissions related to diesel consumption). It is based on the amounts of produced diesel fuel and emissions of harmful substances by the Slovnaft refinery, which is the monopoly producer of diesel fuel in the Slovak Republic.

2.1.1. Indirect Emissions Related to the Production of Harmful Substances (Well-to-Tank Approach, WtT)

Step 1: Indirect emissions from the production of electricity necessary for the refinery (as a producer of fuels, WtT).

Step 1 consists of two consecutive sub-steps:

- From the calculation of emission factors related to electricity production,

- From the calculation of the amount of indirect emissions from electricity consumption in the production of diesel fuel,

Step 1.1 Emission factors for the production of electricity.

Based on the application of Equation (1), there were identified the emission factors of specific harmful substances from electricity production in the period 2005–2017 in the Slovak Republic. The values of the emission factors (EF) are listed in Table 1. The time series of the emission factors are used to define one-criterion regression functions of specific harmful substances emissions from electricity production in the Slovak Republic.

$$EF = \frac{QS}{QE}\ [\mathrm{g/kWh}] \tag{1}$$

where:

QS—produced amount of specific harmful substance [g]
QE—produced amount of electricity [kWh]

Table 1. Emission factors for the production of electricity in the Slovak Republic in g/kWh.

Harmful Substance	2005	2006	2007	2008	2009	2010	2011	2012	2013	2014	2015	2016	2017
CO_2	158.711	154.446	154.830	146.332	136.490	117.961	119.477	113.333	102.392	97.149	113.341	107.209	109.812
PM	0.367	0.239	0.028	0.022	0.022	0.015	0.018	0.013	0.012	0.012	0.024	0.008	0.005
SO_2	1.484	1.335	1.252	1.289	1.342	1.449	1.615	1.333	1.201	0.996	2.112	0.297	0.330
NO_x	0.324	0.258	0.247	0.204	0.213	0.178	0.195	0.161	0.132	0.134	0.174	0.088	0.083
CO	0.037	0.041	0.045	0.044	0.033	0.031	0.034	0.030	0.028	0.028	0.032	0.053	0.044

Source: processed by the authors from the annual reports of the Slovenská Energetika.

Figures 2 and 3 illustrate the one-criterion regression functions defining the development of the emission factors in the production of electricity in the Slovak Republic in 2005–2017. The functions are completed with the values of the coefficients of determination of the given models. The specific types of functions were proposed not only according to the values of the coefficients of determination but also according to the logical interpretability and the possibility to use the functions for the estimate of the emission factors in the future.

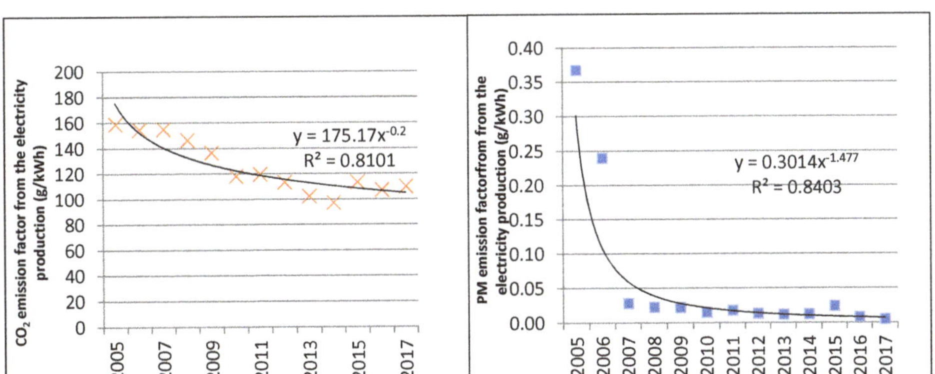

Figure 2. The chain of CO_2 (**left**) and PM (**right**) emissions production in transport with the use of fossil fuels (direct and indirect emissions).

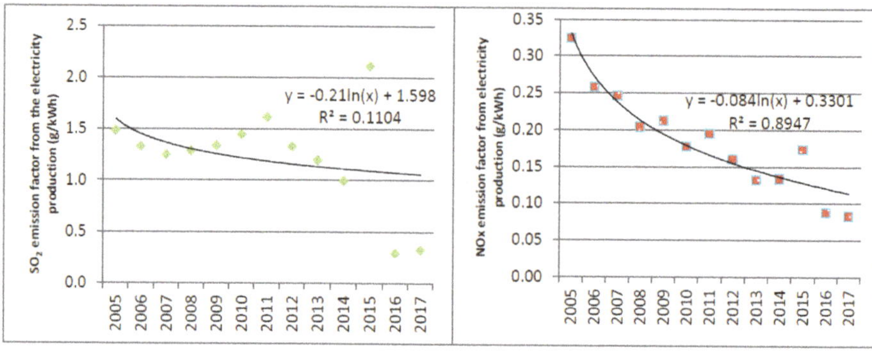

Figure 3. The development of the course of SO_2 (**left**) and NO_x (**right**) emission factor in the production of electricity in the period 2005–2017 in the Slovak Republic.

Table 2 illustrates trend equations together with the R^2 coefficient of determination. Based on these functions, it is possible to calculate the development of each harmful substance in the coming years separately. Table 2 also contains the values of emission factors of the selected harmful substances from the electricity production in the Slovak Republic in the years 2018 and 2019 calculated following the one-criterion regression functions. The regression functions can be updated on the basis of the updated time series of the values of electricity production and the related harmful substances emissions. In this way, it is possible to ensure the timeliness of the used emission factors in the calculation of harmful substances emissions.

Table 2. Emission factors for the production of electricity in the Slovak Republic.

Harmful Substance	Equasions	R^2	2018	2019
CO_2	$Y = 175.17.x^{-0.2}$	0.8101	103.33 g/kWh	101.92 g/kWh
PM	$Y = 0.3014{\cdot}x^{-1.477}$	0.8403	0.00611 g/kWh	0.00552 g/kWh
SO_2	$Y = 1.9268{\cdot}x^{-0.325}$	0.1823	0.817 g/kWh	0.799 g/kWh
NO_x	$Y = 0.3781{\cdot}x^{-0.455}$	0.7641	0.114 g/kWh	0.110 g/kWh
CO	$Y = 0.0395{\cdot}x^{-0.049}$	0.0329	0.035 g/kWh	0.035 g/kWh

Step 1.2 The calculation of harmful substances emissions related to the use of electricity in a refinery (per 1 L of diesel fuel).

Below is the calculation of the values of specific harmful substances that originate from the production of electricity necessary for the production of 1 L of diesel fuel (ESN). Table 2 lists all the values of specific harmful substances during the reporting period. The calculation was based on the Equation (2).

$$ESN = ESE \cdot SEVN\ [\mathrm{g/L}] \tag{2}$$

where:

ESE—specific harmful substance emissions from a power plant in the production of diesel fuel [g/Wh];
SEVN—consumption of electricity per production of 1 liter of diesel fuel [Wh/l]

The specific amount of indirect pollutant emissions produced from the operation of the bus related to the electricity consumption in the production of diesel is calculated as the product of the values of indirect emissions per 1 L of diesel from Table 3 and the consumption of diesel in the monitored period.

Table 3. The values of indirect emissions from electricity consumption in the production of 1000 L of diesel fuel.

Harmful Substance	2005	2006	2007	2008	2009	2010	2011	2012	2013	2014	2015	2016	2017	2018
CO_2	77.133	75.061	75.247	71.117	66.334	57.329	58.066	55.080	49.763	47.214	55.084	52.104	53.369	50.218
PM	0.178	0.116	0.014	0.011	0.011	0.007	0.009	0.006	0.006	0.006	0.012	0.004	0.002	0.003
SO_2	0.721	0.649	0.608	0.626	0.652	0.704	0.785	0.648	0.584	0.484	1.027	0.145	0.161	0.397
NO_x	0.158	0.125	0.120	0.099	0.103	0.086	0.095	0.078	0.064	0.065	0.084	0,043	0.040	0.055
CO	0.018	0.020	0.022	0.021	0.016	0.015	0.016	0.015	0.013	0.014	0.015	0,026	0.022	0.017

We do not consider petrol as fuel since the bus transport vehicles in the Slovak Republic do not use this fuel.

Step 2: Indirect emissions of the refinery from the production of diesel fuel (WtT).

The calculation is based on the production of diesel fuel and the related emissions produced by the Slovnaft refinery as a monopoly producer of diesel fuel in the Slovak Republic. We used the available data of the refinery from 2014 to 2017. Based on the Equation (3), the amounts of specific harmful substances per 1 kg of the produced diesel fuel (*ERN*) were calculated. The values are listed in Table 4.

$$ERN = \frac{QES}{QN}\ [\text{g/kg}] \quad (3)$$

where:

QES—the amount of emissions of specific harmful substance [g],
QN—the amount of produced diesel fuel [kg].

Table 4. Harmful substances calculated in grams per 1 kg of produced diesel fuel.

Type of Emission	2014	2015	2016	2017
CO_2 (g)	420.725	430.185	458.548	452.776
SO_2 (g)	0.483109	0.414994	0.581073	0.683114
CO (g)	0.102530	0.105744	0.095237	0.098991
PM (g)	0.014886	0.013200	0.016660	0.019399
NO_x (g)	0.333581	0.328416	0.428931	0.432166

Table 4 presents the results of the calculations of the amounts of specific harmful substances emissions per 1 kg of diesel fuel produced by the Slovnaft refinery in the Slovak Republic. This year 4,312,900 tons of diesel fuel were produced which resulted in producing 1,814,547 tons of CO_2.

Calculation of CO_2 emissions per 1 L of produced diesel fuel:

As the diesel fuel consumption in road transport is tracked and reported in units per liter, it is necessary to transform the emission factors of indirect emissions from units of kg of emissions per kg of diesel fuel to the units of kg of emissions per liter of diesel fuel. The density of diesel fuel at 15 °C is 0.82 to 0.86 kg/dm^3, the median is 0.84 kg/dm^3.

The volume of 1 kg of diesel fuel

$$V = \frac{m}{\rho} = \frac{1\ \text{kg}}{0.84\ \text{kg/dm}^3} = 1.19\ \text{dm}^3 = 1.19\ \text{L} \quad (4)$$

where:

V—volume [dm^3],
m—mass [kg],
ρ—density [kg/dm^3].

CO_2 emissions per 1 L of produced diesel fuel in 2014:

$$ECO_2 = \frac{420.725\ \text{kg}}{1.19\ \text{L}} = 0.353551\frac{\text{kg}}{\text{L}} = 353.551\ \text{g}\ \text{CO}_2\ \text{on 1 L of produced diesel fuel} \quad (5)$$

The same calculation method as for CO_2 is used according to Equation (5) also for the conversion of other emissions, the results for all types of emissions per 1 L of diesel produced in the period from 2014 to 2017 are shown in Table 5.

Table 5. Calculated harmful substances per 1 L of produced diesel fuel.

Type of Emission	2014	2015	2016	2017
CO_2 (g/L)	353.551	361.500	385.335	380.484
SO_2 (g/L)	0.405974	0.348734	0.488297	0.574045
CO (g/L)	0.086159	0.08886	0.080031	0.083186
PM (g/L)	0.012509	0.011093	0.014	0.016301
NO_x (g/L)	0.28032	0.27598	0.360446	0.363165

The specific amount of indirect pollutant emissions produced from the operation of the bus related to the production of diesel by the refinery is calculated as the product of the indirect emission values per 1 L of diesel from Table 5 and the consumption of diesel by bus in the monitored period.

2.1.2. Direct Emissions from Transport Services

Direct emissions from the operation of a vehicle are related to the fuel consumption of the vehicle during its operation, direct emissions from traffic operation are also referred to as "tank-to-wheel" (TtW).

To calculate the amount of emissions it is possible to use computer programs called emission calculators. There is a wide range of emission calculators on the market, from free versions to prepaid applications. Using them, it is possible to calculate the influence of a vehicle operation during transport on the environment. In other words, they can calculate only direct emissions from vehicle exhaust. As regards indirect emissions, emission calculators can express only the emissions of harmful substance which is equivalent to CO_2 according to the standard EN 16258. Among such emission calculators, there are EcoTransit, Map&Guide or the calculator of DHL company. This principle uses the emission factor for calculating the amount of CO_2 according to the amount of consumed fuel. This methodology is simple and does not take in to account the differences in the fuel production in different regions.

The indirect emissions from the operation of diesel buses and trolleybuses like CO_2, CO, NOx and PM were estimated according to our proposed methodology as presented in this section. Emission factors for the production of diesel fuel are listed in Tables 4 and 5.

HBEFA database was used to get results about direct emissions production of the diesel buses fleet. Any measurements of the exhaust emissions were not done during the operation. The real fleet operation data were available as inputs, like fuel consumption from all vehicles, driven distances, elevation profiles of lines, number of passengers, reached velocities (speed profiles). According to this data and the vehicle technical data were set the amounts of direct emissions with considering and comparing the real fuel consumption and the HBEFA calculated fuel consumption.

Table 6 presents indirect emissions values (WtT) in 2017 for different types of fuels, which were obtained from the research based on regression functions. The WtT emission factors together with the existing TtW emission factors can be used for objectively evaluating the environmental impact of specific types of vehicles with respect to their fuel and type of transport.

Table 6. Emission factors (EF) of direct and indirect harmful substances emissions for different types of fuel or electricity in 2017 in the Slovak Republic.

Diesel—The Amount of Emissions (g/L)			Electricity—The Amount of Emissions (g/kWh)		
Type of Harmful Substance	**WtT**	**TtW**	**Type of Harmful Substance**	**WtT**	**TtW**
CO_2	380.537	*	CO_2	109.81	0
PM	0.016302	*	PM	0.0046	0
SO_2	0.574161	*	SO_2	0.330	0
NO_x	0.36324	*	NOx	0.083	0
CO	0.083222	*	CO	0.044	0

* TtW—The amount of direct emissions from diesel fuel and petrol depends on the specific type of vehicle and its consumption. To calculate the emissions, it is possible to use a suitable calculator of direct emissions using the emission factors obtained from the certified measurements, e.g., Map&Guide or EcoTransit.

3. Results and Discussion

This section of the paper deals with the application of the proposed methodology of the emission calculation, and it also presents the comparison of the amount of direct and indirect harmful substances emissions produced by the fleet of vehicles of the public city transport company in Žilina between 2012 and 2019. The result is the assessment of the change of the fleet of vehicles for newer types of vehicles, which meet stricter emission limits. Such a change can significantly influence the decrease of the direct and indirect emissions produced in the transport operation. The input data represent the number of vehicles, their emission limits, average fuel consumption or electricity consumption, and the number of kilometers made during the operation period. The city transport company uses various types of vehicles, such as diesel buses, trolleybuses, hybrid engine vehicles and electric vehicles. The provided data on the daily composition of the vehicles cover a work day, a vacation day and a weekend day.

In the results, we dealt with the values of CO_2, CO, NOx and PM emissions which were produced directly during the vehicle operation and indirectly during the production of electricity necessary for the refinery. Direct CO_2 emissions are calculated with the use of the formulas presented in the standard EN 16,258. Other direct pollutants such as CO, NOx and PM are calculated with the use of HBEFA simulation methodology.

General relations for the calculation of direct emissions for a particular vehicle:

The amount of CO_2 according to the standard EN 16258

$$QCO_2 = CS \cdot g_t \ \text{[kg]} \tag{6}$$

where:

CS—total fuel consumption [liters],

g_t—tank-to-wheels factor of greenhouse gases for the fuel used (e.g., for diesel fuel, g_t = 2.67 $kgCO_2/l$)

Simulation of direct emissions production using HBEFA 3.3 database.

General emissions models are suitable primary for calculation of emissions in air quality studies, and the framework of integrated assessment studies. Models deliver the factors of emissions and the methodology usable for estimating total and partial emissions at a fleet or unit vehicle level. The most widespread models in the EU include COPERT, HBEFA and VERSIT+ [45–47].

The HBEFA database application estimates the emission factors of several pollutants per vehicle category, selected EURO standard, year of fleet operation or production and for a wide variety of traffic situations [46,48]. The traffic scenarios are mainly represented by four parameters: region type (rural, urban), road type, actual speed limit and traffic flow density (free flow, heavy, saturated, and stop and go) [49].

HBEFA (Handbook of Emission Factors for Road Transport), provides emission factors for vehicle categories like: PC (passenger car), LDV (light duty vehicles), HDV (heavy-duty vehicles), buses, motorcycles. The HBEFA allows users to choose different emission factors (EFs). Values of these EFs are influenced by more vehicle variables such as weight, size, type, engine cylinder capacity, fuel type consumed of the vehicle (gasoline, diesel, others), principle of exhaust treatment technology (with or without catalytic converter), driving style (acceleration and speed) and road longitude slope [50,51].

Output data are results of previous measurement real vehicles from selectable categories in laboratory conditions on vehicle dynamometers and in real driving tests.

Based on the above, HBEFA software was chosen as the most suitable simulation software to simulate the amount of emissions produced by vehicles. It provides sufficiently accurate data and allows the user to select the accurate values of factors according to immediate conditions of vehicle operation. The advantage of this software is possible to get partial results of vehicle fuel consumption and emissions production per each route section according to every change of driving parameters (velocity, slope, ambient temperature, etc.). We used HBEFA such like a database, not a software at all—it is possible to compare short sections, not only whole evaluated route or line and then we can simulate and adjust the input data to real vehicle operation.

The input parameters for the simulation and selection of precise conditions of the vehicle operation were the composition of the vehicle fleet, emission limits of vehicles, speed profile of vehicles operated on lines within the city of Žilina, vertical alignments of the lines, average air temperature reached during the measurement period, and the most important—average fuel consumption (diesel) during the selected spring month.

The measurement period was April (most representative season of the average air temperature and other ambient conditions affecting the vehicle energy consumption). This is a sufficiently long period of time to obtain an accurate long-term fuel consumption. This was one of the input data, on the basis of which the exact value of the pollutants produced by the vehicle was selected. The HBEFA software provides an interval of the resulting values of the production of emissions. From this interval, the values were selected on the basis of the actual fuel consumption of the bus.

The calculation of the rate of emission production was made for a diesel-powered bus, with engines with selected emission limits (see below), with a total weight of 18 t, with a longitudinal slope of the line +/− 2%, on city expressways, service and collector roads with an average speed of vehicle of 24.57–37.3 km/h. The results of the simulation are shown in Table 9. The gradual renewal of the outdated vehicle fleet of the transport enterprise of the City of Zilina, which provides urban public transport in the territory of the regional city of Žilina, started only after 2012 (see Table 7). In 2012, only four Irisbus buses of the EURO 4 emission class were operated in Žilina. On the other hand, seven buses of the outdated concept Karosa B 732 met only the EURO 1 emission class and six Karosa B 732 buses met the EURO 2 emission standard. In 2019, all buses met the highest emission class EURO 6. These are mainly 16 Iveco Urbanway 12 Hybrid buses and 14 Solaris Urbino 12 4th generation buses, which were first put into operation in the Slovak Republic in urban public transport in Žilina.

Table 7. Vehicle fleet before the renewal (2012).

Vehicles 2012				
Make and Type of Vehicle	Emission Class EURO	Consumption (diesel/el—l/100 km; kWh/100 km)	Number (Vehicle)	Capacity (Persons)
Buses				
Karosa B952	EURO 3	30.43	17	100
Renault PS09D1 City bus	EURO 3	33.88	3	100
Karosa B 932	EURO 2	33.74	6	95
Karosa B732	EURO 1	33.33	7	95
Irisbus Citelis Line	EURO 4	44.47	1	96
Irisbus Citelis	EURO 4	34.27	3	117
Total (average-consumption)	-	35.02	37	3682
Trolleybuses				
Škoda 15 Tr	-	217.21	29	150
Škoda 14Tr	-	141.69	13	82
Total (average-consumption)	-	179.45	42	5416
Total buses+trolleybuses	-	-	79	9098

Trolleybuses were renewed to a smaller extent. The biggest renewal, mainly with the help of the EU Structural Funds started in 2016 and was completed in April 2019 with the acquisition of two Škoda Perun electric buses and Škoda Solaris trolleybuses (see Table 8).

Table 8. Parameters reflecting the state of the vehicle fleet after the fleet renewal (2019).

Vehicles 2019				
Make and Type of Vehicle	**Emission Class EURO**	**Consumption (diesel/el—l/100 km; kWh/100 km)**	**Number (Vehicle)**	**Capacity (Persons)**
		Buses		
Solaris Urbino 12 (IV. generation)	EURO 6	36.27	14	98
Solaris Urbino 12 (III. generation)	EURO 6	36.7	5	98
Iveco Urbanway 12 Hybrid	EURO 6	29.73	16	80
Škoda Perun (electro)	-	195.3	2	73
Total (average-consumption)	-	34.23	37	3,288
		Trolleybuses		
Škoda 31Tr SOR	-	205.43	8	166
Škoda 30Tr SOR	-	139.05	7	94
Škoda 27Tr Solaris	-	202.31	18	131
Škoda 26Tr Solaris	-	139.03	9	91
Total (average-consumption)	-	176.224	42	5163
Total buses+trolleybuses	-	-	79	8451

Tables 7 and 8 present a comparison of selected parameters of vehicles of the Transport Enterprise of the City of Žilina in 2012 and 2019. The number of buses and trolleybuses did not change but the capacity (occupancy) of buses slightly decreased. The total average diesel consumption decreased by 0.79 l/100 km, from 35.02 l/100 km in April 2012 to 34.23 l/100 km in April 2019.

The renewal of the Žilina public transport fleet was completed at the beginning of 2019. In April 2019, new vehicles were already deployed in the real operation of the Žilina public transport and it was possible to find out the necessary data from their operation for the research. Due to the fact that public transport is operated according to the timetable during working days with differences over Saturdays, Sundays and public holidays, which was taken into account in the calculations according to the number of these days in 2019 and the same procedure was used for 2012 for vehicles used at that time. Two diesel buses were replaced by electric buses. The fuel consumption of buses was determined in 2012 in the transport enterprise of the City of Žilina by internal calculations on the basis of internal guidelines from data on the amount of diesel refueled and the km performed on individual buses. Each vehicle and each driver currently has an ID card, after insertion of which the stand allows to start refueling diesel. Each refueling is electronically recorded in the registration system. The actual mileage is recorded from the daily vehicle performance record and then the calculations of the average consumption for each vehicle are performed separately. An example of a worksheets of these calculations can be found in Table S2 (Supplementary Materials) and Table S3 (Supplementary Materials) in the Supplement. The authors of the paper are planning to measure the consumption of fuel in real traffic using measuring equipment in the urban public transport Žilina on a selected type of diesel and hybrid bus, as well as on a gas bus, but in another transport company. It will be a new research project funded by SPP (Slovenský plynárenský priemysel—Slovak Gas Industry), a.s.

The electricity consumption of trolleybuses decreased from 179.45 kWh/100 km to 176.224 kWh/100 km. The overall average decrease in the energy intensity of the vehicle fleet is approximately 2%. The renewal of the vehicle fleet was also possible on the basis of the establishment of the organization Integrated Transport of the Žilina Region (Integrovaná doprava Žilinského kraja, s.r.o) and the start of work on the integrated transport system of public passenger transport in the Žilina Region [52].

More detailed outputs of the calculation of WtT and TtW emissions from the operation of the urban public transport vehicle fleet in the city of Žilina in 2012 and 2019 according to the proposed methodology in Section 2 are for CO_2 in Table S1 (Supplementary Materials) in the Supplement. Given that the total driving performance of vehicles remained unchanged in that period, it was possible to calculate a comparison of the production of direct and indirect emissions from vehicle operations in 2012) and 2019 (Table 9).

Table 9. Outputs from the calculation of direct and indirect emissions in 2012 and in 2019.

Summary Table with Emissions Production Comparison Before and After Renewal-Emission Produced at Working Day											
Year	Type of Vehicle	Number of Vehicles in Circulation (pcs)	Number of Kilometer Performed (km)	Direct Emissions (kg)				Indirect Emissions (kg)			
				CO_2	CO	NO_X	PM	CO_2	CO	NO_X	PM
2012	Bus	31	5890	5,056.632	4.992	44.686	0.502	884.217	0.221	0.835	0.049
	trolleybus	30	5889	0.000	0.000	0.000	0.000	1,288.967	0.347	1.826	0.152
	Total	61	11,779	5,056.632	4.992	44.686	0.502	2,173.184	0.568	2.661	0.201
2019	Bus	31	5607	4,540.718	1.063	3.107	0.026	883.243	0.234	0.817	0.038
	electric bus	2	269	0.000	0.000	0.000	0.000	53.941	0.068	0.236	0.011
	trolleybus	29	5855	0.000	0.000	0.000	0.000	1,071.991	0.368	1157	0.058
	Total	62	11,731	4,540.718	1.063	3.107	0.026	2,009.175	0.670	2.210	0.107

In 2012, no hybrid buses or electric buses were deployed.

When comparing the values of emissions from Table 9, a significant decrease (on average up to 88% depending on the emission constituent, see Supplement Table S1 (Supplementary Materials)) of emissions is evident in 2019. The main reason is that in this year the transport enterprise of the City of Žilina operated all vehicles with the Euro VI emission limit. In contrast, in 2012 vehicles with lower emission limits such as Euro I, II, III and IV were deployed. Another crucial factor, especially for the reduction of CO_2 production is the combustion of diesel with a 7% biocomponent, while in 2012, diesel was refueled without a biocomponent. Thanks to the biocomponent in diesel, it was possible to significantly reduce CO_2 emissions while reducing the average consumption of vehicles by only less than 1 l/100 km. The deployment of electric buses that replaced the original diesel buses also contributes to the overall improvement of the CO_2 production for the entire vehicle fleet.

In Figure 4, the total savings in CO_2 production in 2019 (completely renewed vehicle fleet) up to 2023 are calculated. The graph illustrates the difference between the CO_2 production of the renewed vehicle fleet compared to the state before the complete renewal. It is important to mention the growing difference in CO_2 production if the vehicles of the original fleet would operate in the next five years. Over that period, in terms of renewal of vehicle fleet, CO_2 emissions in absolute terms will be reduced by more than 1.3 million tons compared to the operation of the original vehicle fleet, while maintaining the same driving performance as in the current period.

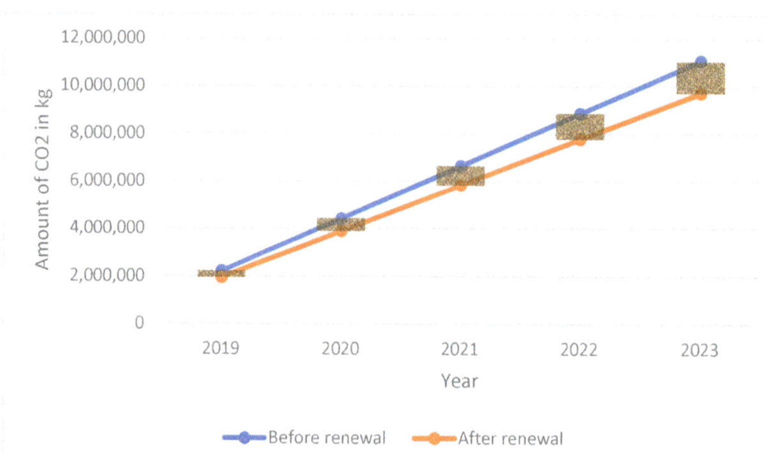

Figure 4. Cumulated annual CO_2 production before and after the renewal of the vehicle fleet.

It is also necessary to point out the planned shutdown of the Nováky Thermal Power Plant in Zemianske Kostoľany in 2023, which should have a positive impact on the energy mix in the Slovak Republic and by analogy also on indirect emissions from electricity generation.

Figure 5 shows the absolute savings in CO_2 production over the years 2019–2023 compared to the base period (before the renewal of vehicle fleet), when annual CO_2 production (including indirect emissions) amounted to more than 2.2 million tons of CO_2. After the renewal of the vehicle fleet, annual production (including indirect emissions) represents more than 1.9 million tons of CO_2. This represents an annual savings of almost 270,000 tons of CO_2. An important fact is an increase in savings (cumulation can be seen in the first and second graph) over the years compared to the base period (before the renewal of vehicle fleet). These are values that will significantly contribute to the improvement of air quality in the city of Žilina.

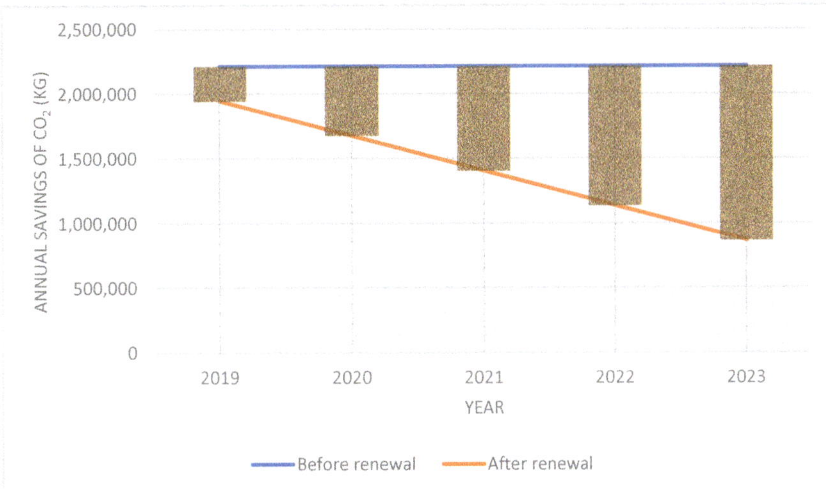

Figure 5. Absolute CO_2 savings over the years 2019–2023.

Figures 4 and 5 take into account the savings in CO_2 emissions in connection with the operation of the renewed vehicle fleet in urban public transport Žilina on the basis of data valid in 2019 without considering changes, the impact of which is currently difficult to predict. The real development of CO_2 emissions savings will depend on several factors, such as the rate of increase in the share of energy from renewable sources in the energy mix. The completion of Unit 3 in 2020 and Unit 4 in 2021 of the nuclear power plant in Mochovce in the Slovak Republic will allow the gradual shutdown of thermal power plants. Based on the information of Slovenské elektrárne, a.s. the capacity of each unit should be 471 MWe. One unit of the power plant will cover 13% of the electricity consumption of the Slovak Republic. According to current calculations, the annual production of completed units will save more than seven million tons of CO_2 emissions. We can then assume the progressive development of CO_2 savings. Changes related to electricity generation may affect the course of CO_2 savings in future years, as the amount of CO_2 savings calculated in Figure 6 does not take into account changes in electricity generation in the following years. An increase in the number of transported passengers is expected, as well as the introduction of an integrated transport system and the transition to a periodic timetable will bring changes in driving performance. The aim of the graphic representation in Figures 4 and 5 is not to predict the real development of emissions production in individual years, but to point out the cumulative nature of the development of savings in emissions production.

Savings in CO_2 production are not the only or greatest benefit in reducing emissions production. The reduction of production of emissions of toxic gases CO and NO_x as well as PM particles has a significantly higher positive impact on human health. Relative savings of the individual components of harmful substances during the calendar year are expressed in Figure 6. The total average savings in CO production are 66%, savings for NO_x gases are up to 88%. PM production decreased by an average of 79% due to vehicle fleet renewal. The absolute savings in the individual months of the calendar year are shown in Figure 7. In addition to the savings in PM production, it is important to pay attention to another fact—the amount of produced particulate matter after the vehicle fleet renewal is less dependent on the change of driving performance which vary from month to month depending on the number of holidays and working days. This is caused by the fact that individual vehicles produce very small amounts of particulate matter. With an increase in driving performance, the change in the total production of particulate matter is smaller than in the case of an increase in the driving performance of the vehicle fleet whose individual vehicles have a higher rate of PM production. This fact is particularly positive for the future, when increasing supply (by analogy with increasing driving performance) in public passenger transport may not result in a dramatic increase in pollutant production. This means that it is possible to provide the public with better transport services with a low negative impact on the health of the population.

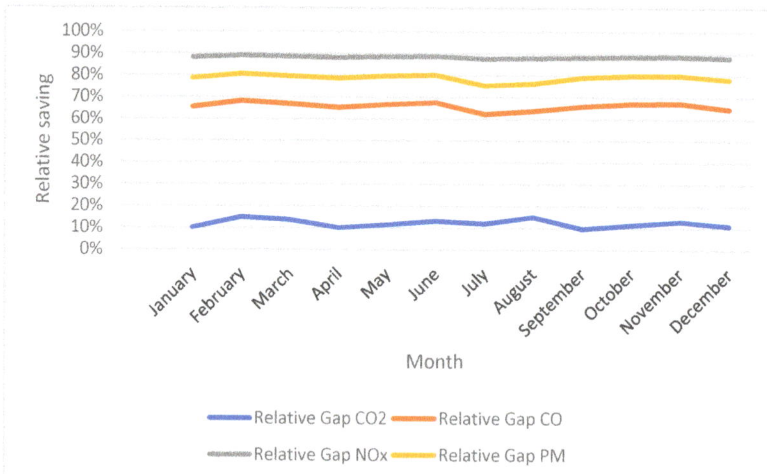

Figure 6. Relative savings in the harmful substances production of public transport in 2019 compared to 2012.

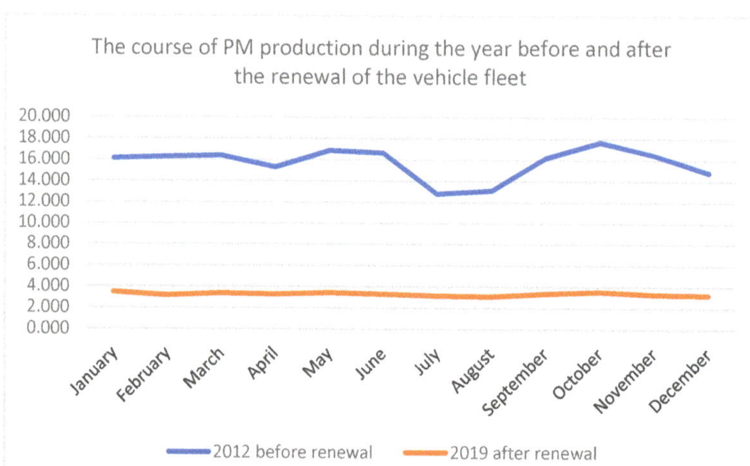

Figure 7. Comparison of PM production during the year before (2012) and after fleet renewal.

The proposed methodology for the calculation of direct and indirect emissions is the result of own research. The values of indirect emissions (WtT) for various types of fuels were obtained by research based on regression functions. These WtT emission factors, together with existing TtW emission factors (direct emissions) can be used for a more objective assessment of the environmental impacts of specific types of vehicles with respect to their fuel as well as the mode of transport.

Many scientific studies were solved in this problem of emissions intensity of transport. Most of them were mentioned in the section Introduction and Literature review. Problematic of urban transport and its impact on the environment and the point of view of used fuel or traction Methodology of this studies often use simulation software to estimate or calculate the energy consumption and emissions production [53], or the results are considering just some prognosis and common vehicle environmental impact declared by producers [54]. These two paths are used mostly just for the calculation of TtW emissions and other detailed statistical data must be used to reach realistic values of WtW approach [55–57]. The academic novelty of this manuscript is the connection of: 1. Data from real vehicle operation characteristics (mainly fuel and energy consumption), 2. Scientific evaluation of WtT emissions from fuel and energy production (statistical data and prognosis), 3. Estimating the production of TtW vehicle emissions by using simulation software. Using these different methods brings a unique result taking in to account holistic approaches from whole problematics in one scientific work.

The methodology and approach used are unique on two levels. Firstly, the existing emission calculators and related standards (EN 16258) in the field of calculation of direct and indirect emissions consider emission factors of indirect greenhouse gas emissions. The proposed methodology and performed calculations also take into account indirect emissions of other pollutants (CO, NO_X, PM). This makes the approach to the calculation of direct and indirect emissions more comprehensive. Secondly, the emission factors of indirect emissions from electricity production determined on the basis of regression functions can be dynamically updated on an annual basis depending on the statistical data from electricity industry. The calculated indirect emissions using updated emission factors respect the structure of the energy mix and current emissions from electricity industry in the conditions of the Slovak Republic. The methodology considers the energy self-sufficiency of the Slovak Republic, after 2020 the Slovak Republic will be an exporter of electricity.

The weakness of our proposed methodology is the fact that it does not consider the amount, origin and emission intensity of imported electricity. Nowadays Slovakia trades electricity with neighboring countries according to current price and consumption. All neighbors, except of for Austria, produce more carbon intensive electricity than Slovakia. Studies dealing with carbon intensity of electricity production declare that the county of origin of the electricity plays a different role in the final carbon intensity of EV operation [37].

As a result, electric driven vehicles can produce more secondary actual emissions than the methodology estimates. This weakness can be removed after considering the data about electricity production and distribution from the countries of its origin. Application of this methodology will reach higher accuracy after 2020 when Slovakia becomes a pure electricity exporting country.

Indirect emissions (WtT) related to the distribution and storage of fuels are only taken into account in the calculation of indirect greenhouse gas emissions. Emission factors according to EN 16,258 were used. For example, in the paper focused on the use of hydrogen as a renewable energy source, Ozawa et al. (2017) in the methodology considered also WtT greenhouse gas emissions in the supply chain [41]. Khan (2017) dealt with the same issue of the calculation of indirect emissions, but only greenhouse gases. For WtT emissions, he also considered greenhouse gas emissions from the transport and distribution of fuels [42].

In the calculation of other pollutants, indirect emissions related to the distribution of fuels and their storage were not considered in our methodology (except for greenhouse gases). This can be considered a shortcoming of the above procedure. The reason is the unavailability of relevant information and data from distribution companies. In the future, we would like to focus our research on this part of the methodology for calculation of indirect emissions in cooperation with fuel manufacturers and distribution companies in the Slovak Republic. The determinants of the values of indirect emission factors from the distribution of fuels are the location of production and storage of fuels, the location of petrol stations, used trucks for distribution as well as transport routes and their level of optimization.

The results are based on the average fuel consumption of buses and the average electricity consumption of trolleybuses and electric buses, which affects the accuracy of the results. Elements like traffic conditions, age of vehicles, number of passengers on board, etc. were not considered in the estimation directly but they were included in the average vehicle fuel/electricity consumption per investigated time period. The final amount of consumed fuel/energy reflected all vehicle operation conditions. So the obtained results can be different from the short time point of view in the comparison of actual emission production. If we consider that bus lines, driving performance and number of passengers, average vehicle speed and ambient conditions, the results from 2012 and 2019 should be suitable for determination of the vehicle fleet renewal on the air pollution.

The above outputs do not take into account the fact that the consumption and content of pollutants in exhaust gases especially of buses with conventional internal combustion engines can change after several years of urban operation. Therefore, the authors of the paper prepared measurements of direct emissions from the operation of buses of urban public transport in the city of Žilina. They are connected with the research into the emission impact of introducing the preference of public passenger transport vehicles at controlled junctions [58,59].

The results of calculations of the impacts of the complex renewal of the public transport vehicle fleet in the city of Žilina on the basis of the developed methodology need to be verified by long-term continuous measurement of air quality in places near urban public transport lines. In this area, it is possible to use the experience of the city of Umeå in Sweden. It takes a number of measures every year to improve air quality not only in the field of public passenger transport. It evaluates the effectiveness of these measures based on the results of air quality measurements and other indicators such as the road cleaning system, increasing use of urban public transport, etc. [60]. A procedure was proposed to evaluate the impacts of the operation of the entire urban public transport fleet for a calendar year. The procedure is based on the actual mileage of specific vehicles operating on urban public transport lines on weekdays, weekends and public holidays and thus provides a tool for assessment of further changes in the structure of the vehicle fleet in future years. It is also possible to examine how the change in the deployment of specific vehicles, for example with different consumption of diesel or electricity, will affect the impacts on air quality in the city. The possibilities of specifying the outputs are through the measurement and control of fuel consumption and electricity consumption in real operation on specific lines where urban public transport vehicles are deployed.

In terms of evaluation of benefits of renewal of vehicle fleet in urban public transport for a modern vehicle fleet equipped with air conditioning, information system and Wi-Fi connection, it is necessary to investigate whether it will also have the effect of increasing the number of passengers and the demand for public passenger transport. This is another direction of research that the authors are planning to address in the next period, given that they already have experience in this field [61].

Increase in the use of public passenger transport, especially in large cities, may in the future influence the changes in the work system, for example by an increase in work from home that does not generate any traffic or change of behavior of young people in cities with reliable public passenger transport where owning a car is not a necessity [7].

4. Conclusions

On the basis of the research outcomes presented in this paper, we recommend that the EU should continue to support the renewal of the urban public transport vehicle fleet in smaller cities. This support can significantly help reduce the negative impacts of urban public transport on the environment. Other synergy effects are in improving the quality of public passenger transport and increasing its use. This has positive impacts on the change in the ratio of the division of transport work between public passenger transport and individual car transport.

According to the EU White Paper on Transport, urban public transport is suitable for the introduction of alternative propulsion systems and fuels, given the large vehicle fleets of city buses. On 5 March 2020, the Government of the Slovak Republic approved the National program for reduction of pollutant emissions for the Slovak Republic pursuant to Article 6 of Directive 2016/2284 of the European Parliament and of the Council on the reduction of national emissions of certain atmospheric pollutants. It includes support for increasing the share of public passenger transport, in particular for electric drive, more massive support for electric vehicles and the introduction of low-emission zones in cities. According to the approved low-carbon strategy of the development of the Slovak Republic by 2030 with a 2050 perspective, carbon neutrality will not be achieved without significant support of public passenger transport, for example by measures to support the development of rail transport (trams and trolleybuses), bus public transport powered by alternative fuels (bio CNG electrification, liquid fuels, hydrogen), etc. It is important that local governments and public administrations support the development of mobility using alternative energy sources. They thus create a positive example and impact on the ecological behavior of the population. Given the size of vehicle fleets and mileage, the nationwide greening of vehicle fleets can also contribute to meeting global climate change targets. There are more ways how to meet the emission targets till 2030 or 2050 but the most effective are for example transition of energy sources used in transport (alternative drive systems, alternative fuels), driving vehicles with higher energy efficiency (new and effective vehicles), higher usage of vehicle / transport system capacity (higher occupancy of vehicles, eliminate "empty" drives).

The paper did not cover all the impacts of the renewal of vehicle fleet in urban public transport but it was focused on the design of the methodology and its application to the operation of vehicles for the production of direct and indirect emissions. It is also necessary to deal with the issue of long-term financing of public passenger transport if the EU Structural Funds are reduced so that the impacts of the operation of public passenger transport on the environment continue to decrease. Although the renewal of the public passenger transport fleet is associated with other indirect emissions related to the production of new vehicles, we did not consider the indirect emissions related to the production of new vehicles. The renewal of vehicle fleets also pursues other objectives, not only the reduction of emissions from transport but is also connected with an increase in the accessibility of public passenger transport for people with reduced mobility, the quality of services provided in public transport and also safety. More modern vehicles are equipped with several active and passive safety systems which can have an impact on reducing accidents and thus also contribute to the protection of human health and life.

Supplementary Materials: The following are available online at http://www.mdpi.com/1996-1073/13/15/3869/s1: Table S1: Comparison of harmful substances produced before and after fleet renewal; Table S2: Worksheets for calculation harmful substances-2012; Table S3: Worksheets for calculation harmful substances-2019.

Author Contributions: Introduction J.G.; literature review V.K., J.G. and. T.F.; material and methods V.K., J.G., F.P., T.S. (Tomáš Settey) and T.S. (Tomáš Skrúcaný); data curation F.P., J.G., T.S. (Tomáš Settey) and V.K.; result: J.G., T.F. and T.S. (Tomáš Skrúcaný); writing—original draft, T.S. (Tomáš Settey) and J.G.; visualization, F.P., V.K., J.G., T.S. (Tomáš Settey) and T.S. (Tomáš Skrúcaný). All authors have read and agreed to the published version of the manuscript.

Funding: This research was funded by the Center of Excellence for Systems and Services of Intelligent Transport II., ITMS 26220120050 supported by the Research & Development Operational Program funded by the ERDF and also thanks to the project of institutional research of the Faculty of Operation and Economics of Transport and Communications no. 8/PEDAS/2019 Assessment of energy intensity and production of gaseous emissions by the operation of vehicles of various transport modes with different types of propulsion in the conditions of selected areas of Central Europe and the Danube region and no. 7/PEDAS/2019 plans and standards of transport services of cities and regions as part of sustainable mobility plans.

Conflicts of Interest: The authors declare no conflict of interest.

Appendix A

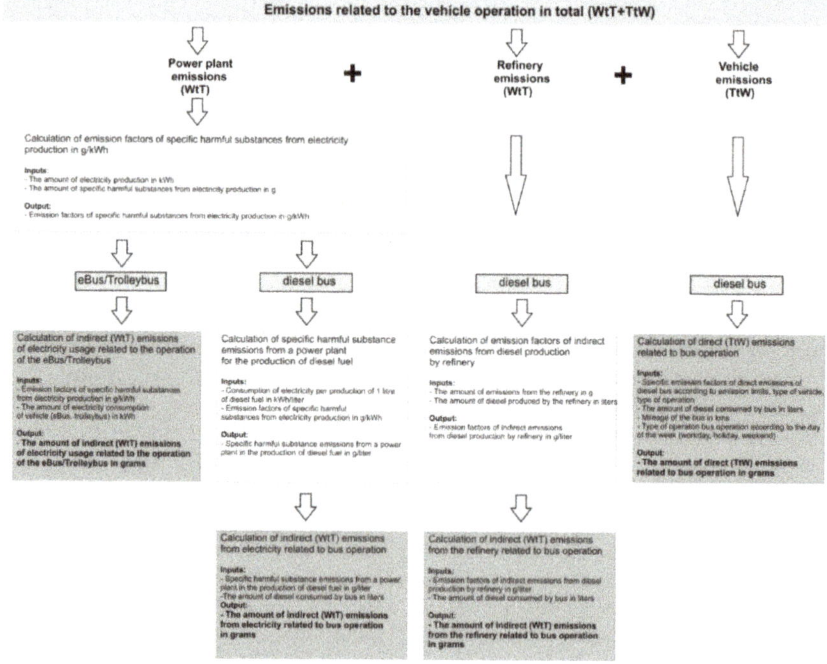

Figure A1. Emissions related to the vehicle operation in total (WtT + TtW).

References

1. Communication from the Commission to the European Parliament, the Council, the European Economic and Social Committee and the Committee of the Regions, Energy Efficiency Plan 2011. Available online: https://eur-lex.europa.eu/legal-content/EN/TXT/PDF/?uri=CELEX:52011DC0109&qid=1586252976389&from=SK (accessed on 27 March 2020).
2. *Air Pollution: Our Health still does not Have Sufficient Protection; Special Report Pursuant to Article 287 (4), Second Subparagraph*; TFEU/European Court of Auditors: Luxembourg, 2018.
3. White PAPER—Roadmap to a Single European Transport Area—Towards a Competitive and Resource Efficient Transport System. Available online: https://eur-lex.europa.eu/legal-content/SK/TXT/PDF/?uri=CELEX:52011DC0144&from=SK (accessed on 27 March 2020).
4. *Strategic Plan for the Development of Public Passenger Transport of the Slovak Republic until 2020*; Delloitte for the Ministry of Transport and Construction of the Slovak Republic: Bratislava, Slovak, 2013.
5. *Strategic Plan for the Development of Transport of the Slovak Republic until 2030*; The Ministry of Transport and Construction of the Slovak Republic: Bratislava, Slovak, 2017.
6. Dydkowski, G. The Application of Just Distribution Theories to Financing Integrated Systems of Regional and Urban Public Transport. *Zesz. Naukowe Transp. Politech. Śl.* **2018**, *100*, 23–33. [CrossRef]
7. Poliak, M.; Poliaková, A.; Mrníková, M.; Šimurková, P.; Jaśkiewicz, M.; Rafał, J. The Competitiveness of Public Transport. *J. Compet.* **2017**, *9*, 81. [CrossRef]
8. Stopka, O.; Zitricky, V.; Abramovic, A.; Marinov, M.; Ricci, S. Innovative Technologies for Sustainable Passenger Transport. *Hindawi J. Adv. Transp.* **2019**, *2019*, 4197246. [CrossRef]
9. L'upták, V.; Droździel, P.; Stopka, O.; Stopková, M.; Rybicka, I. Approach Methodology for Comprehensive Assessing the Public Passenger Transport Timetable Performances at a Regional Scale. *Sustainability* **2019**, *11*, 3532. [CrossRef]

10. Tang, C.; Ceder, A.; Ge, Y.-E. Optimal public-transport operational strategies to reduce cost and vehicle's emission. *PLoS ONE* **2018**, *13*, e021138. [CrossRef] [PubMed]
11. Majumdar, M.; Majhi, B.-K.; Dutta, A.; Randal, M.; Jash, T. Study on possible economic and environmental impacts of electric vehicle infrastructure in public road transport in Kolkata. *Clean Technol. Environ. Policy* **2015**, *17*, 1093–1101. [CrossRef]
12. Bristow, A.L.; Tight, M.; Pridmore, A.; May, A.D. Developing pathways to low carbon land-based passenger transport in Great Britain by 2050. *Energy Policy* **2008**, *36*, 3427–3435. [CrossRef]
13. Suna, S.; Zhao, G.; Wang, T.; Jiaxin, J.; Wang, P.; Lin, Y.; Li, H.; Ying, O.; Mao, H. Past and future trends of vehicle emissions in Tianjin, China, from 2000 to 2030. *Atmos. Environ.* **2019**, *209*, 182–191. [CrossRef]
14. Choudhary, A.; Gokhale, S. Evaluation of emission reduction benefits of traffic flow management and technology upgrade in a congested urban traffic corridor. *Clean Technol. Environ. Policy* **2019**, *21*, 257–273. [CrossRef]
15. LI, P.; Zhao, P.; Brand, C.H. Future energy use and CO_2 emissions of urban passenger transport in China: A travel behavior and urban form based approach. *Appl. Energy* **2018**, *211*, 820–842. [CrossRef]
16. Roșu, L.; Istrate, M.; Bănică, A. Passenger Car Dependency and Consequent Air Pollutants Emissions in Iasi Metropolitan Area (Romania). *Environ. Engineering Manag. J.* **2018**, *17*, 865–875.
17. Khomenko, S.; Nieuwenhuijsen, M.; Ambrós, A.; Wegener, S.; Mueller, N. Is a liveable city a healthy city? Health impacts of urban and transport planning in Vienna, Austria. *Environ. Res.* **2020**, *183*, 109238. [CrossRef] [PubMed]
18. Catalano, M.; Galatioto, F.; Bell, M.; Namdeo, A.; Bergantino, A. Improving the prediction of air pollution peak episodes generated by urban transport networks. *Environ. Sci. Policy* **2016**, *60*, 69–83. [CrossRef]
19. Elbir, T.; Mangir, N.; Kara, M.; Simsir, S.; Eren, T.; Ozdemir, S. Development of a GISbased decision support system for urban air quality management in city of Istanbul. *Atmos. Environ.* **2010**, *44*, 441–454. [CrossRef]
20. Kuranc, A.; Słowik, T.; Wasilewski, J.; Szyszlak-Bargłowicz, J.; Stoma, M.; Šarkan, B. Emission of Particulates and Chosen Gaseous Exhausts Components During a Diesel Engine Starting Process. In Proceedings of the IX International Scientific Symposium "Farm Machinery and Processes Management in Sustainable Agriculture", Lublin, Poland, 22–24 November 2017; pp. 201–215.
21. Stojanovic, D.; Veličkovic, M. The impact of freight transport on greenhouse gases emissions in Serbian cities—The case of Novi Sad. *Metal. Int.* **2012**, *17*, 196–202.
22. Ližbetin, J.; Hlatká, M.; Bartuška, L. Issues concerning declared energy consumption and greenhouse gas emissions of FAME biofuels. *Sustainability* **2018**, *10*, 3025. [CrossRef]
23. Ivkovic, I.; Čokorilo, O.; Kaplanovic, S. The estimation of GHG emission costs in road and air transport sector: Case study of Serbia. *Transport* **2018**, *33*, 260–267. [CrossRef]
24. Petro, F.; Konečný, V. Calculation of Emissions from Transport Services and their use for the Internalisation of External Costs in Road Transport. *Procedia Eng.* **2017**, *192*, 677–682. [CrossRef]
25. Skrúcaný, T.; Kendra, M.; Kalina, T.; Jurkovič, M.; Vojtek, M.; Synák, F. Environmental comparison of different transport modes. *NAŠE MORE: Znan.-Stručni Čas. More Pomor.* **2018**, *4*, 192–196. [CrossRef]
26. Skrucany, T.; Kendra, M.; Gnap, J.; Sarkan, B. Software Simulation of an Energy Consumption and GHG Production in Transport. In *International Conference on Transport Systems Telematics*; Springer: Cham, Swizerland, 2015; pp. 151–160.
27. L'upták, V.; Stopková, M.; Jeřábek, K. Comparative analysis in terms of environmental impact assessment between railway and road passenger transport operation: A case study. *Int. J. Sustain. Aviat.* **2020**, *6*, 21–35.
28. Hlatká, M.; Bartuška, L. Comparing the calculations of energy consumption and greenhouse gases emissions of passenger transport service. *Nase More* **2018**, *65*, 224–229. [CrossRef]
29. Petro, F.; Konečný, V. Calculation of External Costs from Production of Direct and Indirect Emissions from Traffic Operation. *Transp. Res. Procedia* **2019**, *40*, 1162–1167. [CrossRef]
30. Thiel, C.; Schmidt, J.; Zyl, A.-V.; Schmid, E. Cost and well-to-wheel implications of the vehicle fleet CO_2 emission regulation in the European Union. *Transp. Research Part A* **2014**, *63*, 25–42. [CrossRef]
31. Keskisaari, V.; Ottelin, J.; Heinonen, J. Greenhouse gas impacts of different modality style classes using latent class travel behavior model. *J. Transp. Geogr.* **2017**, *65*, 155–164. [CrossRef]
32. Ottelin, J.; Heinonen, J.; Junnila, S. Greenhouse gas emissions from flying can offset the gain from reduced driving in dense urban areas. *J. Transp. Geogr.* **2014**, *41*, 1–9. [CrossRef]

33. Sarkan, B.; Stopka, O.; Gnap, J.; Caban, J. Investigation of Exhaust Emissions of Vehicles with the Spark Ignition Engine within Emission Control; 10th International Scientific Conference Transbaltica 2017. *Transp. Sci. Technol. Procedia Eng.* **2017**, *187*, 775–782. [CrossRef]
34. Milosavljević, B.L.; Pesic, R.; Taranović, D.S.; Davinić, A.L.; Milojević, S. Measurements and modeling pollution from traffic in a street canyon: Assessing and ranking the influences. *Therm. Sci.* **2015**, *19*, 2093–2104. [CrossRef]
35. Blaž, J.; Zupan, S.; Ambrož, M. Study on the Eligibility of Introducing Hybrid-Drive Buses into the Public Passenger Transport. *Stroj. Vestn. J. Mech. Eng.* **2019**, *65*, 12–20. [CrossRef]
36. Napoli, G.; Micari, S.; Dispenza, G.; Di Novo, S.; Antonucci, V.; Andaloro, L. Development of a fuel cell hybrid electric powertrain: A real case study on a Minibus application. *Int. J. Hydrogen Energy* **2017**, *42*, 28034–28047. [CrossRef]
37. Łebkowski, A. Studies of Energy Consumption by a City Bus Powered by a Hybrid Energy Storage System in Variable Road Conditions. *Energies* **2019**, *12*, 951. [CrossRef]
38. Kivekäs, K.; Vepsäläinen, J.; Tammi, K. Stochastic Driving Cycle Synthesis for Analyzing the Energy Consumption of a Battery Electric Bus. *IEEE Access* **2018**, *6*, 55586–55598. [CrossRef]
39. Pietrzak, K.; Pietrzak, O. Environmental Effects of Electromobility in a Sustainable Urban Public Transport. *Sustainabilty* **2020**, *12*, 1052. [CrossRef]
40. Csiszár, C.; Csonka, B.; Földes, D.; Wirth, E.; Lovas, T. Urban public charging station locating method for electric vehicles based on land use approach. *J. Transp. Geogr.* **2019**, *74*, 173–180. [CrossRef]
41. Ozawa, A.; Inoue, M.; Kitagawa, N.; Muramatsu, R.; Anzai, Y.; Genchi, Y.; Kudoh, Y. Assessing Uncertainties of Well-To-Tank Greenhouse Gas Emissions from Hydrogen Supply Chains. *Sustainability* **2017**, *9*, 1101. [CrossRef]
42. Khan, M.I. Comparative Well-to-Tank energy use and greenhouse gas assessment of natural gas as a transportation fuel in Pakistan. *Energy Sustain. Dev.* **2018**, *43*, 38–59. [CrossRef]
43. Konečný, V. Analysis of CO_2 production of transport sector in the Slovak Republic. In Proceedings of the Politransportnyje sistemy: Materialy IV Vserossijskoj naučno-techničeskoj konferencii, Krasnojarsk, Russia, 22–24 November 2006; pp. 167–172.
44. Gnap, J.; Konečný, V.; Poliak, M. *Economic and ecological comparison of trolleybus and bus transport in the conditions of urban public transport in Banská Bystrica, Expert study*; University of Žilina: Žilina, Slovakia, 2005.
45. Ntziachristos, L.; Gkatzoflias, D.; Kouridis, C.; Samaras, Z. COPERT: A European Road Transport Emission Inventory Model. In Proceedings of the 4th International ICSC Symposium on Information Technologies in Environmental Engineering, Thessaloniki, Greece, 28–29 May 2009.
46. Hausberger, S.; Rexeis, M.; Zallinger, M.; Luz, R. *EFs from the Model PHEM for the HBEFA*; Version 3; Report Nr. I-20a/2009 Haus-Em 33a/08/679.; Graz University of Technology: Styria, Austria, 2009.
47. Smit, R.; Smokers, R.; Rabé, E. A new modelling approach for road traffic emissions: VERSIT+. *Transp. Res. Part D: Transp. Environ.* **2007**, *12*, 414–422. [CrossRef]
48. Rexeis, M.; Hausberger, S.; Kuehlwein, J.; Luz, R. *Update of EFs for EURO 5 and EURO 6 Vehicles for the HBEFA*; Version 3.2; Final Report; Technical University Graz: Styria, Austria, 2013.
49. Krecl, P.; Johansson, C.; Targino, A.C.; Ström, J.; Burman, L.; Ström, J. Trends in black carbon and size-resolved particle number concentrations and vehicle emission factors under real-world conditions. *Atmos. Environ.* **2017**, *165*, 155–168. [CrossRef]
50. Colberg, C.A.; Tona, B.; Stahel, W.A.; Meier, M.; Staehelin, J. Comparison of a road traffic emission model (HBEFA) with emissions derived from measurements in the Gubrist road tunnel. *Switz. Atmos. Environ.* **2005**, *39*, 4703–4714. [CrossRef]
51. Olivera, A.C.; García-Nieto, J.M.; Alba, E. Reducing vehicle emissions and fuel consumption in the city by using particle swarm optimization. *Appl. Intell.* **2015**, *42*, 389–405. [CrossRef]
52. Mrnikova, M.; Poliak, M.; Šimurková, P.; Reuter, N. Why is important establishment of the organizer in integrated transport system in Slovak Republic. In Proceedings of the 11th International Scientific and Technical Conference on Automotive Safety, Casta-Papiernicka, Slovakia, 18–20 April 2018; pp. 1–6.
53. Dreier, D.; Silveira, S.; Khatiwanda, D.; Fonseca, K.V.O.; Nieweglowski, R.; Schepanski, R. Well-to-Wheel analysis of fossil energy use and greenhouse gas emissions for conventional hybrid-electric and plug-in hybrid-electric city buses in the BRT system in Curitiba, Brazil. *Transp. Res. Part D Transp. Environ.* **2018**, *58*, 122–138. [CrossRef]

54. Krause, J.; Thiel, C.; Tsokolis, D.; Samaras, Z.; Rota, C.; Ward, A.; Prenniger, P.; Coosemans, T.; Neugebauer, S.; Verhoeve, W. EU road vehicle energy consumption and CO_2 emissions by 2050—Expert-based scenarios. *Energy Policy* **2020**, *138*, 111224. [CrossRef]
55. Dimoula, V.; Kehagia, F.; Tsakalidis, A. A holistic approach for estimating carbon emissions of road and rail transport systems. *Aerosol Air Qual. Res.* **2016**, *16*, 61–68. [CrossRef]
56. Ashtineh, H.; Pishvaee, M.S. Alternative fuel vehicle-rounting problem: A life cycle analysis of transportation fuels. *J. Clean. Prod.* **2019**, *219*, 166–182. [CrossRef]
57. Sleep, S.; Gou, J.; Laurenzi, I.J.; Bergerson, J.A.; MacLean, H.L. Quantifying variability in well-to-wheel greenhouse gas emission intensities of transportation fuels derived from Canadian oil sands mining operations. *J. Clean. Prod.* **2020**, *258*, 120639. [CrossRef]
58. Kalasova, A.; Kupculjakova, J.; Kubikova, S.; Pal'o, J. Recent Advances in Traffic Engineering for Transport Networks and Systems. *Book Ser. Lect. Notes Netw. Syst.* **2018**, *21*, 203–212.
59. Kalasova, A.; Cernicky, L.; Kupculjakova, J. The Impact of Public Transport Priority on the Traffic in the Chosen Part of the City of Zilina. *Transp. Probl.* **2014**, *9*, 19–26.
60. Air Quality in Umeå, Evaluation of the Air Quality Measurement in 2019. Available online: www.umea.se/luft (accessed on 15 June 2020).
61. Gnap, J.; Konecny, V.; Poliak, M. Demand elasticity of public transport. *Ekon. Cas.* **2006**, *54*, 667–684.

energies

MDPI

Article

Which Institutional Conditions Lead to a Successful Local Energy Transition? Applying Fuzzy-Set Qualitative Comparative Analysis to Solar PV Cases in South Korea

Youhyun Lee [1], Bomi Kim [2] and Heeju Hwang [3,*]

1 Department of Public Administration and Police Science, Hannam University, Daejeon 34430, Korea; valerie315@hnu.kr

2 Loan & Deposit Planning Department, Korea Development Bank, Seoul 07242, Korea; spring43@kdb.co.kr

3 Public-Private Infrastructure Investment Management Center, Korea Development Institute, Sejong 30149, Korea

* Correspondence: heeju.hwang@kdi.re.kr

Received: 13 May 2020; Accepted: 13 July 2020; Published: 17 July 2020

Abstract: To explore the most desirable pathway for a successful local energy transition, a fuzzy-set qualitative comparative analysis was conducted on 16 regional cases in South Korea. We developed four propositions based on previous studies and theories as a causal set. Based on the South Korean context, we selected the solar photovoltaic (PV) generation and solar PV expansion rate as barometers for measuring the success of a local energy transition. Our analysis highlights the importance of the International Council for Local Environmental Initiatives (ICLEI) membership (network), local legislation, and the environmental surveillance of locally-based non-governmental organizations (NGOs). The implications of this study will provide insights for developing or newly industrialized countries where an energy transition is underway.

Keywords: local energy transition; fs-QCA; qualitative analysis; South Korean energy policy; solar PV

---

1. Introduction

Can only regions with excellent geographical conditions produce renewable energy? If the answer to this question is "yes," countries such as South Korea (which has high population density and mountainous topography that occupies 70% of the country) can hardly expect to be capable of utilizing renewable energy. Fortunately, the answer to this question is not always "yes." Through the creation and implementation of public policies and various institutional arrangements, our society has overcome its underlying limitations. Natural conditions are unalterable, but institutional conditions are flexible and variable depending on human effort and ingenuity. The purpose of conducting this study is to highlight the importance of institutional efforts in developing renewable energy (Hereinafter "RE") and encouraging an energy transition that can even overcome natural conditions through an analysis of local energy transition efforts in South Korea.

Given the ceaseless increase in energy consumption and the growing concern about climate change, the most likely means of successfully reducing Greenhouse gas (Hereinafter "GHG") emissions is to achieve an energy transition, or a rapid change from traditional to renewable energy sources [1,2]. Germany pioneered the concept of energy transition, "*Energiewende*" [3], but energy transition has recently become the prevailing paradigm, especially at the local level [4]. With the emergence of energy transition phenomena, local actors have become newly powerful entities capable of leading local energy transitions [5,6]. Communities have become influential local authorities [7,8] aligned with grassroots approaches, and local governments have also been driving institutional initiatives for localized energy agendas [9]. Local agencies are also participating in local energy transition efforts [10]. In the case of Japan, citizens have played a crucial role by sharing their views during local policy deliberations [11,12].

Many attempts to achieve more-sustainable development and energy transitions can be found at the local level [6]. Inspired by "*Energiewende*" in Germany, local governments throughout the world, such as those of Barcelona, London, and Paris, are seeking to promote energy transitions in the regions where they are located. An energy transition is not only a transition in an energy system but also an effort to meet a societal challenge that includes institutional changes in a region. Borrowing from several scholars' definitions, a sustainable local energy transition can be defined as a process of transformation that leads to low-carbon patterns of energy production and consumption [13] and includes changes in behavior and cultural discourse [14,15].

Energy transition has now emerged in the South Korean policy context. The current Moon Jae-In administration declared energy transition the basic national energy policy goal. The 3020 energy roadmap promotes increasing the proportion of renewable energy to 20% by 2030. This aim is an ambitious policy goal based on current energy production and consumption trends in South Korea. The goal is to decrease the proportions of coal and nuclear energy to 22 and 21.6 percent, respectively, by 2030 [16]. The South Korean government has several reasons for promoting this ambitious energy transition policy.

The first reason is the need to participate in the fight against climate change. High dependence on the manufacturing industry makes it difficult for South Korea to reduce GHG emissions. Furthermore, achieving the goal of the NDC (Nationally Determined Contributions) has become a major challenge in energy and environmental policy for the South Korean government. The second reason is to improve sustainability and increase energy security. The last reason is the need to improve safety. Since South Korea has a dense population and mountains constitute 70 percent of its territory, the NIMBY (not in my backyard) regarding energy facilities has increased.

Over the past few years, South Korea has experienced a new phase in national energy policy. First, the policy instrument for boosting renewable energy has changed. South Korea switched from the "Feed in Tariff" system to the "Renewable Portfolio Standard (Hereinafter "RPS")" in 2012. This choice has apparently led to a visible increase in domestic renewable energy production [17]. Second, the renewable electricity target was set at 20% in 2030 [18]. In reality, this goal is a highly ambitious target for South Korea regarding the industrial structure and energy mix. Third, the decision-making process of the government regarding energy policy was shifted [19]. The central government-oriented policy-making system is not sustainable for achieving an ultimate energy transition. Throughout history, high dependence on nuclear and fossil fuels has hindered various actors from participating in the establishment of a basic energy plan [4,20].

In addition to waste-based and large-scale hydro energy, solar energy is the dominant renewable resource in South Korea [21]. Indeed, the adoption of the RPS led to an increase in the total amount (or proportion) of renewable energy production for industrial use [22]. As a holistic approach, the RPS system was an effective policy instrument in South Korea [23,24]. However, a problem remains at the local scale. RPS did not affect small-scale renewable energy facilities that service houses [25]. The growing demands of households are a critical issue affecting the energy situation in the region [26]. Households need energy for heating, lighting, and cooking daily. RE for household use usually depends on small-scale solar PV in South Korea. Small-scale solar PV facilities have numerous advantages; they do not occupy large spaces, nor do they aggressively harm the environment [27].

To gauge the progress of local energy transitions, we focus on solar PV facilities that are representative of the actual RE facilities governed by local authorities. As explained above, the South Korean RE support system is designed around quotas for select major electric power producers through the RPS [28]. This system affects overall RE generation and the domestic energy transition for mainly non-household, industrial uses. However, the number of small-scale RE facilities and their potential output (maximum ability to produce electricity in the region) have also been growing, regardless of the system. Among small-scale RE facilities, solar PV is dominant [29].

In general, the development of renewable energy is highly correlated with the geographic potential of renewable resources. However, from Figure 1 below, it can be seen that this assumption does not match the South Korean context exactly. This mismatch implies that there are attractive factors present in addition to the region's environmental and geographical resources. Before we present our research question, two assumptions must be articulated. First, RE production for industrial use is beyond our research scope. Second, the non-institutional factors that might influence an energy transition, such as population density, infrastructure (grid), and energy resource potential, are not our concerns. To concentrate on institutional conditions, we derived several propositions from previous studies. Then, we inquired about the other conditions that can increase local renewable energy generation. Our research question can be formulated as follows.

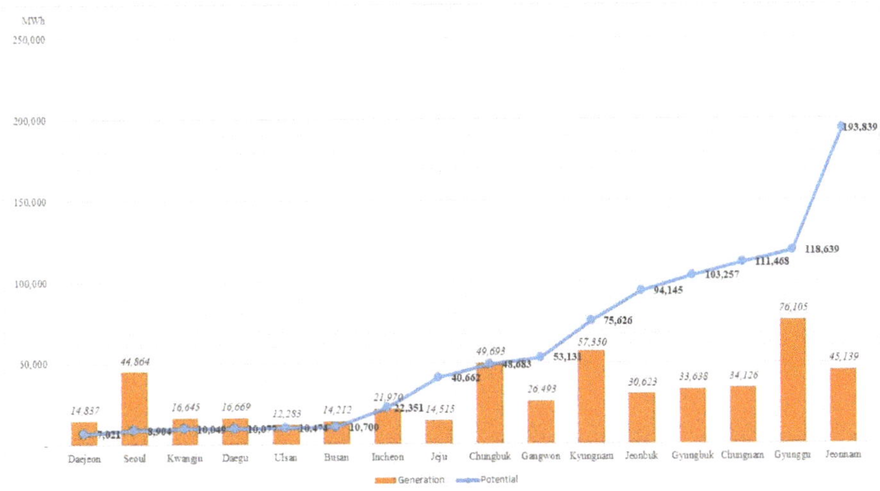

Figure 1. Solar resource potential compared to solar power generation in 16 regions.

Q. What is the most desirable pathway to accomplish a successful local energy transition?

Our study's objective is to provide an explanation of the pathways that lead to successful (meeting the necessary or sufficient conditions) local energy development by analyzing 16 regional cases in South Korea. In our study, a successful energy transition indicates a "comparatively successful" outcome. Considering its methodological limitations, our study cannot provide an imperative numerical outcome. However, it is possible to compare 16 regions' accomplishments in encouraging solar PV generation. Some regions demonstrate lower performance than average, while others demonstrate higher performance. In our study, we focus on solar PV generation when measuring the progress of a local energy transition.

A case-oriented study seeks to utilize the in-depth findings that best fit the specific socio-environmental culture. To explore the possible pathways, we use fuzzy-set Qualitative Comparative Analysis (Hereinafter "fs-QCA") [30]. The remainder of our study is presented as follows. Sections 2 and 3 consists of a literature review, which includes causal conditions and outcomes. Sections 4 and 5 highlights the research design, including the case selection and methodology. Section 6 consists of an analysis of the conditions of successful pathways applying fs-QCA. A summary, and the study's limitations are discussed in Section 7.

2. Local Energy Transition: Solar PV as a Key Measure in the South Korean Context

Local energy transitions cannot be separated from the national one. However, as in most other contexts, unique factors can be found in successful local energy transitions, which are differentiated from the national-scale energy transition. For example, local commitments to ensuring climate protection can provide a good illustration of this phenomenon.

In South Korea, a local energy transition can be a critical factor in achieving national targets, especially at small-scale renewable energy facilities. In the context of industrial use, RPS systems impose generation quotas on selected electricity companies. Accordingly, these electricity companies prefer to establish large-scale RE facilities instead of scattered, small-scale ones that are appropriate for supplying power to households.

Solar PV energy is the most cost-effective renewable energy source for most developed countries. As a result, previous studies have focused on solar PV generation for household use when measuring the expansion of RE [31–34]. The level of technology has already reached the break-even point in the market. In the South Korean energy-generation context, solar PV accounted for 7% of the total RE generation as of 2016 [35].

Moreover, solar PV is a core component of the energy transition in the South Korea [36]. The outcomes of small-scale solar PV implementation for household use can be considered a suitable barometer of the progress of an energy transition for two reasons. First, solar PV is the most feasible source of renewable energy given the population density and geographical features of the nation. South Korea's population grew to 50 million as of 2015, and more than 80% of its population lives in urban areas and cities. The geographical area of South Korea comprises 38,691 square miles, which is equivalent to 1/3 the size of Germany or 1/4 the size of Japan. Thus, it is evident that South Korea does not have appropriate natural terrain for implementing some renewable energy generation methods [37]. A small-scale solar PV facility can be a useful option and can be installed regardless of space constraints (panels can be installed on a rooftop or hung on a balcony in an apartment or multiplex).

Second, only solar PV energy is systematically affordable in the social economy in South Korea. Although the wind power generation accounts for a high proportion of renewable energy globally, in Asia, apart from China, wind power supply is progressing slowly [38]. Since the enactment of the framework act on cooperatives in 2012, the social economy of energy has experienced rapid growth in Korean society. Over 60% of South Korea's energy cooperatives are related to solar PV [39]. Thus, solar PV may be the only renewable resource that can facilitate the energy prosumer trend in renewable resources.

Based on these rationales, we used the amount of regional solar PV energy generation and the regional solar PV expansion rate per year to measure the level of local energy transition that had been achieved. The amount of solar PV generation suggests the mechanism for the current outcome, while the solar PV expansion rate overlooks the mechanism for the acceleration of the energy transition and path.

3. Drivers of Local Energy Transition and Operative Propositions

Numerous studies have illuminated the correlation between local energy transitions and the expansion of RE generation. Zahran et al. [31] focused on using environmental, economic, and socio-political factors to explain the expansion of solar PV for household use. In their study, a political preference for the democratic party and ICLEI membership turned out to be significant factors in the expansion of solar PV in the US. In Kwan [32] and Crago and Chernyakhovskiy [33] studies, incentive programs supported by local financing were found to be one of the most critical factors influencing the prevalence of solar PV facilities.

Recent studies have highlighted the important role of ICLEI membership in the local commitment to sustainable development and awareness of climate change issues [40]. The International Council for Local Environmental Initiatives (ICLEI)-Local Governments for Sustainability was founded in 1990 as a global network of cities and regions dedicated to creating a sustainable future. To date, the ICLEI network includes more than 1700 local governments in more than 100 countries. The critical tasks of the ICLEI consist of 5 pathways: low emission development, nature-based development, circular development, resilient development, and equitable people-centered development.

Although a local energy transition can be achieved without membership in the ICLEI, previous studies have found that membership in the organization is a significant factor in local climate change action [31,41,42]. Moreover, ICLEI membership shows the willingness of local governments to tackle climate change and energy transition in a cleaner and more ecologically friendly way [43]. Seoul city, for example, pioneered the solar city project in accordance with the ICLEI's policy and invested in and implemented solar PV methods to comply with a 100% RE initiative and achieve its zero-carbon goal [44].

The amount of attention paid to local-level legislation has accelerated in recent years [45]. Local ordinances and regulations provide the institutional foundations for local governments to implement their actual climate change actions in a region. Establishing an energy ordinance is not mandatory for every local government in South Korea [19]. In South Korea, a local government has the power to enact local ordinances and regulations regarding energy issues. Accordingly, the level of legislation in this area differs among local governments. However, in areas with local governments that have enacted the energy ordinance, local renewable energy projects are being established based on such ordinances. Blanchet [46] analyzed the two local energy initiatives that were implemented in Berlin city. Although the details varied, the regulations, especially local-scale citizen-led laws, were considered important components of the initiatives.

Financial capacity was considered an influential factor in previous studies [32,33,47]. Unsurprisingly, the most active countries in terms of climate change actions and energy transitions are primarily those in the North. Indeed, "Green (cleaner production) needs green (money)." South Korea has experienced rapid economic growth during the past decade [48]. However, financial strategies for coping with rapid growth can lead to the presence of relatively underdeveloped regions. This regional discrimination has led to differences in the level of financial abundance between regions in South Korea.

Many of the previous studies highlighted the importance of green parties, especially in EU member countries [47,49]. In South Korea, local-scale environmental actions were largely influenced and initiated by grassroots NGOs on behalf of political parties. In South Korea, environmental activism was suppressed by authoritarian regimes in the mid-1980s [50]. Despite the country's late start, NGOs in South Korea have become increasingly active in issues related to the promotion of the public welfare, such as the environment and other social issues [51]. As NGOs in South Korea have become more institutionalized and issue-specific, their social roles have extended their policy influence, making their role equivalent to that of green parties [52].

Environmental NGOs in South Korea grew up in a context of ecologically destructive regional development [19]. As industrialization accelerated, pollution worsened and more frequent local protests by environmental NGOs became prevalent [53]. The role of NGOs in environmental policy is currently apparent at several levels [54,55]. As Korean society does not have empowered political parties focusing on environmental issues, environmental NGOs are assuming the role of green parties. Thus, in our study, environmental NGOs possess the capacity for environmental surveillance and expertise instead of the green parties that were discussed in previous studies. As our study relies on qualitative methodology, we cannot empirically verify its hypotheses. However, it is possible to explore the best combination of several conditions in a qualitative way. Based on our literature review and the logic of qualitative comparative analysis, we formulated four propositions, as shown below.

Proposition 1. *Local Commitment is a necessary or/and sufficient condition for the success of a local energy transition.*

Here, local commitment can be measured by the level of ICLEI membership. Previous studies have highlighted ICLEI membership as a critical factor in the expansion of RE generation [31,40–42]. As ICLEI membership represents the willingness of local governments to promote energy transitions [43], it best indicates the level of local commitment to ensuring the success of an energy transition.

Proposition 2. *Local legislation is a necessary or/and sufficient condition for the success of a local energy transition.*

The importance of local legislation is growing along with the increase in climate change action. Energy-related law is the basis of local energy policies. As mentioned in previous studies, growing numbers of newly enacted energy-related decrees have contributed to the success of local energy transitions [45,46,56]. Therefore, the indispensability of local legislation concretized by the enactment of energy decrees by regional governments should be considered our second proposition.

Proposition 3. *Local financing is a necessary or/and sufficient condition for the success of a local energy transition.* In the past, regional infrastructure development in South Korea has been aligned with the financial

capacity of regions. As in other previous studies [32,33,47], in this study, local financing is considered an important factor in local energy transitions. Local financing can be measured as the GRDP (gross regional domestic product) of 16 regions in South Korea.

Proposition 4. *Environmental surveillance/expertise is a necessary or/and sufficient condition for the success of a local energy transition.*

In Korean elections, green parties are not influential in decision-making even at the national level. On behalf of the green party, environmental NGOs have showed strong leadership in local environmental policy since the 1990s. Environmental NGOs in South Korea have replaced the green parties of Europe, and they are considered "political" actors in the Korean energy policy-making system in reality [56]. Thus, in our study, our last proposition is related to environmental NGOs; Korean environmental NGOs are participating in making and implementing local energy policies with their surveillance capacity and expertise in environmental issues. As a result, environmental surveillance and expertise at the local scale can be measured by the number of environmental NGOs in a region.

4. Methodology: Fuzzy-Set QCA for Intermediate Number Cases

Fuzzy-set qualitative comparative analysis strives to explain complex issues in the social science fields. Developed by Ragin [30], this case-oriented methodology focuses on a combination of specific conditions. A conventional crip-set is comparable to binary variables of 1 and 0. Thus, a crip-set is either inside or outside of a set that is dichotomous [57]. For example, the object can be a Protestant (membership = 1) or non-Protestant (membership = 0) in a survey [57]. By contrast, a fuzzy set provides a level of membership between 0 and 1, which can eventually imply the existence of more qualitative information in transforming each variable. For example, the fuzzy set of "Protestant" includes objects who are fully inside the set (membership = 1); those who are almost fully inside it (membership = 0.90); those who are at the cross-over point in it (membership = 0.50); those who are slightly more outside than inside it (membership = 0.45); and even those who are fully outside of the set (membership = 0), or "non-Protestant" [57]. Thus, a fuzzy set is more empirically grounded and more powerful as a means of explanation than a conventional crip-set [57]. Fs-QCA methodology has been applied in various fields, including energy policy [47]. Fs-QCA has some powerful advantages in research; first, it is an appropriate tool for explaining causal conditions for particular pathways. Specifically, fs-QCA can demonstrate joint causal relations by considering interactive effects between variables (which regression models cannot do because of multi-collinearity).

Second, fs-QCA provides a comprehensive understanding of middle-number cases (15 to 30 examples). Research in social science has mainly been focused on small-N studies and large-N studies. For small-N studies, we use traditional qualitative methodologies, such as in-depth interviews.

For large-N studies, we mostly use empirical methodologies based on statistical models. However, no clear solutions have been suggested for conducting intermediate-sized-N studies, such as, for example, those including 15 to 30 examples [57]. Fs-QCA provides an apt solution to social scientists who have suffered from a "limited repertoire of methods" for studies of intermediate size [57]. In our study, the number of cases is 16, thus fs-QCA is an appropriate methodology.

Third, the causal complexity considered by fs-QCA provides a realistic description of social phenomena [30,58]. When using variable-oriented methods, it is possible to identify broad and theory-related patterns, but case-oriented methods can provide a realistic picture of the specific social phenomena [57] that are actually affecting our society.

Analyzing necessary or sufficient conditions is the key task when implementing fs-QCA methodology. Necessary conditions are usually simple and singular conditions with a high level of consistency. When using an fs-QCA program, we can analyze the necessary conditions before identifying sufficient conditions [59]. The biggest difference between the two types of conditions is the level of consistency. The level of consistency indicates how strongly a condition relates to the results of the analysis [60]. In general, a cut-off point of 0.90 is suggested as the minimum acceptable level of consistency [59]. In the next step, it is possible to analyze the sufficient conditions for a successful energy transition. In general, the threshold for consistency is set at 0.80 for raw consistency [61]. Thus, when analyzing sufficient conditions, we can suggest more complicated combinations of conditions.

The goal of social scientists when using fs-QCA is to find specific patterns [30,62] of solutions that are not unique or universal. Thus, the results of fs-QCA cannot be considered a panacea for every context. However, the results can suggest a plausible solution regardless of the unique context of each region and country.

5. Research Design: Fuzzy-Set QCA Model for Energy Transition Study

Fs-QCA provides a better understanding of middle-number cases, such as those with 15–30 examples according to Ragin, and it also has a powerful explanatory power for "joint causal relations" that regression models do not support because of multicollinearity. First, 16 upper-level local government cases in South Korea were selected for this study which perfectly fits for conducting fuzzy set QCA. Among 16 regions, Sejong was not included because of its short history (Sejong city was created in 2012 to establish a new administrative capital in South Korea). The 16 regional cases selected for this study were Seoul, Busan, Daegu, Incheon, Gwangju, Daejeon, Ulsan, Gyunggi, Gangwon, Chungbuk, Chungnam, Jeonbuk, Jeonnam, Gyungbuk, Gyungnam, and Jeju.

First, solar PV generation and the solar PV expansion rate per year for residential use in 16 regions were selected as dependent variables. The data were collected through Korea Energy Statistical Information System (Hereinafter "KESIS", URL: http://www.kesis.net/main/mainEng.jsp) With the goal of using the most recent data available, we used data from 2015, which is the most recent year for which there is a dataset in KESIS. To measure the solar PV expansion rate, we used the compound annual growth rate of solar PV for each region. The formula was constructed as below.

$$\mathrm{CAGR}(t_0, t_n) = \left[\frac{V(t_n)}{V(t_0)}\right]^{\frac{1}{t_n - t_0}} - 1$$

where $V(t_0)$: start value, $V(t_n)$: finish value, and $t_n - t_0$: number of years.

Second, four causal variables were selected: network, local legislation, financial abundance, and environmental surveillance. A region's level of ICLEI membership was used to determine the local commitment to climate change actions. The data were collected from the ICLEI South Korea Homepage. We calculated the engagement rate of the sub-regions of 16 upper-level local governments from 2015. To measure local legislation power, the number of energy ordinances in each region was compiled from 16 local energy plans. We divided the aggregated numbers of ordinances by the total number of local governments.

To determine a region's financial capacity, we employed the GRDP (gross regional domestic product) of the 16 regions. GRDP can be used to measure the size of a region's economy. GRDP includes the agriculture, fishery, forestry, industry, and service sectors.

Lastly, environmental surveillance was calculated according to the number of environmental NGOs in a region. We compiled data from the list of NGOs registered with the Ministry of Environment (see Appendix A).

Before the calibrations were conducted, descriptive statistics were selected to identify the minimum, the maximum, the mean, and the median. To calibrate the raw data, we used the median to avoid overestimating the dispersion. MS Excel ver.2016 was used to obtain the descriptive statistics from the raw data (see Appendix B for summary of descriptive statistics).

The compiled data from 16 regions were converted to fuzzy-set scores for each case according to the various levels of fuzzy membership. To determine the fuzzy membership classification for each case, we used the maximum, median, and minimum to convert the raw data into fuzzy scores that varied from 0 to 1. For example, regarding local commitment, the minimum raw data score was 1 percent, and the maximum was 13 percent. We used the median of 2 percent as the cross-over point and set 0.95 as the maximum and 0.05 as the minimum for the raw data. The other digits were calculated accordingly, considering the ratio mentioned above. For calibration, we used Microsoft Excel software and fs-QCA 3.0 software by Ragin. The converted fuzzy scores are shown in Table 1 below.

Table 1. Fuzzy scores (calibrated variables).

City (N)	Outcome		Causal Conditions			
	Solar_g	Solar_Exp	ICLEI	Decree	Financing	NGO
Seoul	0.74	0.84	0.87	0.72	0.95	0.95
Busan	0.07	0.95	0.05	0.58	0.7	0.51
Daegu	0.1	0.52	0.05	0.05	0.63	0.5
Incheon	0.23	0.60	0.5	0.75	0.81	0.5
Gwangju	0.1	0.05	0.05	0.05	0.56	0.5
Daejeon	0.07	0.38	0.05	0.05	0.66	0.53
Ulsan	0.05	0.35	0.05	0.85	0.88	0.05
Gyunggi	0.95	0.80	0.95	0.95	0.83	0.71
Gangwon	0.41	0.77	0.8	0.51	0.07	0.53
Chungbuk	0.79	0.62	0.5	0.8	0.19	0.05
Chungnam	0.61	0.45	0.84	0.75	0.2	0.5
Jeonbuk	0.53	0.35	0.05	0.41	0.08	0.12
Jeonnam	0.74	0.20	0.63	0.48	0.05	0.05
Gyungbuk	0.58	0.71	0.5	0.2	0.11	0.51
Gyungnam	0.86	0.23	0.57	0.28	0.41	0.05
Jeju	0.07	0.48	0.5	0.05	0.22	0.05

6. Analysis and Results

Analyses of the necessary and sufficient conditions for a successful local energy transition were performed using fs-QCA software version 3.0. The necessary conditions are those that are directly theoretically relevant to the outcomes and are present all the time. The main reason for studying the necessary conditions is that they have powerful policy implications [57]. How to identify necessary conditions depends on consistency and coverage. Consistency indicates the level of sharing conditions for each case, and it helps the researcher to determine "how well a finding reflects the empirical data" [60]. Coverage is the level of the explanatory capacity covered by the cases. The consistency and coverage scores for the necessary conditions are 0.9 and 0.3, respectively. The objective of investigating the sufficient conditions is to support diversity-oriented research [57]. The sufficient conditions are meaningful when causes are combined. In accordance with Ragin [57], the sufficiency consistency threshold was set at 0.80.

6.1. Necessary and Singular Conditions for a Successful Local Energy Transition

The converted calibrated data were first analyzed to determine the necessary conditions for a successful local energy transition, as shown in Table 2 below. The analysis of necessary conditions was conducted with a two-outcome set: solar PV generation and the solar PV expansion rate. The 0.9-point consistency score is the recommended threshold for determining necessary conditions. As can be explored from Table 2, the environmental surveillance (0.88) for the solar PV expansion rate nearly meets the required consistency score of 0.9. If we adopt a strict interpretation of the textbook definition of fs-QCA, a score of 0.88 cannot indicate a necessary condition because it does not exceed 0.9. However, if we employ the more flexible interpretation of Ragin [57], 0.88 can be considered a quasi-necessary condition for the outcome set. Given the restricted number of cases, this finding indicates that the role of environmental NGOs in a region can be assumed to be a necessary condition for a successful local energy transition.

Table 2. Analysis of necessary conditions.

Conditions	Solar PV Generation		Solar PV Expansion Rate	
	Consistency	**Coverage**	**Consistency**	**Coverage**
Local Commitment (ICLEI)	0.79	0.80	0.80	0.67
Local Legislation Power (DECREE)	0.72	0.78	0.78	0.7
Local Financing (GRDP)	0.76	0.76	0.82	0.68
Environmental Surveillance & Expertise (NGO)	0.62	0.55	**0.88**	**0.64**

6.2. Sufficient Conditions for a Successful Local Energy Transition

Table 3 presents the possible combinations of sufficient conditions for solar PV generation and the solar PV expansion rate. The number of logically possible combinations is $16(2^4) + 16(2^4)$. Analysis of the sufficient conditions includes three steps: constructing the model, editing the rows, and setting the truth table. Truth table analysis is conducted to examine the logical combinations of all possible causes and conditions to explain the resulting conditions. The construction of a truth table begins with an algorithm, which is first inputted into qualitative comparative analysis software [63]; the current version of fs-QCA software also offers a truth table, which is essentially the result of conducting fs-QCA. A truth table shows the possible combinations of conditions and their consistency. In general, a consistency score of 0.8 is the cross-over point.

In this process, setting a truth table is the key procedure for verifying causal conditions [64]. The fs-QCA software shows intermediate, parsimonious, and complex solutions. These three types of solutions provide the combinations of causal conditions. In parsimonious solutions, the model introduces the elimination of existing or absent causes and minimizes additional causes except those of core origin. On the other hand, complex solutions show the maximum amount of combinations of causal conditions. Intermediate solutions are suggested automatically through the process of deriving the most complex result and inversely analyzing the causal conditions. In this study, we will suggest both parsimonious and intermediate solutions and their pathways.

As shown in Table 3, of 16 possible configurations, 8 were empirically observed. A score of "1" implies the partial or full presence of a condition or outcome. Conversely, a score of "0" represents the partial or full absence of a condition or outcome [60].

If we closely examine Table 3, an overall policy evaluation can be conducted according to the truth table. An outcome score of "1" indicates the success of solar PV generation and expansion. Regions such as Jeonbuk had a score of "0" for both of their outcomes in solar PV generation and expansion, signifying the failure of their local energy transitions. However, regions such as Gangwon, Seoul, Gyunggi, and Jeonnam obtained a score of "1" for their outcomes in both solar PV generation and expansion, indicating the success of their local energy transitions.

Regions such as Busan, Ulsan, and Daejeon obtained a score of "0" for their outcomes in solar PV generation; however, they obtained a score of "1" for their outcomes in solar PV expansion. This study indicates that these three regions continue to lack the required absolute production output in solar PV; however, they are constantly endeavoring to promote solar PV generation and to achieve an energy transition. By contrast, regions such as Gyungnam obtained a score of "1" for the outcome of solar PV generation but obtained a score of "0" for the expansion outcome, indicating the opposite result of the above-mentioned cases.

Table 3. Truth table of sufficient conditions for the outcome of solar PV generation.

Conditions	Row	ICLEI	DECREE	GRDP	NGO	Outcome	Consistency	Cases
Solar PV Generation	1	1	0	1	0	1	0.99	Gyungnam
	2	1	1	1	1	1	0.92	Seoul Gyunggi
	3	1	0	0	0	1	0.85	Jeonnam
	4	1	1	0	1	1	0.84	Gangwon
	5	0	1	1	1	0	0.68	Busan
	6	0	1	1	0	0	0.67	Ulsan
	7	0	0	0	0	0	0.58	Jeonbuk
	8	0	0	0	1	0	0.53	Daejeon
	9~16	Logical Remainders						
Solar PV Expansion	1	1	0	0	1	1	1	Gangwon
	2	1	1	1	1	1	1	Busan
	3	1	1	1	1	1	0.98	Seoul Gyunggi
	4	1	1	1	0	1	0.89	Ulsan
	5	0	0	0	0	1	0.84	Jeonnam
	6	0	0	0	1	1	0.82	Daejeon
	7	0	1	1	0	0	0.78	Gyungnam
	8	0	0	0	0	0	0.78	Jeonbuk
	9~16	Logical Remainders						

The overall cut-off for sufficient conditions was constructed using two criteria: a consistency score of 0.8 and a minimum of two cases. Accordingly, the models that could not achieve a consistency score of 0.8 were excluded in the first step. In the second step, the models with a single case were excluded from the final analysis.

An analysis of the combinations of conditions that were sufficient for Solar PV generation was conducted based on three kinds of models as shown in Table 4. The first model, which is a parsimonious solution, includes a single "iclei" condition with a coverage score of 0.80 and a consistency score of 0.79. The Gyunggi (0.95, 0.95), Seoul (0.87, 0.74), Chungnam (0.84, 0.61), Gangwon (0.8, 0.41), Jeonnam (0.63, 0,74), and Gyungnam (0.57, 0.86) cases fit Model I.

The second model is an intermediate solution that includes "iclei*~decree*~ngo." The Gyungnam (0.57, 0.86) and Jeonnam (0.52, 0.74) cases fit Model II. The third model, which is also an intermediate solution, includes "iclei*decree*ngo." Model III includes Seoul (0.72, 0.74), Gyunggi (0.71, 0.95), and Gangwon (0.51, 0.41).

Table 4. Pathways for solar PV generation.

Solution	Model	Pathway	Coverage	Consistency	Number of Cases
Parsimonious	I	iclei	0.80	0.79	6
Intermediate	II	iclei*~decree*~ngo	0.46	0.86	2
	III	iclei*decree*ngo	0.47	0.89	3

An analysis of the combinations of conditions that were sufficient for solar PV expansion was conducted based on six kinds of models as shown in Table 5. Two of the models are based on parsimonious solutions, and the other four models are based on intermediate ones.

The first model, which is a parsimonious solution, includes a single "ngo" condition with a coverage score of 0.64 and consistency score of 0.88. Cases such as Seoul (0.95, 0.84), Gyunggi (0.71, 0.8), Daejeon (0.53, 0.38), Gangwon (0.53, 0.77), Busan (0.51, 0.95), and Gyungbuk (0.51, 0.71) fit Model I.

The second model, which is also a parsimonious solution, includes "iclei*~grdp," a with coverage score of 0.46 and a consistency score of 0.87. Two cases, Gangwon (0.8, 0.77) and Jeonnam (0.53, 0.2), fit Model II.

The third model, which is also a parsimonious solution, includes a "decree" with a coverage score of 0.70 and a consistency score of 0.78. Model III has the most powerful explanatory pathway among the parsimonious solutions for solar PV expansion. Half of the cases fit model III: Gyunggi (0.95, 0.8), Ulsan (0.85, 0.35), Chungbuk (0.8, 0.62), Incheon (0.75, 0.6), Chungnam (0.75, 0.45), Seoul (0.72, 0.84), Busan (0.58, 0.95), and Gangwon (0.51, 0.77). The fourth model, which is also a parsimonious solution, includes "~iclei*grdp" combinations. The Busan (0.53, 0.95) and Ulsan (0.51, 0.35) cases fit Model IV.

Table 5. Pathways for solar PV expansion.

Solution	Model	Pathway	Coverage	Consistency	Number of Cases
Parsimonious	I	ngo	0.64	0.88	6
	II	iclei*~grdp	0.46	0.87	2
	III	decree	0.70	0.78	8
	IV	~iclei*grdp	0.46	0.86	2
Intermediate	V	~iclei*decree*grdp	0.39	0.89	2
	VI	iclei*decree*ngo	0.43	0.98	3

Note: "iclei*~decree*~grdp*~ngo" (0.35, 0.84) and "~iclei*~decree*~grdp*ngo" (0.34, 0.82) were excluded because only a single case of each exists.

The fifth and sixth models were constructed through intermediate solutions. The fifth model includes "~iclei*decree*grdp" combinations with a coverage score of 0.39 and consistency score of 0.89. The Busan (0.53, 0.95) and Ulsan (0.51, 0.35) cases fit Model V. The sixth model consists of "iclei*decree*ngo," with a coverage score of 0.43 and consistency score of 0.98. A score of 0.98 is exceptionally high and indicates how strongly the combinations of "iclei," "decree," and "ngo" relate to solar PB expansion. The Seoul (0.72, 0.84), Gyunggi (0.71, 0.8), and Gangwon (0.51, 0.77) cases fit the powerful Model VI. Based on the results presented above, an optimal pathway for achieving a local energy transition can be suggested by the simple formula below.

Solar PV generation = iclei

+iclei*~decree*~ngo

+iclei*decree*ngo(redundancy)

Solar PV expansion = ngo

+iclei*~grdp

+decree

+~iclei*grdp

+~iclei*decree*grdp

+iclei*decree*ngo(redundancy)

Local energy transition = solar PV generation + solar PV expansion

From the results presented above, the ideal types for a local energy transition can be simplified and summarized as follows (see Figure 2). To suggest a persuasive solution for a local energy transition, we sorted the models that exhibited more than two cases (see the blocks in Tables 4 and 5). In this study, we applied the outcome set through two means: solar PV generation and the solar PV expansion rate. Solar PV generation shows which factors actually facilitated the energy transitions. Below, Type 1 can be used to predict important factors and pathways for other countries.

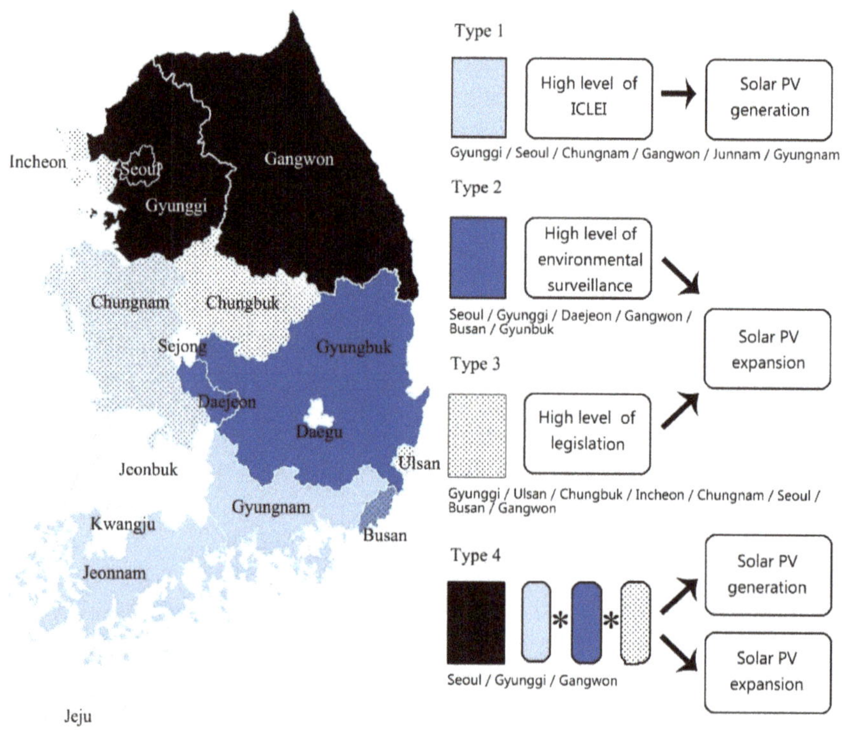

Figure 2. Pathways to a successful local energy transition.

Type 1: High level of ICLEI membership: A high level of ICLEI membership affected the growth of solar PV generation. ICLEI membership singularly proved to be an important condition for the success of a local energy transition. (6 cases observed)

While solar PV generation verified what was important in the future, the solar PV expansion rate showed what will likely be important moving forward. In other words, Types 2 and 3 indicate which factors will promote and accelerate the energy transition henceforth.

Type 2: High level of environmental surveillance: A high level of environmental surveillance obtained from environmental NGOs in a region can be an important factor in expanding and accelerating a local energy transition. (6 cases observed)

Type 3: High level of local legislation: A high level of local legislation power related to the energy decrees in a region can be an important factor in expanding and accelerating a local energy transition. (8 cases observed)

Lastly, Type 4 meets both positive outcomes for solar PV generation and expansion. Type 4 can be summarized as an ideal type of pathway that includes the crucial conditions for accomplishing a local energy transition.

Type 4: A combination of a high level of ICLEI membership, local legislation, and environmental surveillance: The combination of high levels of ICLEI membership, local legislation power, and environmental surveillance can be a successful pathway for promoting a local energy transition. (3 cases observed for each outcome set).

This study presents several meaningful findings as follows. First, ICLEI membership can be a crucial condition for local energy transitions [40,42]. ICLEI membership represents two conditions: a support network and the willingness to achieve sustainability. A previous study highlighted the importance of ICLEI membership. Our findings additionally support those of previous studies and affirm that ICLEI membership can have positive implications for countries such as South Korea, which has unfavorable geographical conditions for renewable energy generation.

Second, environmental surveillance at the local scale has positive effects on local energy transitions. From a global perspective, locally based NGOs are not always pro-renewables [65] since renewable energy facilities contribute to environmental degradation [66]. This situation generally occurs in developed countries where the use of renewable energy is already established. However, countries such as South Korea continue to rely heavily on conventional resources (nuclear and fossil fuel). Accordingly, the positive effect of local NGO involvement in local energy transitions will possibly provide more insight into the situation in developing and newly ascendant countries.

Third, local legislation can positively affect the acceleration of a local energy transition. Since the enactment of an energy ordinance is not mandatory for local governments, the level of legislation also illustrates the amount of effort directed towards a local energy transition. In addition, codified legislation creates an institutional foundation for implementing renewable energy projects in a region. However, the energy ordinance enactment rate remained 39% among lower-level local governments as of 2015 [67].

Lastly, a combination of factors that yields a desirable pathway for a local energy transition can be suggested as follows: a high level of ICLEI membership, a high level of local legislation power, and a high level of environmental surveillance. According to our results, this combination has a positive outcome for both solar PV generation and solar PV expansion. Regarding the strong explanatory power of fs-QCA on social problems, this combination of causal conditions allows us to identify a possible desirable pathway to achieve a successful local energy transition.

In our research, local financing was not a necessary or sufficient condition for a successful local energy transition. This result implies the institutional alignment of a region and the efforts of its local government are more crucial conditions than financial capacity. Another noteworthy finding is that institutional efforts, such as legislation enacted at the local level, the presence of a solid network (e.g., the ICLEI) [68], and the actions of local NGOs, can overcome limitations related to renewable resource potential and geography. This finding provides a rationale for cities or urban areas to organize and promote institutional involvement regardless of the regional environment.

Among the 16 regions that were analyzed, three outstanding ones, Seoul, Gyunggi, and Gangwon, demonstrated the ideal pathway to a successful local energy transition. Especially in the cases of Seoul and Gyunggi, our results support those of previous qualitative studies concerning local energy transitions. Seoul has pioneered an innovative energy strategy involving "One Less Nuclear Power Plant" initiatives, which are focused on achieving energy autonomy and which also highlight the strong leadership of the Seoul city government [4,37]. Mayor Park of Seoul city, who was elected as a chairman of the ICLEI in 2015, announced the "Seoul Declaration, "which clearly indicates the justification for a sustainable energy transition led by a local government.

Gyunggi is also known for its ambitious implementation of solar energy policies, and it claimed a 70% increase in energy independence within the region, promoting a goal of increasing its proportion of renewable energy to 20% in the "Gyunggi Energy Vision 2030" [69]. To accomplish this target, the Gyunggi local government supports a community energy project [70] and established a local energy center, which is the first energy-related public agency in the local level [71]. In the case of Gangwon, no significant in-depth case studies have been conducted qualitatively. The Gangwon region is known for its abundant wind-generation potential in mountainous areas [35], but less is known in terms of its capacity for solar energy generation. However, our analysis implies the rationale for commencing an in-depth case study of the Gangwon region as our next step.

7. Limitation and Conclusions

The main research objective of this study was to explore which combinations of institutional conditions are necessary and/or sufficient for successful local energy transitions and which might be the most desirable pathway to accomplish a local energy transition. To address these issues, 16 regional cases in South Korea were selected for analysis. We examined which conditions affected local energy transitions and which combination of conditions eventually led to successful local energy transitions. To measure the success of local energy transitions, we used solar PV generation for residential use and the solar PV expansion rate for residential use in each region.

In our research, a successful pathway to local energy transition (for both solar PV generation and solar PV expansion) is composed of a high level of ICLEI membership, a high level of local legislation, and a high level of environmental surveillance. This result is quite a convincing result, which can align with the German case. In Wuster and Hagemann's study [48], they found two ways of being successful: green party involvement in governance and high potential, or high potential and a low share of industry in the region. However, the rich federal state (local financing) was not a crucial factor for successful local energy transition. Unlike Germany, the Green party does not play an important role in politics in South Korea. Instead, the South Korean NGOs are much more empowered in local governance systems. Also, their study supports our result that local financing does not affect the success of local energy transition compared to other institutional conditions.

Despite the implications and findings discussed above in Section 6, our study has some limitations. Our study is qualitative in nature. As with other qualitative studies, our study lacks the consideration of control variables for a successful local energy transition. In addition to our four propositions, there might be another, stronger institutional or non-institutional factor that affects local energy transitions. Since the purpose of our study is to explore combination of successful pathway, we failed to consider all factors which might have an effect on local energy transition. Although we did not verify each factor's separate effect, we succeeded to find synergies by combining four institutional conditions derived from previous studies. If we neglect the order of incident of four conditions, we can possibly approach this theme with the regression model using interaction term. And also, it can be a meaningful study to compare four condition's priorities by the utilizing hierarchical regression model if we can update more observations in the future. Nevertheless, the data for solar PV electricity generation for housing first separately collected from 2015. Therefore, the data we used in our study are the best, and it is the most up-to date data collection that we can do. In this regard, the fuzzy set QCA fits perfect for this intermediate sized N.

Additionally, the pathways to a local energy transition may differ according to a country's environment and economic conditions. Our results can provide more useful takeaways for countries where the energy transition process has not yet reached the average level of RE production. Indeed, in terms of energy transition, South Korea remains among the latecomers. In subsequent research, we will expand on the most desirable pathway to a successful local energy transition by examining the cases of other countries, and subsequent research will also provide feedback regarding our results.

Author Contributions: Conceptualization, Y.L., B.K., and H.H.; writing—original draft preparation, Y.L.; formal analysis, Y.L., writing—review and editing, Y.L. and H.H.; visualization, B.K. and H.H.; funding acquisition, B.K. and H.H. All authors have read and agreed to the published version of the manuscript.

Funding: This work was supported by the National Research Foundation of Korea (NRF) for the University Innovation Support Program (UISP).

Acknowledgments: The authors thank the anonymous reviewers for insightful comments that helped us improve the quality of the paper.

Conflicts of Interest: The authors declare no conflicts of interest.

Appendix A

Table A1. Description of the outcome and causal set.

Set	Conditions	Raw Data	Unit	Measurement	Year	Reference
Outcome Set	Solar PV generation	Solar PV electricity generation (for housing)	MWh	-	2015	KESIS (Korea Energy Statistical Information System)
	Solar PV expansion rate	Solar PV expansion rate per year	percentage	-	2015–2017	KESIS (Korea Energy Statistical Information System)
Causal Set	Local commitment	Level of ICLEI membership	percentage	Number of ICLEI members/total number of local governments	2015	ICLEI South Korea
	Local legislation	Enactment of energy decree	percentage	Number of energy decrees in a region/total number of local governments	2015	Local Energy Plan
	Local financing	GRDP	Won	-	2015	Local Finance 365
	Environmental surveillance	Environmental NGO	number	Number of NGOs in a region	2016	Ministry of Environment

Appendix B

Table A2. Descriptive statistics of cases.

Set	Conditions	Unit	Number of N	Minimum	Maximum	Mean	Median
Outcome Set	Solar PV generation	MWh	16	12,273	76,105	28,558	31,919.06
	Solar PV expansion rate	percentage	16	15.3	32.5	24.7	24.7
Causal Set	Local commitment	percentage	16	1	13	3.625	2
	Local legislation	percentage	16	0	80.6	32.11875	32.55
	Local financing	Won	16	15,366	352,857	97,828	67,564
	Environmental surveillance	number	16	0	109	11.1875	3

References

1. Schumacher, K.; Krones, F.; McKenna, R.; Schultmann, F. Public acceptance of renewable energies and energy autonomy: A comparative study in the French, German and Swiss Upper Rhine region. *Energy Policy* **2019**, *126*, 315–332. [CrossRef]
2. Kwon, M.; Jang, H.S.; Feiock, R.C. Climate protection and energy sustainability policy in California cities: What have we learned? *J. Urban Aff.* **2014**, *36*, 905–924. [CrossRef]
3. Renn, O.; Marshall, J.P. Coal, nuclear and renewable energy policies in Germany: From the 1950s to the "Energiewende". *Energy Policy* **2016**, *99*, 224–232. [CrossRef]
4. Lee, Y.; Bae, S. Collaboration and Confucian Reflexivity in Local Energy Governance: The Case of Seoul's One Less Nuclear Power Plant Initiatives. *J. Contemp. East. Asia* **2019**, *18*, 153–174.
5. Walker, G.; Devine-Wright, P.; Hunter, S.; High, H.; Evans, B. Trust and community: Exploring the meanings, contexts and dynamics of community renewable energy. *Energy Policy* **2010**, *38*, 2655–2663. [CrossRef]

6. Seyfang, G.; Haxeltine, A. *Growing Grassroots Innovations: Exploring the Role of Community-Based Initiatives in Governing Sustainable Energy Transitions;* SAGE Publications Sage UK: London, UK, 2012.
7. Devine-Wright, P. *Renewable Energy and the Public: From NIMBY to Participation;* Routledge: Abingdon, UK, 2014.
8. Peters, M.; Fudge, S.; Sinclair, P. Mobilising community action towards a low-carbon future: Opportunities and challenges for local government in the UK. *Energy Policy* **2010**, *38*, 7596–7603. [CrossRef]
9. Fudge, S.; Peters, M.; Woodman, B. Local authorities as niche actors: The case of energy governance in the UK. *Environ. Innov. Soc. Transit.* **2016**, *18*, 1–17. [CrossRef]
10. Jehling, M.; Hitzeroth, M.; Brueckner, M. Applying institutional theory to the analysis of energy transitions: From local agency to multi-scale configurations in Australia and Germany. *Energy Res. Soc. Sci.* **2019**, *53*, 110–120. [CrossRef]
11. Nakamura, H. Disaster experience and participatory energy governance in post-disaster Japan: A survey of citizen willingness to participate in nuclear and energy deliberations. *J. Disaster Res.* **2014**, *9*, 665–672. [CrossRef]
12. Nakamura, H. Local energy governance in post-Fukushima Japan: A survey of citizen willingness to participate in local energy policy deliberations. *Local Environ.* **2015**, *20*, 1000–1017. [CrossRef]
13. Coutard, O.; Rutherford, J. The rise of post-networked cities in Europe? Recombining infrastructural, ecological and urban transformations in low carbon transitions. In *Cities and Low Carbon Transitions;* Routledge: Abingdon, UK, 2010; pp. 123–141.
14. Geels, F.W.; Berkhout, F.; van Vuuren, D.P. Bridging analytical approaches for low-carbon transitions. *Nat. Clim. Chang.* **2016**, *6*, 576–583. [CrossRef]
15. Williams, S.; Doyon, A. Justice in energy transitions. *Environ. Innov. Soc. Transit.* **2019**, *31*, 144–153. [CrossRef]
16. CCC. *South Korea's Energy Transition and Its Implications for Energy Security;* Climate Change Center: Seoul, Korea, 2018.
17. Kwon, T.-H. Policy synergy or conflict for renewable energy support: Case of RPS and auction in South Korea. *Energy Policy* **2018**, *123*, 443–449. [CrossRef]
18. Lee, J. RE target realized by citizen participation. *Electr. Power* **2018**, *12*, 56–57.
19. Lee, Y.H. A Study of Policy Evaluation in Local Climate Change Policy—Case of Incheon Metropolitan City. *Korean J. Local Gov. Stud.* **2018**, *22*, 145–171. [CrossRef]
20. Kim, D.-Y. Energy Governance in South Korea: Long-Term National Energy Master Plans Since 1997. *KDI Sch. Pub Policy Manag. Pap.* **2020**, *20*, 1–22. [CrossRef]
21. Nematollahi, O.; Kim, K.C. A feasibility study of solar energy in South Korea. *Renew. Sustain. Energy Rev.* **2017**, *77*, 566–579. [CrossRef]
22. Kim, J.Y.; Kim, S.B.; Park, S.W. The Effect of the Renewal Portfolio Standards(RPS) on Electric Power Generation in Korea. *Korean Soc. Public Adm.* **2016**, *27*, 131–160.
23. Huh, S.-Y.; Lee, J.; Shin, J. The economic value of South Korea's renewable energy policies (RPS, RFS, and RHO): A contingent valuation study. *Renew. Sustain. Energy Rev.* **2015**, *50*, 64–72. [CrossRef]
24. Chen, W.-M.; Kim, H.; Yamaguchi, H. Renewable energy in eastern Asia: Renewable energy policy review and comparative SWOT analysis for promoting renewable energy in Japan, South Korea, and Taiwan. *Energy Policy* **2014**, *74*, 319–329. [CrossRef]
25. Yoon, J.-H.; Sim, K.-H. Why is South Korea's renewable energy policy failing? A qualitative evaluation. *Energy Policy* **2015**, *86*, 369–379. [CrossRef]
26. Kaygusuz, K. Energy services and energy poverty for sustainable rural development. *Renew. Sustain. Energy Rev.* **2011**, *15*, 936–947. [CrossRef]
27. Lee, H.-J.; Huh, S.-Y.; Yoo, S.-H. Social Preferences for Small-Scale Solar Photovoltaic Power Plants in South Korea: A Choice Experiment Study. *Sustainability* **2018**, *10*, 3589. [CrossRef]
28. Lee, Y.; Seo, I. Sustainability of a policy instrument: Rethinking the renewable portfolio standard in South Korea. *Sustainability* **2019**, *11*, 3082. [CrossRef]
29. Ministry of Trade, Industry and Energy; Korea Energy Agency. White Paper on New and Renewable Energy. 2018. Available online: https://www.knrec.or.kr/download/file_download.aspx?key=291&gubun=pds&div=FILE_NM1 (accessed on 1 November 2019).

30. Ragin, C.C. Qualitative comparative analysis using fuzzy sets (fsQCA). In *Configurational Comparative Methods*; SAGE Publications Inc.: Thousand Oaks, CA, USA, 2009.
31. Zahran, S.; Brody, S.D.; Vedlitz, A.; Lacy, M.G.; Schelly, C.L. Greening local energy: Explaining the geographic distribution of household solar energy use in the United States. *J. Am. Plan. Assoc.* **2008**, *74*, 419–434. [CrossRef]
32. Kwan, C.L. Influence of local environmental, social, economic and political variables on the spatial distribution of residential solar PV arrays across the United States. *Energy Policy* **2012**, *47*, 332–344. [CrossRef]
33. Crago, C.; Chernyakhovskiy, I. Solar PV Technology Adoption in the United States: An Empirical Investigation of State Policy Effectiveness. In Proceedings of the 2014 Annual Meeting, Minneapolis, MN, USA, 27–29 July 2014.
34. Li, H.; Yi, H. Multilevel governance and deployment of solar PV panels in US cities. *Energy Policy* **2014**, *69*, 19–27. [CrossRef]
35. Ministry of Trade, Industry and Energy; Korea Energy Agency. White Paper on New and Renewable Energy. 2016. Available online: https://www.knrec.or.kr/download/file_download.aspx?key=304&gubun=pds&div=FILE_NM1 (accessed on 1 November 2019).
36. Kim, J.H.; Han, J.K. An Experience of Living Lab as Energy Transition Experiment: The Case of Urban Living Lab for Mini-PV System in Seong-Dae-Gol, Seoul, KOREA. *Sci. Technol. Rev.* **2018**, *18*, 219–265.
37. Lee, T.; Lee, T.; Lee, Y. An experiment for urban energy autonomy in Seoul: The one 'less' nuclear power plant policy. *Energy Policy* **2014**, *74*, 311–318. [CrossRef]
38. IEA-PVPS. *Snapshot of Global Photovoltaic Market: A Year Annual PV Installations*; International Energy Agency: Paris, France, 2018.
39. Park, J. The state and tasks of energy cooperatives in Korea: From the perspective of energy citizenship. *ECO* **2015**, *19*, 173–211.
40. Hoppe, T.; van Bueren, E. *Guest Editorial: Governing the Challenges of Climate Change and Energy Transition in Cities*; Springer: Berlin/Heidelberg, Germany, 2015.
41. Brody, S.D.; Zahran, S.; Vedlitz, A.; Grover, H. Examining the relationship between physical vulnerability and public perceptions of global climate change in the United States. *Environ. Behav.* **2008**, *40*, 72–95. [CrossRef]
42. Yi, H.; Krause, R.M.; Feiock, R.C. Back-pedaling or continuing quietly? Assessing the impact of ICLEI membership termination on cities' sustainability actions. *Environ. Politics* **2017**, *26*, 138–160. [CrossRef]
43. Koh, J.K.; Kim, S.W. A Study on Factors Affecting Deployment of Solar PV System. *J. Krda* **2016**, *28*, 109–128.
44. ICLEI. Available online: http://www.icleikorea.org/main (accessed on 1 November 2019).
45. Measham, T.G.; Preston, B.L.; Smith, T.F.; Brooke, C.; Gorddard, R.; Withycombe, G.; Morrison, C. Adapting to climate change through local municipal planning: Barriers and challenges. *Mitig. Adapt. Strateg. Glob. Chang.* **2011**, *16*, 889–909. [CrossRef]
46. Blanchet, T. Struggle over energy transition in Berlin: How do grassroots initiatives affect local energy policy-making? *Energy Policy* **2015**, *78*, 246–254. [CrossRef]
47. Wurster, S.; Hagemann, C. Two ways to success expansion of renewable energies in comparison between Germany's federal states. *Energy Policy* **2018**, *119*, 610–619. [CrossRef]
48. Lee, H.K.; Kim, H.Y. Economic growth for ecological conversions: South Korean case. *Environ. Sci. Eur.* **2018**, *30*, 21. [CrossRef] [PubMed]
49. Kim, E.; Heo, E. Analysis on the Effects of Renewable Policies in OECD Countries Using Dynamic Panel Model. *Environ. Resour. Econ. Rev.* **2016**, *25*, 229–253. [CrossRef]
50. Komori, Y. Evaluating regional environmental governance in Northeast Asia. *Asian Aff. Am. Rev.* **2010**, *37*, 1–25. [CrossRef]
51. Kim, H.-R. The state and civil society in transition: The role of non-governmental organizations in South Korea. *Pac. Rev.* **2000**, *13*, 595–613. [CrossRef]
52. Kim, E. The limits of NGO-government relations in South Korea. *Asian Surv.* **2009**, *49*, 873–894. [CrossRef]
53. Schreurs, M.A. Democratic transition and environmental civil society: Japan and South Korea compared. *Good Soc.* **2002**, *11*, 57–64. [CrossRef]
54. Raustiala, K. States, NGOs, and international environmental institutions. *Int. Stud. Q.* **1997**, *41*, 719–740. [CrossRef]
55. Spiro, P.J. New global communities: Nongovernmental organizations in international decision making institutions. *Wash. Q.* **1995**, *18*, 45–56. [CrossRef]

56. Lee, Y.H. A Study on the Policy Design of Energy Welfare for Solving the Energy Poverty Problem: Case of France and Korea. *J. Comp. Gov.* **2018**, *22*, 43–72.
57. Ragin, C.C. *Fuzzy-Set Social Science*; University of Chicago Press: Chicago, IL, USA, 2000.
58. Vinke-de Kruijf, J. *Research Approach and Preliminary Analysis of a Comparative Sttudy on Multi-Level Learning: The Case of European Climate Change Adaptation Projects*; University of Twente: Enschede, The Netherlands, 2018.
59. Rihoux, B.; Ragin, C.C.; Yamasaki, S.; Bol, D. Conclusions—The way (s) ahead. In *Configurational Comparative Methods: Qualitative Comparative Analysis (QCA) and Related Techniques*; SAGE Publications Inc: Thousand Oaks, CA, USA, 2009; pp. 167–178.
60. Van der Heijden, J. The role of government in voluntary environmental programmes: A fuzzy set qualitative comparative analysis. *Public Adm.* **2015**, *93*, 576–592. [CrossRef]
61. Ragin, C.C. Measurement versus calibration: A set-theoretic approach. In *The Oxford Handbook of Political Methodology*; Oxford University Press: Oxford, UK, 2008.
62. Ragin, C.C.; Amoroso, L.M. *Constructing Social Research: The Unity and Diversity of Method*; Pine Forge Press: Thousand Oaks, CA, USA, 2010.
63. Drass, K.; Ragin, C.C. *Qualitative Comparative Analysis 3.0*; Northwestern University Institute for Policy Research: Evanston, IL, USA, 1992.
64. Verweij, S.; Klijn, E.H.; Edelenbos, J.; Van Buuren, A. What makes governance networks work? A fuzzy set qualitative comparative analysis of 14 Dutch spatial planning projects. *Public Adm.* **2013**, *91*, 1035–1055. [CrossRef]
65. Landeta-Manzano, B.; Arana-Landín, G.; Calvo, P.M.; Heras-Saizarbitoria, I. Wind energy and local communities: A manufacturer's efforts to gain acceptance. *Energy Policy* **2018**, *121*, 314–324. [CrossRef]
66. Dvořák, T. The use of local direct democracy in the Czech Republic: How NIMBY disputes drive protest behaviour. *Local Gov. Stud.* **2018**, *44*, 329–349. [CrossRef]
67. Park, K. *Cooperation between Central Government and Local Government in Energy Project Implementation*; Korea Energy Economic Institute: Ulsan, Korea, 2017.
68. Foljanty-Jost, G. NGOs in environmental networks in Germany and Japan: The question of power and influence. *Soc. Sci. Jpn. J.* **2005**, *8*, 103–117. [CrossRef]
69. Koh, J.; Kwon, O. A Study on Gyeonggi-Do's Community Energy Potential and Policy Implications. *GRI Rev.* **2018**, *21*, 231–258.
70. Choi, S.; Choi, G. A Study on the Energy Cooperative for Energy Transition—A case study of Citizens' Energy Cooperative in metropolitan area. *J. Korean Urban Manag. Assoc.* **2018**, *31*, 65–84. [CrossRef]
71. Koh, J.; Cho, S.; Ye, M. A Progressive Plan for Gyeonggi-Do Energy Center. *GRI Rev.* **2018**, *3*, 1–142.

energies

MDPI

Article

National Energy Transition Framework toward SDG7 with Legal Reforms and Policy Bundles: The Case of Taiwan and Its Comparison with Japan

Mu-Xing Lin [1], Hwa Meei Liou [2] and Kuei Tien Chou [3,*]

1 Graduate Institute of National Development, National Taiwan University, Taipei 10617, Taiwan; d03341012@ntu.edu.tw

2 Graduate Institute of Technology Management, National Taiwan University of Science and Technology, Taipei 10607, Taiwan; liouhm@mail.ntust.edu.tw

3 Graduate Institute of National Development, and Risk Society and Policy Research Center, National Taiwan University, Taipei 10617, Taiwan

* Correspondence: ktchou@ntu.edu.tw; Tel.: +886-(0)2336-633-28

Received: 2 February 2020; Accepted: 12 March 2020; Published: 16 March 2020

Abstract: The main problem explored in this study is how Taiwan and other countries meet the challenges of the United Nations Sustainable Development Goals regarding energy transition by using legal instruments or policy bundles. This study adopts textual analysis and legal policy analysis as its main form of research methodology, and the theory of energy justice, as well as principles of energy management, to correlate with the Sustainable Development Goals. Furthermore, this study aims to construct an analysis structure for national energy transition and to analyze the current situation within Taiwan's electricity sector reforms, while providing evidence of the national experience of electrical industry reforms as an international reference. This study also compares the differences between the seventh Sustainable Development Goal relationship and national energy transitions in Taiwan and Japan, based on the similar initiative of the revised Electricity Act with the policy bundle. This study specifically finds that, firstly, the theory for energy justice is connected with the principles for energy management, owing to the same concepts of "Fair Competition", via the recognition of "Energy Development and Poverty", which correlates with "Environment Protection". Therefore, the concept of energy transition proposed in this study integrates national energy development policy goals and combines them with environmental sustainability, the green economy, and social equity. Secondly, the national energy transition in Taiwan is a response to the Sustainable Development Goals, and electricity sector-related laws could be used as legal tools for national energy transition. This study concludes that Taiwanese and Japanese governments can strengthen their environmental regulations to promote fair competition directly, with fair competition then being able to enhance stable electricity supply, to enable these countries to move towards the seventh Sustainable Development Goal and its indicators. Finally, the analysis structure used in this study could be used as a policy analysis tool for other countries during their own energy transition, when a nation is willing to strategically reform its electricity sector and make sustainable choices regarding transition paths and policy bundles that are suitable for the situation of the individual country. Then, a nation can make revisions to its laws and formulate a policy that is in line with local conditions, while as simultaneously implementing the Sustainable Development Goals.

Keywords: sustainable development goals; energy transition; electricity sector; energy justice; sustainable energy

1. Introduction

1.1. Sustainable Development Goals at the National Level

The United Nations (UN) adopted the "Transforming our World: The 2030 Agenda for Sustainable Development" as its development agenda, which includes a manifesto called the Sustainable Development Goals (SDGs), as well as related targets and follow up and audit protocols [1,2]. The seventh Sustainable Development Goal (SDG7) aims to ensure that, by 2030, all people have access to affordable, reliable, sustainable and modern energy, which, at a national level, has a concrete meaning in terms of the common development of renewable energy and a significant improvement in energy efficiency. The International Bank for Reconstruction and Development et al. [3] have worked together with the UN Environment Programme (UNEP) to publish annual progress reports on the progress made toward sustainable energy at the international, regional and national levels. Moreover, in terms of SDGs, besides climate change, energy transition is the greatest transitional challenge facing developed countries [4]. Sustainable energy is vital in the efforts to attain these SDGs, as renewable energy could be used in the poorest communities to provide modern energy services [5]. Munro et al. [6] pointed out that, in order for SDG7 to be realized, energy justice needs to be considered, with misgivings over the potential for energy transition to marginalize traditional energy producers and consumers. Energy transition, therefore, needs to emphasize the importance of clear procedures and a political agenda for recognition justice. This study seeks to interpret the concepts of procedural justice [7,8] and recognition justice [9,10] in a Taiwanese context and to propose related energy policies within this context, assuming that the SDG7 not only provides the meaning of energy, but also enables individual countries to respond to the SDG7 by setting goals and making their own policy bundles for national energy transitions, while enabling interactions between the national energy transition goals for each individual country to adjust the timing and priority of their policy bundles in order to reach their goals for national energy transition. Furthermore, the international research community has also launched a well-known and flagship project, called "The World in 2050 (TWI2050)", as a follow up to the SDGs [11]. The report entitled "Transformations to achieve the Sustainable Development Goals" was first prepared by The World in 2050 initiative at the High-Level Political Forum 2018 in New York [12]. Although policy-oriented rather than legally binding, this report integrates the seventeen SDGs into six transformation fields, including "decarbonization and energy", extending the time scale from 2030 to 2050, and seeking a national energy transition in the long term, which relates to this study.

1.2. Taiwan's Efforts to Implement SDGs and the Highlights of This Study

As it has progressed into a developed country, Taiwan has made the UN SDGs an important foundational basis for its policymaking. Beginning in 2015, the Executive Yuan's National Council for Sustainable Development has held a series of "Sustainable Development Goals Review Research Meetings", setting clear development and completion times for the SDGs [13,14]. In November 2016 at the 29th meeting of the Executive Yuan's National Council for Sustainable Development, a resolution was passed agreeing to Taiwan's participation in the UN SDGs; the Taiwanese SDGs were established, and a draft was announced, including the proportion of household access to power supply, the electricity generated by clean fuel, and the electricity generated by renewable energy as measurable SDG7 indicators [15]. As Taiwan is currently in the process of an energy transition, there is also a great need to implement reforms related to associated policy and legislation. According to Taiwan's Power Company statistics, in 2018, the percentage of electricity in Taiwan generated by renewable energy, including electricity generated by hydroelectricity power, only accounted for 4.9% of the total electricity generated, with fossil fuels accounting for 82.2% and the electricity generated by nuclear power accounting for 11.4% [16]. These statistics show that greater efforts still need to be made towards increasing renewable energy and carrying out a transition in the electricity structure. This effort of policymakers is not only needed in terms of changes to the electricity structure, as energy transition in

Taiwan faces both vertical international pressure to reduce carbon emissions, and horizontal pressures from domestic energy democracy, with energy carbon emissions, industry transition and air pollution forming a triple helix of challenges. Changes in the use of high carbon emission fossil energy, as well as adjustments to high energy consumption and high air pollution industries, are related to the international and domestic demands for a reduction in carbon dioxide and air pollution [17,18]. Transitioning from fossil fuel to clean forms of energy and low carbon electricity is an important method to resolve the dialectical problems of this spiral-entangled triple helix, as low carbon electricity also produces lower levels of air pollution and CO_2. When the International Energy Agency (IEA) [19] first proposed a market structure analysis report for a low carbon electricity system, the report referred to the environmental regulation of carbon pricing, electricity taxation and de-carbonization as being an important direction for the international energy transition in the post-Paris Treaty climate. Moreover, Fouquet [20] noted that some countries, through institutions, have been able to increase their speed of energy transitions and their use of low pollution energy. Taiwan has made "The Amendment of Electricity Act" into an institutional tool for energy transition to enable electricity industry reforms, as well as to break the monopoly of the electricity market. Among critical reviews of the new Electricity Act in Taiwan, the study of Gao et al. [21] focused on the Taiwanese electricity sector's liberalization and introduced other countries' approaches to energy transition using an inductive method. Going beyond the study of Gao et al. [21], with a concentration on renewable energy development, the main purpose of the present study is to establish an analytical structure with diverse and sustainable perspectives, derived from energy theory and the UN SDGs, to theorize the national energy transition, and to provide empirical Taiwanese evidence in a social context by holding six focus group forums. There are five main areas of significance or foci to this study: (1) the national energy transition in response to the UN SDGs; (2) a transition analytical structure derived from energy theory; (3) evidence of the challenges for national energy transition seen in the current reforms of Taiwan's electricity sector; (4) laws related to the electricity sector acting as legal tools for national energy transition; (5) the concepts that every country should use methods in line with their own local conditions to develop a policy bundle and move toward an energy transition. The main contributions of this study to the existing literature are explained as follows: (1) this pioneering study correlates the national energy transition with SDG7 by adopting energy theory and conducting legal studies. (2) The national energy transition framework features generalized knowledge, deducted from energy justice theory and energy management principles, and this study also provides practicable evidence through case studies of Taiwan and Japan. (3) Energy justice theory is too abstract to be applied, but this study establishes an applicable analytical framework using components from this theory, along with its path, to fill the knowledge gap between theory and practice or the policy gap between ambitious SDGs and their concrete indicators. (4) Outside the particular Taiwanese contexts, this study demonstrates that other countries, like Japan, can adopt a generalized energy transition framework. Regardless of Taiwan and Japan, other nations can develop their own national frameworks to diagnose their energy transition status and to focus on different components or paths in the framework. (5) Dioha and Kumar [22] innovatively studied SDG7's effects, especially on the residential sector, via a quantitative analysis. However, in the present study, the national energy transition, which we approach via a qualitative analysis, is cross-sectional, and a national pathway is necessary to promote sustainable energy comprehensively and efficiently. (6) Marcillo-Delgado et al. [23] adopted a quantitative analysis of electricity access, which is one of the SDG7 indicators and is an aspect of stable supply. Based on stable supply achieved through a path with a diverse supply with renewable energy, Korkovelos et al. [24] provided electrification modelling to support SDG7. However, based on the analytic framework of this study, the energy transition framework could be assumed to consider the comprehension and efficiency of balancing a stable supply, as well as other aspects, such as environmental protection, fair competition, and energy democracy.

The paper proceeds as follows: Section 2 presents preliminary findings on the implications of SDGs for national energy transition and a legal tool to promote the energy transition in Taiwan. Section 3 outlines energy theory. Section 4 mainly presents the methodology and analytical framework. Section 5 compares the status of energy transition between Taiwan and Japan and presents a generalized national energy transition framework with a path through deduction, from energy theory to implementation. Finally, Section 6 offers our conclusions.

2. This Study

2.1. Preliminary Findings on the Relationships Between SDGs and National Energy Transition

The 2030 agenda for Sustainable Development has significant implications when it comes to energy transition (Table 1), with the main goal of SDG7 primarily pursuing the development of sustainable energy and the implementation of an energy transition at the national level [22,25–28]. SDG7 involves building a vision for developing sustainable energy and an energy transition, as well as a new agenda for sustainable energy services, with the aim to steadily provide the public affordable and clean energy or electricity. From this perspective, SDG7 is interrelated with the other goals in the agenda, with SDG8 working towards the goal of decent work and economic growth by focusing on the development of the social economy, such as by increasing energy-related research and development funding and human resources, in order to improve energy use efficiency. SDG9 aims at fostering innovation to promote industrial transition, while at the same time achieving energy efficiency. Like SDG8, this goal is focused on the economic aspect of development, but it is not clear how global energy transition will affect national labor markets [29]. SDG11 is aimed at the development of a sustainable city, there are also energy implications for this goal in terms of improving energy efficiency. Finally, SDG13 focuses on climate change and, according to Sachs et al., [30] the measurable indicators of average CO_2 emissions are the consequences of energy carbon emission, with environmental pollution issues related to high carbon emissions being found in the energy sector, energy industry and high energy consumption, which, in turn, means focusing on the aspect of environmental protection.

Table 1. The meaning of energy and its indicators in the 2030 Agenda for Sustainable Development.

Contents	Levels	SDGs	Meaning of Energy	Indicators
Sustainable Energy	Vision	SDG7	• Affordable, Reliable, Sustainable, and Modern Energy • Joint Development for Renewable Energy • Double the Rate of Improvement in Energy Efficiency	• Access to Electricity • Access to Non-Solid Fuels • CO_2 Emissions from Fuel Combustion, and Electricity Output • Share of Renewable Energy in Total Final Energy Consumption
Energy Transition				
Sustainable Energy Service	Agenda			
Decent Work	Goal	SDG8	Improvement in Energy Efficiency	
Economic Growth	Goal			
Industrial Transition	Goal	SDG9		
Technology Innovation	Opportunity			
	Goal			
Sustainable Growth of Cities	Goal	SDG11		
Climate Change	Goal	SDG13	Energy-related Carbon Emission	Energy-related CO_2 Emissions Per Capita

2.2. New Electricity Act to Implement SDG7 and Promote the National Energy Transition

Taiwan is currently in the early stages of promoting sustainable energy development and needs reforms in both policymaking and legislature, in order to break past its path dependence and construct innovative energy transition mechanisms. This study integrates the trends outlined by the UN SDGs and examines Taiwan's reforms of the Electricity Act and other complementary measures to inspect energy transition. In 2017, the Taiwanese government's amendments to the National Energy Policy, as well as the Electricity Act, revealed a double axis of policymaking and legal reforms, in turn driving the promotion of sustainable energy development. The above developments can be seen as important moments in the concretization of SDG7.

Besides examining an amended draft of the Electricity Act, this paper also adopts an energy justice and energy management perspective to develop related research, thereby carrying out empirical research based on a comprehensive overview of all of the policies related to energy transition or the development of renewable energy. Based on a previous adoption of Section 3.1 "Energy Theory and the Taiwan Status", this study adopts a definition of "energy justice" as "a global energy system that fairly disseminates both the benefits and costs of energy services, and one that has representative and impartial energy decision-making" [7], as well as a "definition of energy management" as a "forward-looking, organized and systematic coordination of energy procurement, conversion, distribution and utilization in order to cover requirements whilst taking into consideration ecological and economic objectives" [31]. That is to say, by an analytical structure for analyzing energy transition, while also carrying out an empirical study of Taiwan's practical transition experience, this study is able to dissect and identify the elements lacking in the current national energy transition policy and the driving force for energy transition, while, in turn, providing a reference for international comparisons.

3. Energy Theory and Taiwan's Status

3.1. Energy Justice and Its Challenges in Taiwan

The core implication of energy justice [32–35] lie in the recognition justice of energy development [9,10], the procedural justice of energy democracy [7,8], and the distributive justice of energy risk [36–38], which takes further steps to structuralize energy justice, using an analysis of other countries' domestic energy development situations, as well as discussing their suitability. Fan [39] discussed the overall international perspective on energy justice, noting that, internationally, there is an emphasis on discussions of fuel poverty, and that this notion of fuel poverty relates to issues of distributive justice and environmental justice. Therefore, the three core concepts of "energy justice" are "recognition justice, procedural justice and distributive justice". Based on a preliminary review, Taiwan's government has yet to integrate recognition justice in a way that provides a fully open and friendly environment for either natural or legal persons to operate electricity-generating enterprises. As a result, there is yet to be a recognition of public self-powered power generation or the right to buy, sell or dispose of electricity. In terms of procedural justice, there is a need for the government to adhere to legal due process in the formulation of policies, plans, individual cases, or development plans, including the opening of the policymaking process to public participation. For example, besides the formulation of policy, legally speaking, an environmental impact assessment should be carried out in order to enable the public to participate earlier in the process [40]. Quitzow et al. [41] described how, in the case of Germany's energy transition, there has not only been the development of renewable energy technology, the implementation of electricity feed-in tariffs, and the covering of related costs, but also a model of public participation and process of such developments, a fact that has earned itself attention in its own right in a climate where the impact of many countries' energy policies has produced social and economic change accompanied by tension and conflict.

Distributive justice could extend its influence to simultaneously address the social injustice of energy subsidies and energy poverty to bring about greater social equity [42,43]. Firstly, in terms of the social injustice of energy subsidies, in 2013, the Taiwan Control Yuan [44] noted that civilians having to fund subsidies for industrial-use electricity was unreasonable and demanded that the economic sector and the Taiwan Power Company (TaiPower Company) review this policy. Furthermore, in relation to the issues of social equity and energy poverty, the Department of Energy and Climate Change, UK, beginning in 2010, released its first Annual Fuel Poverty Statistics Report, which showed the statistics related to fuel poverty [45]. Moreover, the Health and Public Services Committee of London Assembly redefined fuel poverty [46]. Furthermore, the delineation of South Korea's energy poverty line must be taken into account in assessments of governmental energy welfare policies [47]. In conclusion, the issue of distributive justice is not limited to public services, but also relates to energy subsidies, fuel poverty and other energy-related welfare policies.

3.2. Energy Management and Its Challenges in Taiwan

Energy management is multifaceted, and there must be a mediation of the conflicts that are produced between its various facets. Liou [48] organized the theory and practical opinions on this subject from various countries and, based on this information, developed energy management principles including stable supply, fair competition and environmental protection.

In terms of stable supply, Taiwan has low energy self-sufficiency, and lacks energy autonomy, with its energy security being affected by Taiwan's diplomatic relations and energy value. The energy supply is affected by the technological ability or system integration of electricity generation, transmission, and distribution; therefore, these factors are addressed in the Bureau of Energy's Electricity Act. The Bureau of Energy [49] announced that a stable supply of electricity is the premise for promoting energy transition and electricity reforms, including opening electricity-generating enterprises and electricity-retailing enterprises to market competition, allowing consumer plants the free purchase of electricity and promoting the localization and decentralization of electricity use to improve the operational efficiency of electricity enterprises, to increase user rights and to create an electricity development environment that is decentralization friendly.

The root cause of unfair competition relates to whether the environmental costs are internalized for nuclear power, thermal-generated electricity, and renewable energy. In response to its problem of insufficient power supply, in 1995, Taiwan gradually opened up its market to independent power producer (IPPs) and signed 25-year contracts with them. Altogether, there are nine IPPs, all of which feature coal-fired or gas-fired power plants [50]. The establishment of these IPPs meant an end to the TaiPower Company's monopoly of the electricity generation industry. IPPs must negotiate their electricity prices with the TaiPower Company [51]. Even with the new Electricity Act's opening up of the electricity-generating market for renewable energy electricity-generating enterprises, the Green Citizens' Action Alliance, the Terminator of Nuclear Heresies and the Taiwan Youth Climate Coalition all questioned the decision to open up renewable energy-generated electricity to market competition for fear that this action would limit the development of renewable energy [52]. Opening up renewable energy-generated electricity to market competition could very easily result in obstructing the development of renewable energy, because the true cost of nuclear power and thermal-generated electricity has yet to be realistically internalized. As a result, traditional forms of energy and renewable energy are not yet on an equal footing and need management through environmental regulations to enable the internalization of the relevant environmental costs to become a reality, in turn promoting fair competition between electricity-generating enterprises.

From the perspective of environmental protection, Liou [53] studied the reduction of carbon dioxide emissions in Taiwan's electricity enterprises, proposing that Taiwan's carbon dioxide emissions currently account for 1% of the total world emissions, while, in 2013, the carbon dioxide emissions caused by Taiwan's electricity enterprises accounted for 58% of Taiwan's overall emissions. This study reviewed the legislation design related to reducing carbon dioxide emissions in electricity enterprises and provided relevant recommendations. Furthermore, the revisions made to the Electricity Act in terms of how it relates to environmental protection were focused on the internalization of external costs. However, the Environmental Protection Administration (EPA) expressed that, at the present time, there was no discussion of a stipulation for energy tax [54]. The EPA also studied and compared electricity enterprises' power generation equipment with that of renewable energy in terms of the Regulation of the Fund for the Development of Renewable Energy, proposing that this form of funding could replace energy taxes, while achieving the same function. In this way, future electricity generation enterprises could pay different amounts into the fund depending on the form of fossil fuel they use [55].

4. Methodology, Analysis Structure, and Data

4.1. Methodology

This study adopted textual analysis and legal policy analysis as its main form of research methodology. For textual analysis, this study analyzed papers and materials (please refer to Section 4.3, "Data"). For legal policy analysis, the following steps were taken: Firstly, we referenced the contents of each draft made of the revisions to the legislation [49,56–59], as well as the content of the recommendations made by the stakeholders at each public hearing [52]. Secondly, this study classified the contents of the drafts or revisions mentioned above to present the legal policy of the Electricity Act. This study's analytical structure also made use of legislative policy to analyze current challenges within the national energy transition. Moreover, this study attempted to organize related theories, and use them to deconstruct the complex content of the revisions to the Electricity Act, as well as legislative policy, developing a discourse for reforms and liberalization of electricity enterprises. For example, energy theory and Taiwan's status were used to summarize the role of energy justice and energy management discourse, which relate to the challenges Taiwan currently faces, in turn incorporating theoretical implications as a default aim of revision and adopting the systematic foundational principles and analytical structure of the national energy transition in order to discuss the related problems or the aims of the revisions to the Electricity Act. Finally, this study compiled information for a follow-up policy evaluation of the Electricity Act amendments to fulfill the requirements of these theoretical elements.

4.2. Energy Transition Analysis Structure

In order to provide a complete analysis structure for the amendments being made to the Electricity Act, this study attempted to fuse various theories and the components of energy management and energy justice with the development goals cited in Taiwan's Policy Guidelines for Energy Development, in order to organize them into an energy transition analysis structure (Figure 1). The outside half circle in Figure 1 represents the three components of the principles of energy management. The components in the circle were arranged in temporal order in the preliminary stage of this study, but the connected paths between these components are justified later and shown in Figure 2. In more detail (Figure 1), the theory for energy justice is connected with the principles for energy management, which are based on the same contents of "Fair Competition" and recognition of "Energy Development and Poverty" as it correlates with "Environment Protection". That is to say, the concept of energy transition proposed in this study fuses national energy development policy goals and combines them with environmental sustainability, the green economy, and the concepts of sustainability, and social equity. This kind of analysis structure agrees with the international trend established by the UN SDGs. Moreover, the main legal foci of this study are the revisions made to the Electricity Act; the Electricity Act itself is the key component, with other policies arranged accordingly. In terms of research methodology, this study considered the legislation before and after the 2017 amendments to the Electricity Act. We adopted a national energy transition analysis structure, alongside the legislative policy (after revisions), to dissect the current status and challenges facing Taiwan's energy transition. Finally, we interrogated the revision process and situation after the revisions were completed to be able to identify legislation and policies that support the 2017 new Electricity Act, in order to promote a national energy transition policy bundle.

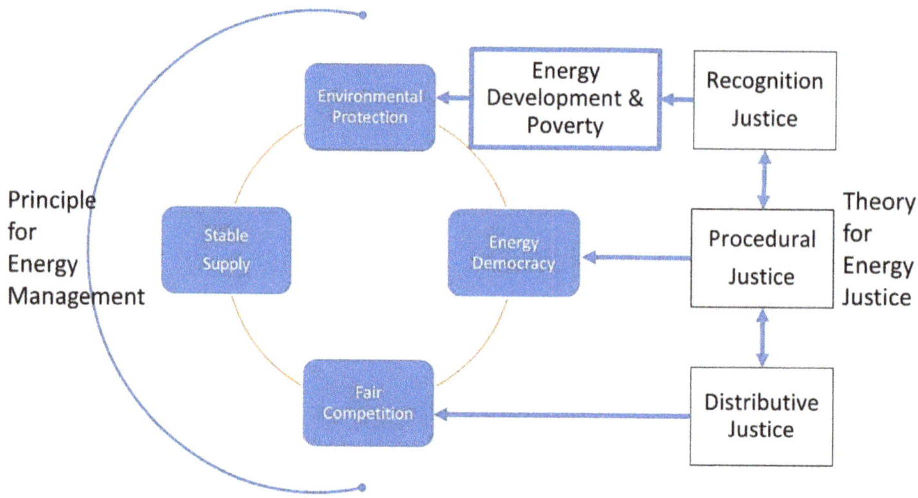

Figure 1. Analytical framework for national energy transition.

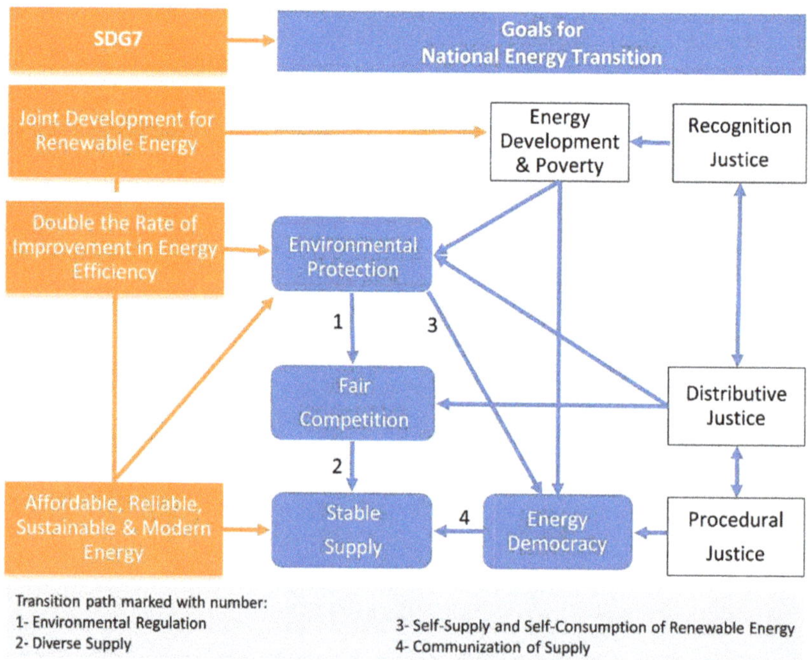

Figure 2. Response to Sustainable Development Goals (SDGs) with the national energy transition framework and path.

4.3. Data

The texts analyzed for this study not only included national energy policies, national energy laws, the Bureau of Energy's meeting minutes and the focus group's forum transcript, but also periodicals, books and other articles, civil group statements, media reports, and so on. The studied national energy policies include the Policy Guidelines for Sustainable Energy, Policy Guidelines for Energy

Development, and the New Energy Policy. National energy laws included the Basic Environment Act, the Greenhouse Gas Reduction and Management Act, and the Renewable Energy Development Act. The focus group held six forums, attended by 45 people comprised of stakeholders in the revisions to the Electricity Act, academics and experts. Attendees included Bureau of Energy officials and thinktanks (administrative offices), the TaiPower Company (a single integrated electricity enterprise), a government-owned electricity enterprise, a renewable energy-based electricity-generating enterprise (Energy Cooperatives), social enterprises, legislators or administrator officers (Legislative Yuan), energy technology service industries, academic experts and environmental groups, IPPs, electricity consumers, and self-use power generator equipment users. This focus group, besides recognizing stakeholders, was able to produce an early assessment of the effects of the bill on those stakeholders, thereby aiding legal development and policy analysis. In relation to the UN SDGs, the textual analysis in this paper also included minutes from the National Council for Sustainable Development [13–15] meetings and press releases; in terms of the distributive injustice of energy poverty [42,43], this paper mainly focuses on a textual analysis of the Taiwan Control Yuan Survey report in 2013 [44].

5. Results and Discussion

5.1. The Main Revision to the Electricity Act and Its Policy Analysis

By analyzing the revisions made to the Electricity Act, we explored whether these revisions have the ability to drive energy transition, while also considering the policy packages that support this process. The newly passed revised version of the Act is more conservative than the ruling party or the "original" Executive Yuan version, adopting a go-ahead for the Green Energy two-stage revision policy. Currently, the most important aspects of the first stage of revisions to this Act are the liberalization of green electricity, involving the opening of the electricity market to renewable energy-generated enterprises. These electricity-generating enterprises should be able to distribute electricity to or buy and sell electricity directly from the end consumer. In order to prevent thermal power plants from monopolizing the market, the market has only been opened up to electricity-generating enterprises of renewable energy, implementing direct supply and self-consumption, or through electricity transmission and distribution enterprises, implementing the sale of electricity. Other main aspects of these revisions include the resource-sharing of the electricity grid and the establishment of an electricity control mechanism. As a singular integrated electricity enterprise or government-owned electricity enterprise, the TaiPower Company should establish subsidiary companies under the parent holding company to manage the generation of electricity, the distribution of electricity and electricity sales. According to a 2017 report by the Executive Yuan, the right of liberalization will be granted to other traditional thermal power plants only during the second stage of the revisions to the Act at some point in the future.

Under this structure, the new Electricity Act has yet to strengthen environmental regulations to ensure a transition away from high carbon electricity enterprises, which makes it difficult to promote healthy competition between traditional electricity-generating enterprises and renewable energy generators. Internationally, the Intergovernmental Panel on Climate Change (IPCC)'s Emissions Factor mainly uses carbon emission controls in electricity-generating enterprises, in particular, adopting source controls for carbon emissions in electricity enterprises to control electricity-generating enterprises and the carbon emissions released in the production process [59]. In contrast, an environmental regulation tool called the "Electricity Emission Factor" was adopted in Taiwan's newest revised version of the Electricity Act and was adapted from the IPCC's "Electricity Carbon Emission Factor". However, this factor only attempts to introduce controls for electricity-retailing utility enterprises and fails to introduce controls for the levels of carbon emissions produced. Thus, rather than introducing controls for electricity enterprises' energy sources, Taiwan has, instead, introduced changes in the form of the end controls for electricity retailing enterprises. As a result, Taiwan's adoption of the electricity carbon emission factor has made public electricity-retailing utility enterprises work from the demand side and downstream from the electricity market, toward the supply side and upstream from the electricity market to reversely require electricity generation enterprises to supply a certain percentage of renewable energy to generate electricity. However, if the regulations for electricity retailing utility enterprises, according to the electricity carbon emission factors, are not stringent enough, it will be difficult for such carbon emission controls to work effectively.

5.2. The New Electricity Act Is Not Enough to Create an Electricity Market

The IEA stresses that the electricity market needs reviving through innovative system design and forward-looking environmental regulations [19]. However, Taiwan's newly revised version of the Electricity Act neglects the environmental protection aspect, persisting with a traditional energy fund payment method that places little burden on electricity generating enterprises, while having yet to formulate an energy tax act or give electricity-generating enterprises matching obligations. Moreover, while the new Electricity Act both opens up renewable energy-generating enterprises to direct supply and authorizes the wheeling of electricity, it fails to provide economic incentives or other incentives, thereby compelling renewable energy electricity-generating enterprises, energy cooperatives or social enterprises to sell wholesale to the state, with the subsidy costs being footed by the people, thus increasing the state's financial burden. With electricity prices yet to rise, and before competitive pricing is included in electricity generation costs, the only way for renewable energy to be developed is through government subsidies, which lack market mechanisms [60]. Although Taiwan has passed amendments to the Electricity Act, it is important to implement the development of the "Renewable Energy Certificate (REC)" or other market mechanisms, such as the certification of the exchange market with the carbon trading market to build an active exchange market.

5.3. Taiwan's Energy Transition Framework in the New Electricity Act Era

We used policy analysis of the Electricity Act to study the driving force behind, and the possible transition paths toward, achieving an energy transition and then analyzed whether each potential transition path would be strong, moderate, or weak. The following discussion focuses on Taiwan's national energy transition path, along with the way that the new Electricity Act works with legislative policy, by analyzing Taiwan's energy transition status in the new Electricity Act era (Table 2). This study also provides objectively quantitative data (Table 3) to support whether a transition path is strong, moderate, or weak.

Table 2. Taiwan's energy transition framework and its inter-connected policy bundles in new era of the Electricity Act.

Key Points in the Revision of the Electricity Act (2017)	Aspect	Degree of Aspect Development	Path	Degree of Path Development	Policy Bundle
Electricity Carbon Emission Factor; Legalized Agenda for Nuclear-Free Homeland	Environment-al Protection	Weak	Environment-al Regulation	Weak	Policy Guidelines for Energy Development (2017)/ Energy White Paper (2018; Draft)/ Energy Tax (on Planning)
			Distribution of Interests and Risk	Weak	REC (2017); Obligation for Consumer Plant to Setup Renewable Energy (2019)
Electricity Industry Regulatory Authority; Revolution in Electricity Enterprise; Dispute Mediation Between Enterprises	Fair Competition	Weak	Diverse Supply	Moderate	The Second Revision of Electricity Act (An Unclear Policy)
			Public Services	Weak	
Mechanism for the Stabilization of Pricing; Obligation to Provide Backup Capacity; Obligation for Electricity Retailing Utility Enterprise to Supply	Stable Supply	Strong	Obligation to Supply	Strong	Policy Guidelines for Energy Development (2017)/ Energy Security (A Developing Goal)
Flexible and Legalized Organization of Electricity Enterprise	Energy Democracy	Weak	Communization of Supply	Weak	Policy Guidelines for Energy Development (2017)/ Energy White Paper (2018; Draft)/ Action plan for People's Power Plant (2018; Draft)
			Self-Supply and Self-Consumption of Renewable Energy	Weak	

The influence of the 2017 amendment to the Electricity Act on Taiwan's energy transition is weak, as seen from the content of Table 2. From an analytical perspective of energy transition, besides being strong in stabilizing the supply side of development, other aspects of development are weak. In terms of path development, besides the electricity supply obligation path development being directly linked to the stable supply aspect (and, therefore, being strong), the fair competition aspect and the path development towards diversification of the related electricity supply is moderate, while other aspects linked to path development are all weak. Therefore, three quarters of the individual aspects of development are currently weak. However, among the various path developments, five out of seven are weak. Below is a clear explanation of the various aspects and paths:

(1) The stable supply aspect is strong because Taiwan's government has not only designed an electricity pricing stabilizing mechanism (which uses more than 15% of the electricity reserve capacity rate, mandated by the Regulation Governing Electricity Reserve Capacity originating in the 27th article of the Electricity Act), but it has also entrusted the obligation for supplying electricity to public retailing electricity enterprises. Moreover, the TaiPower Company [61] has launched a long-term Power Supply Development Plan extending from 2018 to 2028. Furthermore, Taiwan's Executive Yuan [62] reviewed the energy policy to confirm the percentage of more than 15% of the electricity reserve capacity rate after a referendum at the end of 2018. As a result, the electricity supply obligation (path development) is strong. However, the path development of the diversification of electricity supply, owing to the fair competition aspect or the go-ahead policy for green energy, is moderate. Moreover, there is a lack of a driving force for the communization of supply (path development: weak) from the energy democracy aspect without the mechanism for supporting community power plants or people's power plants;

(2) The fair competition aspect is weak, as the reforms taking place in the electricity enterprises have yet to see sweeping changes, and the independency and transitional capacity building of the control mechanisms for electricity enterprises are not yet strong enough. Moreover, the environmental regulations applied to the electricity carbon emission factor have yet to apply source management to electricity enterprises, which, in turn, means that in areas related to environmental protection environmental regulations (path development) remain weak and are incapable of promoting the transition of electricity enterprises and other fair competition aspects;

(3) Developments in the area of environmental protection are weak, as Taiwan has yet to implement either a carbon trading or energy tax system. The environmental cost has also has yet to be suitably internationalized, which, in turn, affects the distributive justice of environmental risk [36,37], with a weak path development in the distribution of interests and risk. The Taiwanese government is in the process of working toward greater cooperation between the public and private sectors through the "Energy Transition's white paper (at plan's level under a policy)", which identifies the formulation of an energy tax as vitally important in enabling the public to participate in the agenda;

(4) The development of the environmental protection aspect is weak. Nevertheless, Taiwan began to implement the REC form of market subsidies in 2017, which have gradually replaced feed-in tariffs and enabled an improved distribution of interests. However, the market for RECs is still in its nascent stages, and the government is still in the process of establishing a more fully comprehensive market mechanism, which directly relates to weak path development in interest and risk distribution. Furthermore, the goal of a nuclear-free homeland was mandated to be reached by 2025 in the 95th Article of the Electricity Act. Although the referendum in the end of 2018 resulted in the revocation of this article [63], the goal of the nuclear-free homeland is still mandated by the Basic Environment Act;

(5) The development of the energy democracy aspect is also weak. However, Taiwan was able to make the overall organization of electricity enterprises more flexible through the revisions made to the Electricity Act (i.e., the administrative authorities agreed to greater decentralization of energy management—for example, by allowing local government authorities to run private

energy companies or energy cooperatives). However, due to the fact that the communization of the power supply (development aspect: weak) and the legal construction of the peoples power plants are both lacking, the principal recipients of the electricity feed-in tariff system are private enterprises with a certain level of capital rather than energy cooperatives aimed at developing the whole community, community power plants or social enterprises invested in by normal civilians.

Since the 2017, amendments to the Electricity Act have failed to provide a driving force for energy transition in Taiwan, there is a need for other support measures to be established alongside the Electricity Act, in order to develop a national energy transition policy bundle. Currently, in terms of policy packages, there has been a few areas of development, such as the new REC system brought into effect by the Taiwanese government in 2017, and attempts to allow the REC market subsidy system to work alongside state-funded feed-in tariffs in order to reduce the number of casualties of pollution from power generation or the amount of Taiwanese people's money spent on subsidy systems. Thus, the government's REC system offers a response to the problem of distributive injustice by giving consumer plants energy-rationing obligations. This makes it mandatory for consumer plants to be set up with renewable-energy-based power generation equipment. Furthermore, rather than making it compulsory to establish the equipment themselves, plants can purchase an REC as an alternative plan, enabling them to comply with the agreement of self-generation of power through the obligation to set up renewable-energy-based power generation equipment. In this way, Taiwan can avoid an increase in electricity prices caused by internalizing the environmental costs of traditional thermal-generated energy, while, at the same time, enabling market subsidies in the form of RECs to replace state subsidies in the form of feed-in tariffs [60].

5.4. National Energy Transition Correlated with SDG7

In terms of national energy development and poverty (Figure 2), a response to SDG7's proposal for "Renewable Energy Joint Development" can be seen in the concept of recognition justice [9,10]. Beginning with the recognition justice aspect of energy justice, the amendment made to the Electricity Act involves electricity retailing sales discounts for vulnerable parties, a concept that is part of the fuel poverty or energy poverty issue as a whole. At the same time, this recognizes that the public can be involved in generating and selling electricity not only through energy cooperatives but also through other electricity enterprise organizations. As a result, in the amendments made to the Electricity Act public participation in energy development or the installation of renewable-energy-based power generation equipment is recognized, while, in terms of renewable energy, the low cost of installing of solar-generated electricity or on-land wind-generated electricity means that the potential for mutual public participation in renewable energy development is higher. Although public mutual participation is possible, renewable energy development in Taiwan needs more stakeholder communication on the demand for land and encourages relevant energy development on rooftops prior to ground solar photovoltaic power. In Figure 2, this study frames SDGs seven, eight, nine, 11 and 13 alongside the correlation between national energy transition and energy justice. Therefore, the amendment to the Electricity Act not only relates to environmental protection, fair competition, stable supply and energy democracy (the four axes of energy development), but also closely connects and responds to the various aspects of the SDGs as well as the issues of recognition justice, distributive justice and procedural justice (which are connected to energy justice). In this way, we view the revised Electricity Act as an important step in the national energy transition and consider whether it conforms to the various important international SDG trends and energy justice, and if it adequately handles the demands of energy poverty.

One of the main legal tools in national energy transition is electricity sector laws. The amendments made to these laws embody the SDGs. The interim structural problems evident within the amendment to the Electricity Act point to the challenges facing Taiwan in this stage of national energy transition, including returning the right to generate electricity to the people, the limits of a two-stage process to amend the law and (green energy first, unbundling of the power and the grid, and the liberalization of electricity retailing), encourage fair competition, the externalization of environmental cost, the lack of environmental regulations to drive reforms in electricity generating enterprises, the weakening of electricity emission factor source management, interests and risk distribution, energy poverty, etc. On the one hand, these challenges fit with the core concepts of the SDG7 affordable energy issue; on the other hand, these challenges reflect the energy justice-related issues of recognition justice, distributive justice and procedural justice, and the other issues of energy democracy and energy poverty that extend from these problems. At the same time, these two main areas could be connected by taking an energy management perspective on environmental protection, fair competition, stable supply, and social justice.

Regarding the aspects of energy transition analyzed here (Table 2), this study assesses the SDG7 indicators (Table 1) with the following polices (more concrete quantitative values are showed in another form in Table 3). Firstly, the indicator of "Access to Electricity" is strongly connected with the development of a stable supply aspect and its path of "Obligation to Supply", which is mandated by the Electricity Act and by the policy bundle of "Policy Guidelines for Energy Development". Secondly, the indicators of "Access to Non-Solid Fuels" and "CO_2 Emissions from Fuel Combustion and Electricity Output" are extremely relevant to the development of the environmental protection aspect and its path of "Environmental Regulation", with the policy bundle of "Electricity Carbon Emission Factor" and "Energy Tax". Furthermore, the indicator of "Share of Renewable Energy in Total Final Energy Consumption" is related to the development of the environmental protection aspect and its path of "Distribution of Interests and Risk", with the policy bundle of "Electricity Carbon Emission Factor", "REC", and "Obligation for Consumer Plant to Setup Renewable Energy" related to the development of the fair competition aspect and its path of "Diverse Supply", with the policy bundle of "Revolution in Electricity Enterprise" and "The Second Revision of Electricity Act", as well as to the development of the energy democracy aspect and its path of "Self-Supply and Self-Consumption of Renewable Energy", with the policy bundle of "Flexible and Legalized Organization of Electricity Enterprise", and "Action Plan for People's Power Plant". Therefore, with their policy tools, the key factors in the revision of the Electricity Act and the policy bundles (Table 2) pave clear paths toward the improvement of the SDGs and their indicators. In more detail (Figure 2), the strong development of a stable supply aspect will enhance SDG7's pursuit of "Affordable, Reliable, Sustainable, and Modern Energy" and its indicators. On the other hand, the weak development of the fair competition aspect will problematize directly promoting the development of a stable supply aspect, which will make it difficult to indirectly enhance SDG7 and its indicators.

5.5. Comparison of National Energy Transition between Taiwan and Japan

This study also compared the differences between the SDG7 relationships and the national energy transitions in Taiwan and Japan, based on the similar initiative of their Electricity Acts and policy bundles (Table 3). This study selected Japan because both Taiwan and Japan are island nations that suffer earthquakes, typhoons and a lack of energy reserves. After the Fukushima Nuclear Accident in 2011, the Japanese Government launched the "Policy on Electricity System Reform" (in 2013) and then revised its "Electricity Business Act". During the period from 2013 to 2015, Japan aimed to stabilize its power supply and lowered electricity prices, and promote the liberalization of its electricity distribution through the policy bundle of "Establishing Organization for Cross-Regional Coordination of Transmission Operators, Full Retail Competition, and Legal Unbundling of Transmission and Distribution Sectors, during the period from 2015 to 2020" [64,65]. This policy bundle for energy transition in Japan was mainly concentrated on fair competition and stable supply, which corresponds with the SDG7 goal of "affordable and reliable energy" and its indicator "access to energy".

Table 3. A comparison of the seventh Sustainable Development Goal (SDG7) relationship and energy transition between Taiwan and Japan, based on the similar initiative of Electricity Act with policy bundles.

SDG7 Relationship	Item	Taiwan	Japan
Energy Transition at the National Level	Amendment	Electricity Act (2017)	Electricity Business Act (2013–2015)
Energy Transition at the National Level	Energy Mix	Electricity Demand(+2%/Year) Nuclear12%→0% (−1%/Year) Renewable5%→20% (+1%/Year) Coal- and Oil-fired 50%→30% (-2%/Year) LNG-fired32%→50%(+2%/Year) (From Year 2016 to 2025) [62,66]	Electricity Demand (+1%/year) Nuclear 1%→20%~22% (+2%/Year) Renewable10%→22-24%(+1%/Year) Coal- and Oil-fired 65%→29% (−3%/Year) LNG-fired 24%→27% (−0%/Year) (From Year 2016 to 2030) [67]
Access to Electricity	Status	100% (2016) [68]	100% (2016) [69]
	Policy Bundle	Dividing into an Electricity Transmission and Distribution Enterprise (2023–2026)	Establishing Organization for Cross-Regional Coordination of Transmission Operators (2015)
		Obligation for Electricity Retailing Utility Enterprise to Supply (2017)	Obligation for Electricity Retailing Utility Enterprise to Supply (2016)
Access to Non-Solid Fuels	Status	Electricity from Coal-fired 46% (2016) [66]	Electricity from Coal-fired 25% (2016) [67]
	Policy Bundle	Project Report of Energy Policy Review for Response to the Result of Referendum (2019)/ National Energy Mix	Retailing from Non-fossil Fuels-based Power Ratio (2016)
CO_2 Emissions from Fuel Combustion and Electricity Output	Status	11.9 t- CO_2 Per Capita (2015) [70]	9.9 t- CO_2 Per Capita (2015) [70]
	Plan	0.529 Kg- CO_2 Per KWh→ 0.394 Kg- CO_2 Per KWh (−3%/Year) (From Year 2016 to 2025) [71]	0.516 Kg- CO_2 Per KWh→ 0.370 Kg- CO_2 Per KWh (−2%/Year) (From Year 2016 to 2030) [72,73]
	Policy Bundle	Electricity Carbon Emission Factor (2017)	Retailing from Non-fossil Fuels-based Power Ratio (2016)
Share of Renewable Energy in Total Final Energy Consumption	Status	5% (2016) [66]	10% (2016) [67]
	Policy Bundle	Project Report of Energy Policy Review for Response to the Result of Referendum (2019)/ National Energy Mix	Long-term Energy Supply and Demand Outlook (2015)/ National Energy Mix
		REC (2017)	Non-Fossil Fuel Energy Certificate Program (2018)

On the other hand, in 2015, the Japanese Government proposed a "Long-Term Energy Supply and Demand Outlook", to confirm its national energy portfolio in 2030 and to promote the development of renewable energy to account for more than 20% of all generated electricity [74], a response to the SDG7 meaning of "joint development for renewable energy" and its indicator of a "Share of Renewable Energy in Total Final Energy Consumption". Moreover, Taiwan and Japan have individually taken the increasing electricity demand and energy savings into a scenario analysis of energy transition [62,67]. Taiwan and Japan also have a common need to make more efforts in their environmental regulation for carbon reductions in the electricity sector, sharing similar policy bundles for creating a market of non-solid fuels via the "Electricity Carbon Emission Factor" in Taiwan and the "Retailing from

Non-Fossil Fuels-Based Power Ratio 44%" in Japan [75], to improve the SDG7 indicators of "CO_2 Emissions from Fuel Combustion and Electricity Output" and "Access to Non-Solid Fuels".

In short, Taiwan and Japan have paid more attention to the SDG7 indicator of "Access to Electricity", with Japan opening full retail competition, albeit in name only and accompanied by the unfair issues of the unbundling of the transmission and distribution sectors, while neglecting the SDG7 indicator of "CO_2 Emissions from Fuel Combustion and Electricity Output" by failing to implement more stringent environmental regulations or related policy bundles, as indicated in the form (Table 2). In particular, Taiwan faces more challenges on the issue of carbon reduction, with the goal of national energy mix by 2030 to replace nuclear power with renewable energy and to reduce coal-fired power by increasing liquefied natural gas (LNG)-fired power. Finally, following their goals and framework for national energy transition (Figure 2), Taiwan and Japan can strengthen their environmental regulations to promote fair competition directly, with fair competition then able to enhance stable electricity supply, thereby allowing these countries to move towards SDG7 and its indicators.

6. Conclusions and Policy Implications

The main problem explored in this study is how a nation should face the challenges of energy transition using legal tools or policy bundles. Responding to SDG7 and adopting a textual analysis of the law as its research methodology alongside legal policy analysis (during which a number of texts were analyzed), alongside the theories of energy justice and energy management, this study also established a national energy transition analysis structure, using Taiwan's amendments to the Electricity Act as a case study. The results of this analysis found that, for the legal reforms made to Taiwan's electricity sector, despite responding to the challenges in the early stages of national energy transition, as well as embodying many aspects of the SDGs, the majority of aspects or paths toward national energy transition are weak and need to be supported by other legislation and policies besides those within the electricity sector in order to achieve national energy transition and reach the SDGs. At the same time, this study also compares the SDG7 relationship and national energy transition between Taiwan and Japan, to justify the theoretical and analytical framework in this work.

Globally speaking, most countries like Taiwan are currently in a crucial period of energy transition. While facing pressure to reduce carbon emissions and air pollution, there is also the challenge of industry transitions; these challenges can all be approached using the structure shown in Figure 2 to ensure further resolution. Whether in the mature stages of energy transition (like Germany) or just starting out (like Taiwan, South Korea and Japan), all countries have responded to the SDGs' emphasis on energy justice, energy democracy and energy management and can be analyzed together. Countries that have just started out can particularly learn from a critical analysis of Taiwan's 2017 amendment to the Electricity Act and the weak driving force for energy transition that it has created. Here, we can observe a corresponding relationship—the more comprehensive the legislative revisions made to the electricity sector, the greater the influence not only on domestic issues of environment, society and the economy in terms of balanced development, but also on issues of international climate change, sustainable economy and social development responsibility.

Particularly worthy of a mention is the fact that, within the goals towards national energy transition, environmental protection involves energy competition, technological innovation, clean energy, energy interests and distributive justice issues that environmental health risks touch upon, which are at the core of the SDGs. This study provides an analytical structure for national energy transition, helping the country by outlining an energy transition agenda, including policy planning and the early stages of risk assessment for carrying out various transition goals or analytical aspects, with policy decisions made accordingly. Moreover, other countries that desire to implement this energy transition process can refer to the national energy transition goals discussed in this study, while, at the same, time taking an inventory of their domestic energy legislations or policies and looking at the different political, economic and social conditions unique to themselves, thereby adopting different legislation or policy bundles depending on the their unique local conditions in order to support or drive such energy transition paths. Furthermore, Renn et al. [76] proposed that sustainable development faces the issue of complex systems and that the science community could find solutions by identifying the pathways of these systems. This study responds to complex systems under the structure of SDG7 and its related goals, by establishing a national transition framework to identify the components, along with the path and the policy bundle, to balance national demands related to technology, the economy, society, and the environment. Finally, this study suggests two areas for future research, the first being comparison of the energy transition in Taiwan with the transitions of other countries, and the second to classify policy bundles into patterns to provide the other countries with references for policymaking. For policymakers, the approach of this study is beneficial not only by accurately diagnosing the status of the national energy transition, but also by efficiently concentrating on a particular component or path of the transition framework to strengthen or weaken related policy bundles based on particular national contexts and on a case-by-case basis. However, the approach of this study has limitations for policymakers, who will find it difficult to identify existing policies and implement related policies, due to the diversity of national policy contexts. This study, ultimately, has the following limitations: (1) beyond comparative case studies in Taiwan, Japan, or other east Asian countries, more national cases should be investigated based on this study to classify types of policy bundles in a particular component or path. (2) Exclusive of Table 3, this study did not focus on a quantitative analysis, as we aimed to develop a generalized analytic framework and to identify qualitative paths and policies among different components. (3) Using a quantitative analysis to study the paths between the components in the national energy transition framework has much greater potential than this study alone, such as using environmental regulation to promote fair competition via electricity carbon emission factors or implementing an energy tax, as mentioned in Table 2.

Author Contributions: Conceptualization, M.-X.L., H.M.L. and K.T.C.; methodology, M.-X.L., H.M.L and K.T.C.; formal analysis, H.M.L; investigation, M.-X.L.; resources, K.T.C.; data curation, M.-X.L.; writing—original draft preparation, M.-X.L.; writing—review and editing, M.-X.L. All authors have read and agreed to the published version of the manuscript.

Funding: This research received no external funding.

Conflicts of Interest: The authors declare no conflict of interest.

References

1. UN. Transforming our World: The 2030 Agenda for Sustainable Development. 2015. Available online: http://www.un.org/ga/search/view_doc.asp?symbol=A/RES/70/1&Lang=E (accessed on 13 January 2020).
2. UN. The Sustainable Development Goals Report. 2016. Available online: http://www.un.org.lb/Library/Assets/The-Sustainable-Development-Goals-Report-2016-Global.pdf (accessed on 13 January 2020).
3. International Bank for Reconstruction and Development; World Bank; IEA. Global Tracking Framework: Progress toward Sustainable Energy. 2017. Available online: http://www.worldbank.org/en/topic/energy/publication/global-tracking-framework-2017 (accessed on 13 January 2020).
4. Osborn, D.; Cutter, A.; Ullah, F. *Universal Sustainable Development Goals: Understanding the Transformational Challenge for Developed Countries*; Report of a Study by Stakeholder Forum; Stakeholder Forum: Kent, UK, 2015.
5. Akinwale, Y.O.; Ogundari, I.O. Exploration of Renewable Energy Resources for Sustainable Development in Nigeria: A Study of the Federal Capital Territory. *Int. J. Energy Econ. Policy* **2017**, *7*, 240–246.
6. Munro, P.G.; Horst, G.V.D.; Healy, S. Energy justice for all? Rethinking Sustainable Development Goal 7 through struggles over traditional energy practices in Sierra Leone. *Energy Policy* **2017**, *105*, 635–641. [CrossRef]
7. Sovacool, B.K.; Dworkin, M.H. Energy justice: Conceptual insights and practical applications. *Appl. Energy* **2015**, *142*, 435–444. [CrossRef]
8. Walker, C.; Baxter, J. Procedural justice in Canadian wind energy development: A Comparison of Community-Based and Technocratic Siting Processes. *Energy Res. Soc. Sci.* **2017**, *29*, 160–169. [CrossRef]
9. Hernández, D. Sacrifice Along the Energy Continuum: A Call for Energy Justice. *Environ. Justice* **2015**, *8*, 151–156. [CrossRef] [PubMed]

10. Bailey, I.; Darkal, H. (Not) Talking about Justice: Justice Self-Recognition and the Integration of Energy and Environmental-Social Justice into Renewable Energy Siting. *Local Environ.* **2018**, *23*, 335–351. [CrossRef]
11. International Institute for Applied Systems Analysis. Transformations to Achieve the Sustainable Development Goals. Report Prepared by the World in 2050 Initiative (TWI2050). 2018. Available online: www.twi2050.org (accessed on 13 January 2020).
12. UN. Launch of Report Transformations to Achieve the Sustainable Development Goals. 2018. Available online: http://www.iiasa.ac.at/web/home/research/twi/180712-TWI2050-report-launch.html (accessed on 13 January 2020).
13. National Council for Sustainable Development. The 27th Recording for Committee Meeting. 2015. Available online: http://nsdn.epa.gov.tw/Files/Meeting/27meeting.pdf (accessed on 6 June 2018).
14. National Council for Sustainable Development. News Press Release. 2017. Available online: https://nsdn.epa.gov.tw/Nsdn_News.aspx?id=1306 (accessed on 13 January 2020).
15. National Council for Sustainable Development. The 39th Recording for Working Group Meeting. 2015. Available online: http://nsdn.epa.gov.tw/Files/WMeeting/39meeting.pdf (accessed on 6 June 2018).
16. TaiPower. Electricity Purchased by TaiPower Over the Years. 2020. Available online: https://www.taipower.com.tw/tc/Chart.aspx?mid=194 (accessed on 13 January 2020).
17. Chou, K.T. Challenge to Taiwan Energy Transition Driven by Climate Change. In *Introduction to the Fourteen Lectures on Taiwan Energy Transition*; Chou, K.T., Lin, Z.L., Eds.; Chuliu: Taipei, Taiwan, 2016.
18. Chou, K.T. Triple Helix Energy Transition Movement in Taiwan. In *Energy Transition in East Asia: A Social Science Perspective*; Chou, K.T., Ed.; Routledge: New York, NY, USA, 2018.
19. International Energy Agency. *Re-Powering Markets: Market Design and Regulation during the Transition to Low-Carbon Power Systems*; IEA: Paris, France, 2016.
20. Fouquet, R. Historical energy transitions: Speed; prices and system transformation. *Energy Res. Soc. Sci.* **2016**, *22*, 7–12. [CrossRef]
21. Gao, A.M.Z.; Fan, C.T.; Liao, C.N. Application of German Eenergy Ttransition in Taiwan: A Critical Review of Unique Electricity Liberalization as a Core Strategy to Achieve Renewable Energy Growth. *Energy Policy* **2018**, *120*, 644–654. [CrossRef]
22. Dioha, M.O.; Kumar, A. Exploring sustainable energy transitions in sub-Saharan Africa residential sector: The case of Nigeria. *Renew. Sustain. Energy Rev.* **2020**, *117*, 109510. [CrossRef]
23. Marcillo-Delgadoa, J.C.; Ortegob, M.I.; Pérez-Fogueta, A. A compositional Approach for Modelling SDG7 indicators: Case study Applied to Electricity Access. *Renew. Sustain. Energy Rev.* **2019**, *107*, 388–398. [CrossRef]
24. Korkovelos, A.; Khavari, B.; Sahlberg, A.; Howells, M.; Arderne, C. The Role of Open Access Data in Geospatial Electrification Planning and the Achievement of SDG7. An OnSSET-Based Case Study for Malawi. *Energies* **2019**, *12*, 1395. [CrossRef]
25. Hillerbrand, R. Why affordable clean energy is not enough. A capability perspective on the sustainable development goals. *Sustainability* **2018**, *10*, 2485. [CrossRef]
26. Acheampong, M.; Yu, Q.; Ertem, F.C.; Ebude, L.D.E.; Tanim, S.; Eduful, M.; Vaziri, M.; Ananga, E. Is Ghana Ready to Attain Sustainable Development Goal (SDG) Number 7? A Comprehensive Assessment of Its Renewable Energy Potential and Pitfalls. *Energies* **2019**, *12*, 408. [CrossRef]
27. Bhat, K.S.; Bachhiesl, U.; Feichtinger, G.; Stigler, H. A techno-economic model-based analysis of the renewable energy transition in the Indian subcontinent region. *Elektrotech. Informationstechnik* **2019**, *136*, 361–367. [CrossRef]
28. Santika, W.G.; Anisuzzaman, M.; Bahri, P.A.; Shafiullah, G.M.; Rupf, G.V.; Urmee, T. From goals to joules: A quantitative approach of interlinkages between energy and the Sustainable Development Goals. *Energy Res. Soc. Sci.* **2019**, *50*, 201–214. [CrossRef]
29. McCollum, D.L.; Echeverri, L.G.; Busch, S.; Pachauri, S.; Parkinson, S.; Rogelj, J.; Krey, V.; Minx, J.C.; Nilsson, M.; Stevance, A.S.; et al. Connecting the sustainable development goals by their energy inter-linkages. *Environ. Res. Lett.* **2018**, *13*, 033006. [CrossRef]
30. Sachs, J.; Schmidt-Traub, G.; Kroll, C.; Durand-Delacre, D.; Teksoz, K. *SDG Index and Dashboards—Global Report*; Bertelsmann Stiftung and Sustainable Development Solutions Network (SDSN): New York, NY, USA, 2016.

31. The Association of German Engineers. VDI-Standard: VDI 4602 Part 2. 2013. Available online: https://www.vdi.de/fileadmin/pages/vdi_de/redakteure/richtlinien/inhaltsverzeichnisse/1929951.pdf (accessed on 13 January 2020).
32. Goldthau, A.; Sovacool, B.K. The uniqueness of the energy security; justice; and governance problem. *Energy Policy* **2012**, *41*, 232–240. [CrossRef]
33. Heffron, R.J.; McCauley, D.; Sovacool, B.K. Resolving society's energy trilemma through the Energy Justice Metric. *Energy Policy* **2015**, *87*, 168–176. [CrossRef]
34. Jenskin, K.; Darren, M.; Raphael, H.; Hannes, S.; Robert, R. Energy justice: A Conceptual Review. *Energy Res. Soc. Sci.* **2016**, *11*, 174–182. [CrossRef]
35. Fuller, S.; McCauley, D. Framing Energy Justice: Perspectives from Activism and Advocacy. *Energy Res. Soc. Sci.* **2016**, *11*, 1–8. [CrossRef]
36. Jenkins, K.; McCauley, D.; Warren, C.R. Attributing Responsibility for Energy Justice: A case Study of the Hinkley Point Nuclear Complex. *Energy Policy* **2017**, *108*, 836–843. [CrossRef]
37. Damgaard, C.; McCauley, D.; Long, J. Assessing the Energy Justice Implications of Bioenergy Development in Nepal. *Energy Sustain. Soc.* **2017**, *7*, 1–16. [CrossRef]
38. Islara, M.; Brogaarda, S.; Lemberg-Pedersen, M. Feasibility of Energy Justice: Exploring National and Local Efforts for Energy Development in Nepal. *Energy Policy* **2017**, *105*, 668–676. [CrossRef]
39. Fan, M.F. The Issues of Justice on Low-Carbon Energy Transition. In *Introduction to the Fourteen Lectures on Taiwan Energy Transition*; Chou, K.T., Lin, Z.L., Eds.; Chuliu: Taipei, Taiwan, 2016.
40. Lin, M.X. Strategic Environmental Assessment on the Meaning of the Constitution. *Tsinghua Discourses Rule Law* **2012**, *16*, 136–150.
41. Quitzow, L.; Canzler, W.; Grundmann, P.; Leibenath, M.; Moss, T.; Rave, T. The German Energiewende What's happening? Introducing the special issue. *Util. Policy* **2016**, *41*, 163–171. [CrossRef]
42. Hiteva, R.P. Fuel poverty and vulnerability in the EU low-carbon transition: The case of renewable electricity. *Local Environ.* **2013**, *18*, 487–505. [CrossRef]
43. Bouzarovski, S.; Herrero, S.T.; Petrova, S.; Ürge-Vorsatz, D. Unpacking the spaces and politics of energy poverty: Path-dependencies; deprivation and fuel switching in post-communist Hungary. *Local Environ.* **2016**, *21*, 1151–1170. [CrossRef]
44. Control Yuan. Survey Report. 2013. Available online: http://www.cy.gov.tw/sp.asp?xdURL=./di/Message/message_1.asp&ctNode=903&msg_id=4471 (accessed on 13 January 2020).
45. UK. Collection: Fuel Poverty Statistics. 2013. Available online: https://www.gov.uk/government/collections/fuel-poverty-statistics (accessed on 13 January 2020).
46. London Assembly. *In from the Cold? Tackling Fuel Poverty in London*; Greater London Authority: London, UK, 2012.
47. Shin, J.S. Estimation of Fuel-Poverty in Korean Households. 2011. Available online: http://www.keei.re.kr/main.nsf/index_en.html?open&p=%2Fweb_keei%2Fen_publish.nsf%2Fby_report_types%2F985C6BB7B56C60C2492579AA001F60E4&s=%3FOpenDocument%26menucode%3DES173%26category%3DResearch%2520Papers (accessed on 13 January 2020).
48. Liou, H.M. Basic Principles and Legal Foundation of Energy Management: Post Kyoto Era. *Nat. Taiwan Univ. Law J.* **2006**, *35*, 45–102.
49. Bureau of Energy. Draft of Amendment to The Electricity Act. 2016. Available online: http://web3.moeaboe.gov.tw/ECW/populace/content/SubMenu.aspx?menu_id=3124 (accessed on 11 December 2018).
50. TaiPower. Status of Purchasing Electricity. 2016. Available online: https://www.taipower.com.tw/tc/page.aspx?mid=207&cid=163&cchk=9c1fa9ec-c80e-4e08-b4e8-be1464b3811c (accessed on 13 January 2020).
51. Liou, H.M. Liberalization of electric Utilities and Competition Law: The Development in European Union; Germany and Taiwan. *Chengchi Law Rev.* **2002**, *72*, 65–139.
52. Bureau of Energy. Public Hearing for Draft of the Amendment to The Electricity Act: Video Recording from the 1st to 3rd Meeting. 2016. Available online: http://web3.moeaboe.gov.tw/ECW/populace/content/SubMenu.aspx?menu_id=3124 (accessed on 11 December 2018).
53. Liou, H.M. Carbon Emission Reduction of Taiwan's Electric Power Industry. *J. Adv. Clean Energy* **2015**, *2*, 18–34.
54. Lin, S.H.; Lee, Y.Y. Currently No Plans to Discuss Energy Tax. 2016. Available online: http://www.chinatimes.com/newspapers/20160617000093-260202 (accessed on 13 January 2020).
55. Huang, L.Y. Pay to Energy Foundation/The Percentage from Renewable Energy Hopes Bring Down. 2016. Available online: http://www.cna.com.tw/news/afe/201607270469-1.aspx (accessed on 13 January 2020).

56. DPP. *Draft of the Amendment to the Electricity Act*; Term 4th; Session 8th; and Conference 12th Legislative Yuan Bill Related Document: Legislator Proposal 660(15654); Legislative Yuan: Taipei, Taiwan, 2013.
57. Executive Yuan. *Draft of the Amendment to the Electricity Act*; Term 4th; Session 8th; and Conference 1st Legislative Yuan Bill Related Document: Legislator Proposal 660(15317); Legislative Yuan: Taipei, Taiwan, 2015.
58. Executive Yuan. *The Electricity Act. Executive Yuan Gazette 23(21)*; Executive Yuan: Taipei, Taiwan, 2017.
59. IPCC NGGIP. Database on Greenhouse Gas Emission Factors (EFDB). 2017. Available online: http://www.ipcc-nggip.iges.or.jp/EFDB/documents.php (accessed on 13 January 2020).
60. Lin, M.X. The Lost Segment of the Electricity Act? Scheme on Renewable Energy Certificates Connected with Renewable Portfolio Standard. 2017. Available online: http://www.storm.mg/article/241479 (accessed on 13 January 2020).
61. TaiPower. Long-termed Power Supply Development Plan. 2018. Available online: https://www.taipower.com.tw/upload/212/106%E5%B9%B4%E9%95%B7%E6%9C%9F%E9%9B%BB%E6%BA%90%E9%96%8B%E7%99%BC%E6%96%B9%E6%A1%88(10610%E6%A1%88-107%E5%B9%B41%E6%9C%88%E4%BF%AE%E6%AD%A3%E6%A1%88).pdf (accessed on 13 January 2020).
62. Executive Yuan. *Project Report of Energy Policy Review for Response to the Result of Referendum*; Term 9th; Session 7th; and Conference 13st Legislative Yuan Bill Related Document: Legislator Proposal 887(16450); Legislative Yuan: Taipei, Taiwan, 2019.
63. Bureau of Energy. MOEA Announced that the Assessment Report on the Response to the Result of Energy Referendum. 2019. Available online: https://www.moea.gov.tw/MNS/populace/news/News.aspx?kind=1&menu_id=40&news_id=82760 (accessed on 4 March 2019).
64. Hiranuma, H. Japan's Energy Policy in a Post-3/11 World: Juggling Safety; Sustainability and Economics. 2014. Available online: https://www.tkfd.or.jp/en/research/detail.php?id=296 (accessed on 13 January 2020).
65. METI. Electricity Market Reform in Japan. 2015. Available online: https://www.meti.go.jp/english/policy/energy_environment/electricity_system_reform/reference.html (accessed on 13 January 2020).
66. RSPRC. Interpreting 2016 Energy Trends in Taiwan. 2017. Available online: https://rsprc.ntu.edu.tw/zh-tw/m01-3/energy-transformation/627-0330_tw-energy-status-2016 (accessed on 13 January 2020).
67. METI. Japan's Energy. 2018. Available online: https://www.enecho.meti.go.jp/en/category/brochures/pdf/japan_energy_2017.pdf (accessed on 13 January 2020).
68. TaiPower. Taiwan Power Company Sustainability Report. 2017. Available online: https://www.taipower.com.tw/upload/85/2018011009015371352.pdf (accessed on 13 January 2020).
69. World Bank Group. *Regulatory Indicators for Sustainable Energy*; RISE, 2017. Available online: https://rise.esmap.org/country/japan (accessed on 13 January 2020).
70. EC. CO_2 Time Series 1990–2015 Per Capita for World Countries. 2017. Available online: https://edgar.jrc.ec.europa.eu/overview.php?v=CO2ts_pc1990-2015 (accessed on 13 January 2020).
71. Executive Yuan. Report for Energy Policy. 2018. Available online: https://www.ey.gov.tw/File/FA89BE915504DBD8/311ed920-9788-499d-87ff-7d665b870a77?A=C (accessed on 13 January 2020).
72. FEPC (Federation of Electric Power Companies). Establishment of the Electricity Business Council for a Low-Carbon Society. 2016. Available online: http://www.fepc.or.jp/about_us/pr/pdf/kaiken_s_e_20160219.pdf (accessed on 13 January 2020).
73. ELCS (Council for a Low Carbon Society). CO_2 Emission in 2017. 2018. Available online: https://e-lcs.jp/news/detail/000048.html (accessed on 13 January 2020).
74. METI. Long-term Energy Supply and Demand Outlook. 2015. Available online: https://www.meti.go.jp/english/press/2015/0716_01.html (accessed on 13 January 2020).
75. Nee, M.T.; Lin, M.X.; Liou, H.M.; Chou, K.T. A Comparative Study of the Carbon Emission Regulations between Taiwan's and Japan's Electricity Sectors. *Sci. Technol. Law Rev.* **2019**, 31, 48–70.
76. Renn, O.; Chabay, I.; van der Leeuw, S.; Droy, S. Beyond the Indicators: Improving Science, Scholarship, Policy and Practice to Meet the Complex Challenges of Sustainability. *Sustainability* **2020**, 12, 578. [CrossRef]

MDPI
St. Alban-Anlage 66
4052 Basel
Switzerland
Tel. +41 61 683 77 34
Fax +41 61 302 89 18
www.mdpi.com

Energies Editorial Office
E-mail: energies@mdpi.com
www.mdpi.com/journal/energies

www.ingramcontent.com/pod-product-compliance
Lightning Source LLC
Chambersburg PA
CBHW061844210926
51850CB00006B/965
9783036571669